Ichnology of the Permian-Triassic Mass Extinction in China

Xueqian Feng • Zhong-Qiang Chen
David J. Bottjer • He Zhao

Ichnology of the Permian-Triassic Mass Extinction in China

Springer

Xueqian Feng
State Key Laboratory of Geomicrobiology and Environmental Changes
China University of Geosciences (Wuhan)
Wuhan, Hubei, China

Zhong-Qiang Chen
State Key Laboratory of Geomicrobiology and Environmental Changes
China University of Geosciences (Wuhan)
Wuhan, Hubei, China

David J. Bottjer
Department of Earth Sciences
University of Southern California
Los Angeles, CA, USA

He Zhao
College of Marine Science and Technology
China University of Geosciences
Wuhan, Hubei, China

ISBN 978-981-96-6638-6 ISBN 978-981-96-6639-3 (eBook)
https://doi.org/10.1007/978-981-96-6639-3

This Springer imprint is published by the registered company Springer Nature Singapore Pte Ltd.
The registered company address is: 152 Beach Road, #21-01/04 Gateway East, Singapore 189721, Singapore

Preface

When you are walking on a beach or river bank, you may be attracted by the holes or trackways left on the ground, and they are usually abundant. You take out your cellphone and take photos, but some questions may puzzle you: what are these and who produced them? Are there any organisms living in the holes? Why are they so densely distributed?

If you are interested in collecting fossils in the field geological investigation, it is not surprising to find similar structures in the sedimentary rocks. These biogenic structures preserved in the rocks are trace fossils. Like body fossils, these trace fossils are relatively common in the rocks, and some of them are spectacular and enigmatic.

What's more, trace fossils have been shown as an independent line of evidence to track the origin of metazoan and understand the evolution of life on earth. The two aspects have been proven to be successful in recent years. Chen et al. (2018) reported the earliest known trackways, preserved in association with burrows, from the terminal Ediacaran Shibantan Member of the Dengying Formation (ca. 551 to ca. 541 million years ago) in the Yangtze Gorges area of South China. These trackways were interpreted to be produced by bilaterian animals with paired appendages, marking the arrival of a new era characterized by bilaterian animals. Buatois and Mángano (2018) quantitatively reconstructed the global ichnodiversity of Phanerozoic based on a comprehensive dataset of bioturbation and bioerosion structures. The results demonstrated that the three marine evolutionary radiations (Cambrian Explosion, Great Ordovician Biodiversification Event, and Mesozoic Marine Revolution) detected in the body-fossil record were also expressed in the trace-fossil records. Sepkoski's logistic model of the diversification of marine invertebrates based on body fossils, the Cambrian evolutionary fauna, Paleozoic evolutionary fauna, and Modern evolutionary fauna was also reflected by the trace-fossil records (Sepkoski, 1984; Buatois and Mángano, 2018).

The Permian-Triassic (P-Tr) mass extinction, the largest biocrisis of Phanerozoic life, displayed severely devastated biodiversity of animals and marked the final transition from the Paleozoic evolutionary fauna to the Modern evolutionary fauna (Sepkoski, 1984). Modern-type marine ecosystems were originated from the

P-Tr mass extinction, and followed by numerous evolutionary innovations during the Mesozoic, termed as the Mesozoic Marine Revolution (MMR; Vermeij 1977). That is to say, the P-Tr mass extinction event is crucial in shaping modern marine ecosystems. The studies of the P-T mass extinction were primarily observed in the body fossil of well-skeletonized animals. Understanding how whole ecosystems were affected and rebuilt following the extinction requires evidences from both skeletonized and soft-bodied animals; the best comprehensive information on soft-bodied animals comes from trace fossils. Thus, trace fossils across the P-T mass extinction are a significant supplement of the body fossils.

This book provides the reader with a great deal of high-quality photographs, figures, and other ecological metrics of trace fossils across the P-T mass extinction. The fossil materials are largely from the collection in the field outcrops by the authors during the past 10 years. The book is designed for undergraduates and graduate students in the earth sciences, and also for professionals who are interested in ichnology and the recovery patterns of marine trace-making organisms in the aftermath of P-T mass extinction. Foremost, Chap. 1 introduces the P-Tr mass extinction event and the recovery in the aftermath, then summarizes the global research status and progress of trace fossils in recent years, and emphasizes trace fossils are crucial tools in assessing the recovery process and pattern of marine trace-making organisms. In particular, bioturbation, a typical ecosystem engineering, is outlined, as more and more studies are now discussing this aspect.

Chapter 2 introduces the geological background of the South and North China during the P-T transition and focuses on the paleogeographic and stratigraphic features of the studied sections. The chapter also provides a number of field photographs of the strata and sedimentary structures. The studied sections are divided into five environmental categories according to the stratigraphic units, two for siliciclastic systems (brackish, shoreface), two for carbonate systems (ramp, platform), and one for mixed siliciclastic-carbonate systems (shelf-slope-basin). Each field section is briefly introduced in aspects of stratigraphic units, lithology, fossil assemblages, biozones and biostratigraphy, and palaeoenvironmental interpretations.

Chapter 3 is the main body of this book, focusing on the systematic ichnology. A total of 39 ichnogenera are systematically described and illustrated, and 60 fossil plates are displayed, at least one plate for a single ichnogenus.

Chapter 4 conducts quantitative analyses on these trace fossils across the P-T mass extinction. It first introduces the ichno-ecological metrics applied in the book, and then displays the results of these four metrics: ichnodiversity, ichnodisparity, ecospace utilization, and ecosystem engineering. Here, we conduct the quantitative analyses of every environmental category.

Chapter 5 discusses the recovery of infaunal ecosystem following the P-T mass extinction. The chapter integrates data of the four ichno-ecological metrics from all environmental categories together to generate general tendencies of each ichno-ecological metric across the P-T mass extinction and the Early Triassic. For better comparison, with body fossils, a dataset of all known marine invertebrate body fossils derived largely from a previously constructed database is compiled spanning the Late Permian to the end of the Early Triassic from South China.

The history of ichnology has been well summarized in many ichnological monographs, and thanks to Baucon et al. (2012), who have done an excellent job on this topic, we won't discuss any more in this book. Ichnology was brought in China since 1980s, and more and more papers have been published in relation to the late Ediacaran biota, Cambrian explosion, and Permian-Triassic (P-Tr) mass extinction, especially the last one. Other studies have also expanded to ichnofacies, ichnofabrics, and neoichnology. When compared with scientific papers, systematic ichnological textbooks or monographs based on trace fossil materials from China are rarely published in English, which do not match the well-exposed sedimentary rocks and abundant trace fossil records. The origination of this book is attributed to extensive field investigations during the PhD project of the first author. The large number of photos and specimens of trace fossils constitute the main body of this book. The identification of trace fossils sometimes is challenging, which requires extensive practice, as well as numbers of specimens one have seen. There is a long way to see, to learn, and to practice, and this book is just a beginning.

This book is supported by three grants of National Natural Science Foundation of China (41930322, 42372136, and 42477215), the "CUG Scholar" Scientific Research Funds at China University of Geosciences (Wuhan) (No. 2023081), State-sponsored Postdoctoral Fellowship Program (GZC20232474, 2024M753028), and the Postdoctor Project of Hubei Province (2004HBBHCXA084).

Wuhan, China Xueqian Feng
December 2024

References

Baucon A, Bordy E, Brustur T, Buatois LA, Cunningham T, De C, Duffin C, Felletti F, Gaillard C, Hu B, Hu L, Jensen S, Knaust D, Lockley M, Lowe P, Mayor A, Mayoral E, Mikuláš R, Muttoni G, Neto de Carvalho C, Pemberton SG, Pollard J, Rindsberg AK, Santos A, Seike K, Song H-BH, Turner S, Uchman A, Wang Y-Y, Yi-Ming G, Zhang L, Zhang W-T (2012) A history of ideas in ichnology. In: Knaust D, Bromley RG (eds) Trace fossils as indicators of sedimentary environments, Developments in sedimentology 64. Elsevier, Amsterdam, pp 3–43

Buatois LA, Mángano MG (2018) The other biodiversity record: innovations in animal- substrate interactions through geologic time. GSA Today 28:4–10

Chen Z, Chen X, Zhou CM, Yuan XL, Xiao SH (2018) Late Ediacaran trackways produced by bilaterian animals with paired appendages. Sci Adv 4:eaao6691

Sepkoski Jr JJ (1984) A kinetic model of Phanerozoic taxonomic diversity, III. Post-Paleozoic families and mass extinctions. Paleobiology 10:246–267

Vermeij GJ (1977) The Mesozoic marine revolution: evidence from snails, predators and grazers. Paleobiology 3:245–258

Contents

Chapter 1
Introduction

1.1 Brief Introduction of the P-Tr Mass Extinction Event

Many biologists believe that the Earth is currently experiencing an unprecedented mass extinction event. According to statistical analysis of the extinction rate of modern organisms, extinction rates of major modern vertebrates are more than 50 times higher than normal extinction rates (Barnosky et al. 2011; Ceballos et al. 2015). Whether the sixth mass extinction is coming, the prediction of its time and its impact on the Earth's ecosystem are a hot area in current research. There have been five mass extinctions in the geological past (Raup and Sepkoski 1982). Studying mass extinction events in the geological past therefore is valuable in dealing with the sixth mass extinction. In fact, research on "Big Five" mass extinctions has long been a hot topic in geosciences (Ravelo et al. 2004; Rong and Huang 2014; Bond and Grasby 2017; Shen and Zhang 2017).

The Permian-Triassic (P-Tr) mass extinction (252 million years ago) was the most serious crisis of life on Earth since the Phanerozoic (Erwin et al. 2002). It has led not only to a dramatic decline in biodiversity but also to a near collapse of the ecosystem (Knoll et al. 1996; Chen and Benton 2012). The P-Tr mass extinction reshaped the structure of marine ecosystems, leading to biological recovery and radiation after the mass extinction. The biological features in the marine ecosystem have changed from Paleozoic evolutionary fauna such as brachiopods, corals and sponges to Modern evolutionary fauna such as bivalves, gastropods and large reptiles (Sepkoski 1984). The Modern evolutionary fauna is dominated by carnivores (Vermeij 1977; Benton et al. 2013; Motani et al. 2015; Brayard et al. 2017; Song et al. 2018; Dai et al. 2023). The earliest Early Triassic oceans were dominated by remnants and opportunistic organisms that were resistant to harsh environments or had the ability to move and swim, such as bivalves, soft-bodied organisms, fish and arthropods. Marine ecosystems did not fully recover until ten million years after the mass extinction (Chen and Benton 2012). The trophic pyramid model has shown

X. Feng et al., *Ichnology of the Permian-Triassic Mass Extinction in China*,
https://doi.org/10.1007/978-981-96-6639-3_1

that the recovery and reconstruction of marine ecosystems occurred from bottom to top. That is to say, the primary producers at the bottom of the trophic pyramid recovered first, and then the functional groups at the middle and upper levels of the pyramid recovered. It was a bottom-up mechanism of nutrients and energy flux in the food web (Chen and Benton 2012; Huang et al. 2023).

However, in the late Early Triassic, even in the Dienerian of the Early Triassic, predators in higher trophic levels, such as fish, ichthyosaurs and crocodiles, have appeared in many areas, especially in Nanzhang-Yuan'an, Chaohu, Guiyang and other areas in South China (Benton et al. 2013; Yuan et al. 2022). Even some close relatives of dinosaurs appeared in the Early Triassic. Therefore, some studies have suggested that dinosaurs may have appeared in the Early Triassic, but the fossil record has not yet been found (Benton and Wu 2022). Most of these top predators are not Paleozoic survivors but newly appeared and further radiated in the Early Triassic. Their morphological structure, functional, and predatory patterns are significantly different from those of top predators that lived in the Paleozoic ocean. Some exceptional preserved fossil lagerstätten have been found in the Early Triassic, such as the Paris Biota in Idaho, USA, which contained at least 7 phyla and 20 metazoan orders. The Paris Biota combined Cambrian-Ordovician leptomitid sponges and Jurassic gladius-bearing coleoid cephalopods, which were previously not found in the Triassic strata. The biota contained diverse and trophically multileveled fauna from primary producers up to top predators, indicating a complex marine ecosystem in the Early Triassic (Brayard et al. 2017). More recently, another well-preserved fossil lagerstätten, the Guiyang Biota, was reported from the Daye Formation of the Guiyang, South China, which was only about one million years after the P-Tr mass extinction (Dai et al. 2023). The Guiyang Biota is characterized by fish fauna and malacostracans, representing a complex marine ecosystem (Dai et al. 2023). Hence, the Paris Biota and Guiyang Biota highlight the evolutionary significance of some Early Triassic fauna in driving the dominance of Modern marine evolutionary fauna (Brayard et al. 2017; Dai et al. 2023).

The emergence of these trophically complex ecosystems in the Early Triassic reflects the rapid recovery of top consumers, which seems to be somewhat contradictory to the traditional bottom-up reconstruction model of the trophic pyramid. Thus, a large number of studies of the patterns and rates of recovery of different clades or on the reconstruction of the whole ecosystem have shown that the recovery after the mass extinction has distinct heterogeneity in time and space. The complex process may not be simply summarized by a simple model. The mechanism and driving force underlying the rapid recovery of these complex ecosystems in the Early Triassic are still unclear.

There are many ways to study biological "recovery," and the simplest way is to count biodiversity. To see if the species diversity in the recovery period has returned to the level before extinction. As mentioned above, some groups or some special communities can quickly return to their pre-extinction level (Brayard et al. 2009; Stanley 2009; Payne et al. 2011; Song et al. 2013). However, the structure of these rapidly recovered communities is very unbalanced and is dominated by "disaster taxa," forming a community that exists for a short time in a specific environment

(Fraiser and Bottjer 2007; Petsios and Bottjer 2016). Such cases make the Early Triassic fauna very special (Chen and Benton 2012). Therefore, statistical biodiversity alone is far from enough to reveal the true pattern of ecosystem recovery. Many researchers are using ecological methods to study recovery.

1.2 Introduction of Trace Fossils and Current Research Status

In modern oceans, apart from those organisms with skeletons and shells, soft-bodied organisms are also abundant. They constitute the complete energy pyramid of the ecosystem, together with skeletal and shelly organisms. Similarly, there were also a large number of soft-bodied organisms in ancient oceans. These organisms were usually difficult to preserve as body fossils, except in lagerstätten. Most studies on biotic recovery have focused on organisms with skeletons and shells. This has resulted in an obvious bias in the studies of global biodiversity and marine ecosystems in the geological past. Studies have shown that soft-bodied organisms, such as worms, annelids, and arthropods, usually live on the seafloor or within sediments, and they can prominently transform sediments, leaving various traces on the seafloor or in the sediments, such as crawling tracks, burrows, dwelling and so on. When these traces were preserved in rocks as a kind of fossils, they were trace fossils (Buatois and Mángano 2011; Mángano and Buatois 2014; Luo et al. 2021). Therefore, trace fossils are an important supplement to the study of marine ecosystems and paleoecology (Bromley 1996; Yang et al. 2004; Qi et al. 2007; Buatois and Mángano 2011, 2013; Knaust and Bromley 2012; Mángano and Buatois 2014; Zhou et al. 2014; Hu et al. 2015; Knaust 2017; Chen et al. 2018; Xiao et al. 2019; Buatois et al. 2020; Feng et al. 2022). Trace fossils reflect the interactions between organisms and sediments or seawater. The sediment-modification process by trace-making organisms may have an impact on the biogeochemical cycle of nutrient elements in the water column (Lohrer et al. 2004; Boyle et al. 2014; Kristensen et al. 2018; Tarhan et al. 2021).

Ecosystem engineering refers to organisms that change the resources of other species directly or indirectly by modifying their habitats in the ecosystem (Jones et al. 1997; Meysman et al. 2006; Hastings et al. 2007). A large number of modern ecological studies have shown that ecosystem engineering has significant positive feedback on the diversity of the whole ecosystem (McIlroy and Logan 1999; Lohrer et al. 2004; Meysman et al. 2006; Erwin 2008). For example, the biodiversity in coral reef ecosystems is significantly higher than that of non-coral reef ecosystems, according to observation and study in the modern ocean. Coral reef is a typical ecosystem engineering that can provide a habitable environment for other organisms. Ecosystem engineering emphasizes that organisms influence other organisms by transforming the physical environment. The time effect of this process may be longer than the life span of the organism itself, such as in the coral reef case mentioned above. The coral reefs may have vanished, but the ecosystem engineering effects still exist.

Marine organisms such as worms, annelids, and arthropods that live on the seafloor and in the sediments can disturb and modify the sediment and substrate. The process is bioturbation. Bioturbation is also an important ecosystem engineering (Kristensen et al. 2012). Bioturbation has a significant impact on the cycling of some nutrient elements such as C, N, and P in seawater and sediments. This process enhances the intensity of geochemical reaction just below the water-sediment interface, which has a positive feedback effect on microbial biomass, expanding the habitable ecospace and ultimately facilitating diversification of the whole ecosystem (McIlroy and Logan 1999; Lohrer et al. 2004; Meysman et al. 2006; Erwin 2008; Boyle et al. 2014; Tarhan et al. 2015, 2021; Gougeon et al. 2018). However, at present, the research on the effect of bioturbation on biogeochemical cycles and its feedback mechanism on the whole ecosystem is poorly understood, especially from an ecological perspective (Cribb and Bottjer 2020; Feng et al. 2022).

There are two main ways of bioturbation: biomixing and bioirrigation. Biomixing mainly refers to the disturbance and transformation of sediment particles, including biodiffusors, upward conveyors, downward conveyors, and regenerators (Kristensen et al. 2012; Minter et al. 2017). The transformation of sediment particles by biomixing can affect the vertical and lateral distribution of organic matter in sediments, as well as the sorting and size of sediment particles. Biomixing is mainly the biological bulldozing of sediment particles. These sediment particles are bulldozed back and forth by trace-making organisms near the oxidation zone-oxygen depletion zone. A large number of reducing substances such as iron sulfides are oxidized, which consumes much dissolved oxygen in the water column and pore water, resulting in an anoxic condition. Bioirrigation is related to burrow ventilation, which means the exchange between sediment pore water and seawater caused by the activities of trace-making organisms, accompanied by a large amount of solute migration. Bioirrigation enhances the exchange between the water column and pore water, resulting in the oxygenation of the sediment (van de Velde and Meysman 2016; van de Velde et al. 2018). The bioirrigation effect of trace-making organisms makes the one-dimensional diffusion of solute in sediment pore water change into three-dimensional diffusion and can quickly exchange the water column and the sediment pore water. A large number of soluble nutrients, meanwhile, are exchanged between the burrow and the seawater and between the pore water and the seawater. Biomixing and bioirrigation seem to have opposite impacts on biogeochemical cycling, as biomixing elevates oxygen consumption and bioirrigation tends to oxygenate the sediment. Therefore, these two ways of bioturbation have opposite effects on benthic biodiversity.

Studies have shown that the peak diversity of soft-bodied trace-makers (trace fossils) in the early Cambrian was 20 million years earlier than the body fossil record (Buatois et al. 2020). These early Cambrian trace-makers are thought to be marine ecosystem engineers; their bioturbation behavior changed the environmental conditions of the seafloor, promoted the circulation of oxygen and nutrients between the seawater and the sediment, and ultimately contributed to the Cambrian explosion (Marenco and Bottjer 2008; Mángano and Buatois 2014; Buatois and Mángano 2018; Buatois et al. 2017, 2020). However, other studies based on marine bioturbation in the early Paleozoic have found that bioturbation, mainly biomixing, resulted in ocean anoxia and global warming (van de Velde et al. 2018). Thus, bioturbation has a negative feedback effect on the ecosystem. Therefore, the ecosystem engineering effects of bioturbation need to consider biomixing and biol-irrigation with opposite mechanisms, respectively. There are several works documenting the bioturbation from the perspective of ecosystem engineering (Erwin 2024; Cribb and Darroch 2024; Mángano et al. 2024).

1.3 Summary of P-Tr Ichnological Study

1.3.1 Recovery Model

Body fossils are rare in the stratigraphic record after the mass extinction, and their preservation is very uneven (Fraiser and Bottjer 2005). In contrast, trace fossils are relatively common and well preserved. Therefore, trace fossils are a useful tool in evaluating the recovery of trace-making organisms and marine ecosystems in the aftermath of P-Tr mass extinction (Twitchett et al. 2004). According to several ecological proxies such as burrow size, complexity, penetration depth and bioturbation index, Twitchett (2006) summarized a semi-quantitative method to evaluate the recovery process and stage of the infaunal ecosystem (Fig. 1.1a). In Twitchett's model, biotic recovery is divided into four stages: trace fossils are small during the initial stage, which are usually less than 5 mm in diameter and have low diversity, mainly simple, horizontal burrows such as *Planolites*. The penetration depth of trace fossils is very shallow. In the second stage, the diversity of ichnofossils began to increase, and the number of ichnodiversity may increase to 5. Some vertical burrows such as *Arenicolites* or *Diplocraterion* began to appear. These vertical burrows directly lead to a significant increase in the ecological tiering of trace fossils. In the third stage, trace fossils are further diversified, and the ichnodiversity can reach 10 or even more. Some complex forms, such as *Rhizocorallium* and *Thalassinoides* start to appear. The burrow size of trace fossils increased significantly, but most of them are less than 20 mm in diameter. *Rhizocorallium* and *Thalassinoides* are the iconic trace fossils in the fourth recovery stage, and the diameter of the tube is more than 20 mm. The diversity is always high, and the number of ichnogenera is more than 10, which usually rebounded to the level of pre-extinction. The application of this semi-quantitative recovery model through the analysis of trace fossils can provide an effective assessment for the restoration of marine ecosystems.

Later, Pietsch and Bottjer (2014) revised the four-stage recovery model proposed by Twitchett (2006) and added some benthic ecological proxies to evaluate the recovery process of benthic organisms and ecosystems. Pietsch and Bottjer (2014) applied these proxies in many Lower Triassic sections around the world. They

A

		Stage 1	Stage 2	Stage 3	Stage 4
Body Fossils	Diversity	<5 sp.	5-10 sp.	10+ sp.	high
	Size	small	increasing	increasing	pre-extinction
	Dominance	>75%	45-75%	25-45%	<25%
	Tiering	lowest	infaunal bivalves	high,epifauna appear	highest, coral reefs return
	Fossil composition	disaster taxa			reefs
Trace Fossils	Fossil composition	simple burrows			
	Tiering	shallowest	vertical traces appear	increased *Thalassinoides*	deep, complex
	Dominance	*Planolites*	45-75%	25-45%	<25%
	Diameter	<5 mm	5-20 mm	<20 mm	pre-extinction
	Diversity	1-2 isp.	2-5 isp.	5+ isp.	high

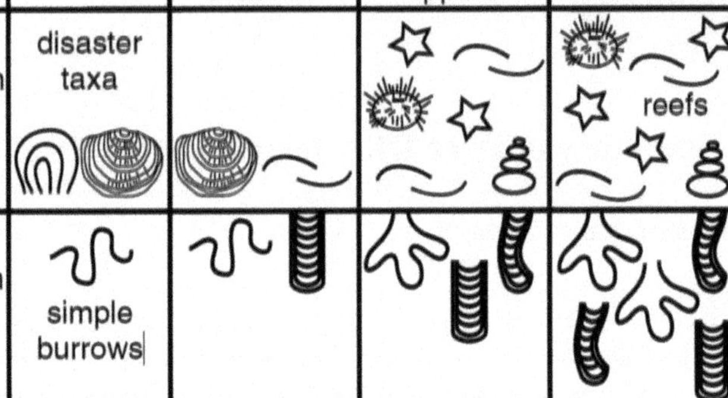

B

EARLY (EARLY TRIASSIC) STAGES OF RECOVERY			LATE (MIDDLE TRIASSIC) STAGES OF RECOVERY	
Stage 1	Stage 2	Stage 3	Stage 4	Stage 5
- very low diversity (a few ichnotaxa) - shallow tiering - small diameters - mainly *Planolites*	- low diversity (<10 ichnotaxa) - shallow tiering - small diameters - vertical domichnia, e.g., *Arenicolites*, *Diplocraterion*, *Skolithos*	- moderate to high diversity (>10 ichnotaxa) - shallow to deep tiering - high variety of ethologies - simple and complex horizontal and vertical forms - common monoichnospecific assemblages - *Rhizocorallium*, *Thalassinoides*		
		- small (<10 mm) and minor medium (10–30 mm) diameters	- common large *Rhizocorallium* - rare to abundant medium-sized *Thalassinoides*	- common large *Thalassinoides* reaching Permian dimensions - rare to common large *Rhizocorallium*

Twitchet & Wignall (1996), Twitchet (1999), Zhao et al. (2015)
Luo et al. (2016), Feng et al. (2018)
Beatty et al. (2008), Fraiser & Bottjer (2009), Knaust (2010), Feng et al. (2017a)
Pruss & Bottjer (2004), Zonneveld et al. (2010), Hofman et al. (2011), Chen et al. (2012)
Mørk & Bromley (2008), Knaust & Costamagna (2012), Feng et al. (2017b)
Stachacz and Matysik (2020)

Fig. 1.1 Stages of the ecological recovery model. (**a**) The four stages are determined by epifaunal diversity, size, dominance ratios, tiering, and fossil composition as well as infaunal diversity, burrow size, burrow depth, and trace fossil composition. (Modified from Twitchett (2006) and Pietsch and Bottjer (2014)). (**b**) Five-stage recovery of burrowing organisms after the P/Tr mass extinction, Stages 1 to 3 correspond to Twitchett's (2006) stages 1 to 3, whereas stages 4 and 5 represent his stage 4. (Modified from Twitchett (2006) and Stachacz and Matysik (2020)) revealed that an oxygenated environment was a key factor in shaping the recovery of benthic invertebrates after the P-Tr mass extinction (Fig. 1.1a).

The complexity of trace fossils may indicate the complex behaviors of trace makers (Miller 2003). The reoccurrence of several complex ichnofauna such as *Rhizocorallium* and *Thalassinoides* in the Early Triassic has been regarded as the final recovery stage of the marine ecosystem. More recently, according to the large number of specimens of the ichnogenus *Rhizocorallium* collected from ten Lower-Middle Triassic sections from South China, Feng et al. (2018) found that *Rhizocorallium* possessed a relatively smaller size, fewer ichnospecies, and limited environmental and geographical distributions during the Griesbachian-Smithian. *Rhizocorallium* began to rise in the number of ichnospecies and morphological complexity, as well as environmental expansion in the Spathian (late Early Triassic), while most specimens remained relatively small. Thus, the occurrence *Rhizocorallium* in the Spathian may indicate stages 2–3 of biotic recovery. Instead, *Rhizocorallium* not only showed the highest ichnospecies richness but also exhibited the greatest morphological complexity and largest burrow sizes during middle-late Anisian, which may indicate the final recovery stage after the P-Tr crisis (Fig. 1.2).

In addition, Stachacz and Matysik (2020) documented abundant trace fossils of the Middle Triassic (Anisian) from the Germanic Basin of southern Poland and proposed a five-stage recovery model. The last stage of the Four-stage model was divided into two phases. The early phase of recovery occurred in the Early Triassic and ended with the appearance of small *Rhizocorallium* and *Thalassinoides*, and the later phase took place in the Middle Triassic and was characterized by large burrows of *Rhizocorallium* and *Thalassinoides*. Large *Rhizocorallium* and medium-sized *Thalassinoides* occurred in the former phase of the last stage, while the widespread occurrence of large *Thalassinoides* with similar size of pre-extinction Permian counterparts may indicate the final recovery of trace-maker communities (Fig. 1.1b).

1.3.2 Review of Global Trace Fossil Research Status from the Late Permian to the Early Triassic in Marine Environments

The Shangsi section in Jiangue County, Sichuan Province, is one of the most well-studied sections that was regarded as the candidate Global Stratotype Section and Point (GSSP) of the P-Tr Boundary (PTB) (Lai et al. 1996). Similar to the distribution of other benthos, ichnodiversity first increased in the upper part of the Changhsingian Talung Formation and then rapidly decreased near the P/Tr boundary (Wignall et al. 1995). *Zoophycos* with backfill structures and *Thalassinoides* with three-dimensional network structures are the main forms in the upper part of the Talung Formation, and the cross-cutting relationship between them indicates a deep ecological tiering level (Zhang et al. 2018, 2020). The abundance and diversity

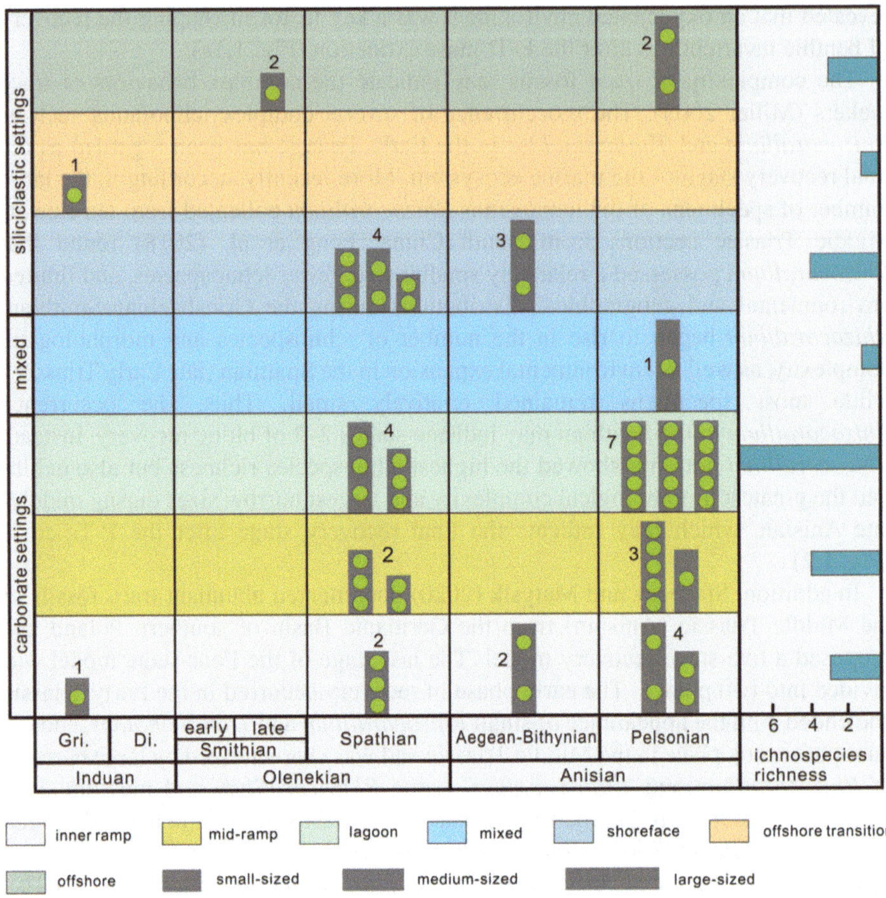

Fig. 1.2 Variations in ichnospecies richness and burrow sizes of key ichnotaxa *Rhizocorallium* ichnospecies/varieties among various environmental settings in the Early–Middle Triassic. Each solid circle in the vertical bar represents an occurrence of *Rhizocorallium*. The numbers next to the bar refer to the number of ichnospecies (or varieties) in each environmental setting of a time bin. Mixed—mixed carbonate/siliciclastic environment. The vertical bars in the plot represent increments of burrow size. The horizontal bar represents the ichnospecies richness of each habitat type. (Modified from Feng et al. (2018))

of the ichnoassemblages significantly decreased; only small-sized *Thalassinoides* can be identified in the uppermost of the Talung Formation (Wignall et al. 1995).

During the global regression in the latest Guadalupian, continuous sedimentation of the G-L boundary was hiatus in Gondwana, North America, northern Eurasia, and much of Paleo-Tethys, but was well preserved in the Laibin area of Guangxi (Penglaitan and Tieqiao sections), which has attracted international attention (Jin et al. 2006; Shen et al. 2013). The Late Permian strata are developed in the northern Penglaitan section of this area, yielding a large number of ammonoids, brachiopods and conodonts. Abundant trace fossils were also preserved in the strata (Ehiro and

Shen 2010). Eighteen ichnogenera and 29 ichnospecies have been documented from the northern Penglaitan section (Ding et al. 2016).

The Dongpan section is located about 80 km southwest of Nanning City, Guangxi Province, and was also deposited in a deep basin during the Late Permian. Trace fossils mainly occurred on two horizons and were more abundant and diverse on the lower horizon. Only 4 ichnospecies occurred in the upper horizon of the ichnofossil-bearing strata. The ichnofauna in the Dongpan section included some facies-crossing ichnotaxa such as *Chondrites*, *Nereites*, *Zoophycos* and *Planolites* (Zhang et al. 2018).

Twelve ichnogenera were found below Bed 27, the P/Tr boundary horizon of the Global Stratotype Section and Point (GSSP) of the Meishan section, Changxing County, Zhejiang Province (Chen et al. 2015). Major turnover in ichnodiversity occurred across Bed 24 of the extinction horizon, indicating the demise of most trace makers in such a short time. Ichnodiversity continued to decrease in the above two horizons, which was very similar to other benthic fauna recorded by body fossils. The ichnotaxa in Bed 27, together with an irregular bedding surface, indicated a firmground condition. All ichnoecological proxies, such as ichnodiversity, complexity, burrow size, tiering, and bioturbation of the strata, confirmed a two-stage model of the P-Tr extinction (Chen et al. 2015).

There are few reports on trace fossils in the Late Permian from other places outside China. Although Twitchett (2006) proposed a four-stage recovery model (Twitchett 1999; Twitchett and Barras 2004), the pre-extinction Permian ichnofauna is largely unknown (Twitchett and Barras 2004). One of the few cases is from Svalbard, Norway; trace fossils are very abundant in the Changhsingian to Early Triassic (Uchman et al. 2016). Twelve ichnogenera were recognized in the Late Permian strata, comprising *Zoophycos*, *Diplocraterion*, *Nereites*, *Phycosiphon* and *Chondrites* (Uchman et al. 2016). *Zoophycos* and *Chondrites* disappeared at the P/Tr boundary (Wignall et al. 1998; Uchman et al. 2016). In addition, Knaust (2010) identified 22 ichnogenera from the Late Permian to Early Triassic in drilling cores of the Persian Gulf of Iran. The Late Permian sequence contains 21 ichnogenera, including *Asterosoma*, *Cylindrichnus*, *Parahaentzschelinia*, *Phycosiphon*, *Rosselia*, *Siphonichnus*, *Teichichnus*, *Thalassinoides*, and *Zoophycos*. The bioturbation degree of fossil-bearing strata was quite different; some were intensely bioturbated, while others were very weakly bioturbated (Knaust 2010).

Early Triassic trace fossils are quite common and have been regarded as one of the most effective materials to assess the recovery of trace-making organisms after the mass extinction. It has been widely reported in many parts of the world (Twitchett and Wignall 1996; Wignall et al. 1998; Twitchett 1999; Pruss and Bottjer 2004; Beatty et al. 2008; Fraiser and Bottjer 2009; Hu et al. 2009, 2015; Knaust 2010; Zonneveld et al. 2010; Hofmann et al. 2011, 2015; Mata and Bottjer 2011; Chen et al. 2011, 2012, 2015; Lovelace and Lovelace 2012; Foster et al. 2015, 2017; Shi et al. 2015, 2019; Zhao et al. 2015; Baucon and De Carvalho 2016; Bordy and Krummeck 2016; Luo et al. 2016, 2020, 2021; Uchman et al. 2016; Feng et al. 2017, 2018, 2019a, b, 2022; Guo et al. 2019; Zhang et al. 2019, 2020). In most cases, ichnofauna did not undergo significant diversification until the late Early Triassic

(Olenekian Stage), particularly during the Spathian substage. Early Griesbachian trace fossils were relatively rare and were predominantly characterized by simple, horizontal *Planolites* (Figs. 1.3 and 1.4). In contrast, Spathian ichnoassemblages typically exhibited greater complexity, often comprising more than ten ichnospecies, which represented a conspicuous increase in ichnodiversity compared to the more depauperate ichnoassemblages of the Induan Stage (Mata and Bottjer 2011; Chen et al. 2011; Foster et al. 2015; Shi et al. 2015; Zhao et al. 2015; Luo et al. 2016).

However, in rare instances, the Griesbachian (or Induan) ichnofauna exhibited significant diversification, reaching levels of ichnotaxic richness comparable to—or even exceeding—those observed in the Smithian and Spathian (Fig. 1.3). These

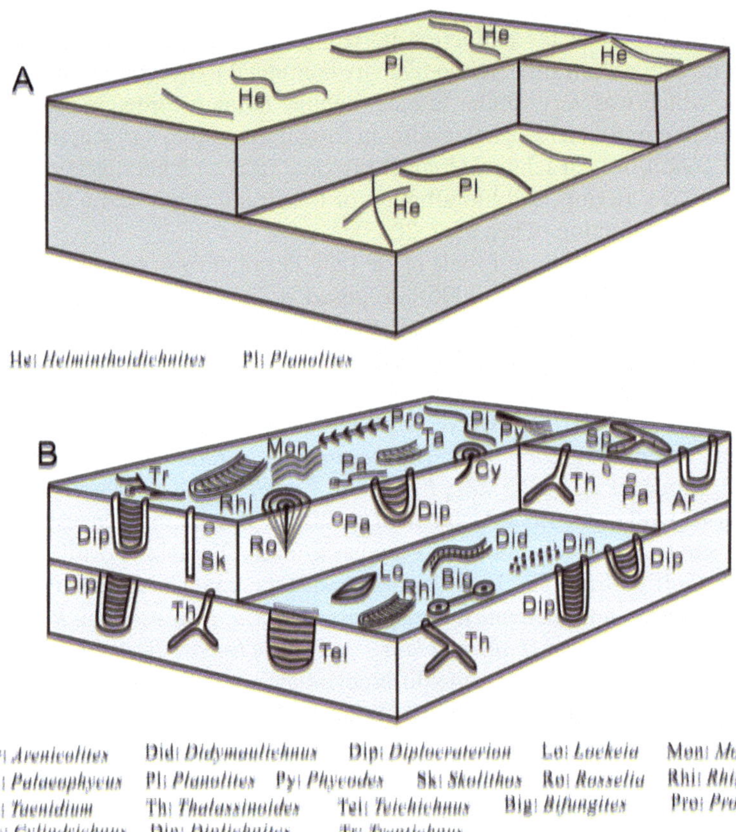

Fig. 1.3 Sketch reconstructions illustrating two distinct types of ichnoassemblages from the Early Triassic. (**a**) Simple, horizontal burrows. This type has been documented in many carbonate systems during the Induan. (**b**) Diverse ichnotaxa, characterized by complex burrow forms such as *Thalassinoides*, *Rhizocorallium*, and *Diplocraterion*. This type has been reported from most Olenekian localities, as well as several Induan localities: Svalbard, western Spitsbergen (Wignall et al. 1998; Foster et al. 2017); western Canada (Beatty et al. 2008; Zonneveld et al. 2010); Persian Gulf, Iran (Knaust 2010), Dolomites, Italy (Twitchett and Barras 2004; Hofmann et al. 2011). (Modified from Feng et al. (2019a))

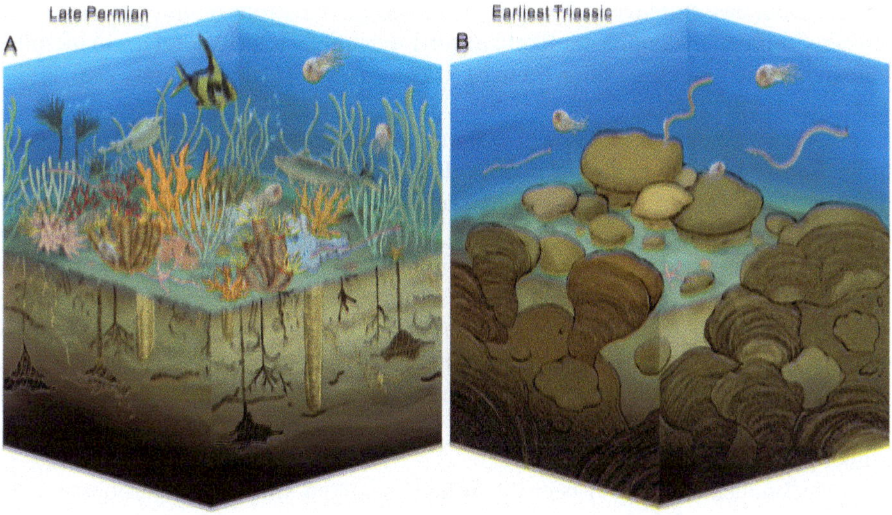

Fig. 1.4 Reconstruction of marine ecosystems before and after the P-Tr mass extinction in China. (**a**) Pre-extinction marine ecosystem in the latest Permian, characterized by diverse nekton, benthic animals, and burrowing organisms. (**b**) Marine ecosystem in the earliest Griesbachian (early Induan) following the mass extinction, featuring limited nekton, simple horizontal burrows, and the proliferation of microbialites. (Modified from Feng et al. (2022))

unusually diverse ichnocoenoses have been documented in several regions, including Svalbard in western Spitsbergen (Wignall et al. 1998), Alberta and British Columbia in western Canada (Beatty et al. 2008; Zonneveld et al. 2010), the Persian Gulf in Iran (Knaust 2010), and the Dolomites in Italy (Hofmann et al. 2011).

In the Boreal sections of Spitsbergen, 12 ichnotaxa have been identified within shoreface sandstone beds of the Vardebukta Formation of the Dienerian age. These include complex burrow forms such as *Thalassinoides* and *Rhizocorallium*, while *Zoophycus* and *Chondrites* are notably absent, which suggests that the recovery in Spitsbergen occurred more rapidly than in the Paleotethys region. Furthermore, these successions also yield a notable diversity of fauna, including bryozoans, calcareous algae, and deep infaunal bivalves, indicating that marine ecosystem recovery may have begun earlier at higher paleolatitudes (Wignall et al. 1998). More recently, the earliest Induan marine assemblage comprising exceptionally diverse invertebrates, alongside a diverse ichnoassemblage, has been documented in this region (Foster et al. 2017), further underscoring the significance of high-latitude ecosystems in the aftermath of the extinction event (Fig. 1.3b).

Diverse and abundant trace fossils comprising over 30 ichnogenera have been documented from Griesbachian successions of the Montney Formation from northwestern Alberta and northeastern British Columbia in western Canada. These marine invertebrate trace fossils were predominantly characterized by arthropod-produced trace fossils, such as *Thalassinoides* and *Rhizocorallium*, which were commonly found in intensely bioturbated (ii 5) sandstone beds. These "anomalously diverse" ichnoassemblages were thought to have been strongly influenced by specific environmental conditions, likely occurring within a lower shoreface to offshore transition zone—a setting often referred to as a "habitable zone" (Beatty et al. 2008; Zonneveld et al. 2010).

Following a detailed analysis of carbonate well cores from the Khuff Formation in the Persian Gulf, Iran, Knaust (2010) identified 13 ichnogenera of the Induan age. Among these, *Diplocraterion* appeared approximately 55 cm above the Permian-Triassic (P-T) boundary, marking the first occurrence of a Triassic trace fossil in the cores. The *Diplocraterion* specimens exhibited a limb diameter of 1 mm, a spreiten width of 5–7 mm, a spreiten length of 15 mm, and a penetration depth of 4 cm. The presence of such a diverse ichnoassemblage in a low-latitude carbonate platform in the Persian Gulf challenged the notion that refugees were confined solely to high latitudes. Instead, it suggested that equatorial settings also served as refuges and that biotic recovery in these regions occurred at a pace comparable to that observed in higher paleolatitudes (Knaust 2010).

Ten ichnotaxa have also been identified in mixed siliciclastic-carbonate shelf sediments from the Werfen Formation in the Dolomites of northern Italy, which was late Griesbachian in age. The ichnoassemblage in this region was distinguished by the presence of large, complex forms such as *Thalassinoides*, *Spongeliomorpha*, and *Rhizocorallium*. These deep-tier ichnotaxa suggested that a significant global recovery of benthic ecosystems may have begun shortly after the P-Tr mass extinction. Furthermore, the presence of these trace fossils indicated that the early recovery of trace-making organisms was not restricted by latitudes (Hofmann et al. 2011).

Although Lower Triassic successions are well-exposed in South China (Benton et al. 2013), ichnofossil-rich intervals are primarily documented from the late Early Triassic and are found in only five key localities: the Yashan (Chen et al. 2011) and Susong (Luo et al. 2016) sections in the Lower Yangtze region, the Daxiakou section (Zhao et al. 2015) in the Middle Yangtze region, and the Gaimao (Shi et al. 2015) and Tianshengqiao (Luo et al. 2016) sections in the Upper Yangtze region. Early Smithian (Olenekian) trace fossils were extremely rare in the Yangtze region, with only *Palaeophycus*, *Planolites*, and *Treptichnus* documented across the entire area. However, by the late Smithian, ichnotaxa had significantly diversified in most regions of South China, with the exception of Daxiakou, where only *Planolites* and *Didymaulichnus* were present (Zhao et al. 2015). At the Yashan section, eight ichnogenera had been identified (Chen et al. 2011). The Susong section yielded a more diverse assemblage of 11 ichnogenera (Luo et al. 2016). Similarly, the Gaimao section contained nine ichnogenera (Shi et al. 2015). In the upper Smithian strata of the Tianshengqiao section, seven ichnogenera were present (Luo et al. 2016).

Trace fossils were relatively scarce during the late Early Triassic in the Dolomites region of northern Italy, with occurrences primarily limited to a few bedding surfaces. The identified ichnogenera included *Planolites*, *Palaeophycus*, *Diplocraterion*, *Cochlichnus*, and *Asteriacites* (Twitchett and Wignall 1996; Twitchett 1999). In northern Hungary, Smithian strata yielded only *Arenicolites* (Foster et al. 2015). In the Western Carpathians of Slovakia, Lower Triassic successions contain *Diplocraterion*, *Skolithos*, and *Arenicolites* (Šimo and Olšavský 2007). In contrast, trace fossils from the Kockatea Shale Formation in the Perth Basin, Western Australia, were exceptionally abundant and included 16 ichnogenera (Chen et al. 2012). Similarly, the Sinbad Limestone Member of the Moenkopi Formation contained eight ichnogenera. However, in the Union Wash Formation of the western United States, only *Planolites* had been documented (Fraiser and Bottjer 2009; Mata and Woods 2008).

Luo et al. (2020) evaluated the applicability and reliability of several ichnological proxies commonly used to assess the recovery of infaunal ecosystems in marine environmental settings following the P-Tr mass extinction. These proxies included ichnodiversity, ichnofabric indices, bedding plane bioturbation index, burrow size, trace complexity, tiering level, key ichnotaxa, and ichnodisparity. Their study suggested that while the bedding plane bioturbation index required further exploration as a recovery indicator, ichnodisparity may show significant potential for assessing biotic recovery in future research. Key ichnotaxa, such as *Rhizocorallium* and *Thalassinoides*—produced only by malacostracan crustaceans can be identified as reliable indicators of ecosystem recovery, as these trace makers were more sensitive to aquatic temperature changes and oxygen concentration.

An analysis of the tendency in global ichnodiversity and ichnodisparity from the late Permian to the Middle Triassic revealed that late Permian levels of ichnodiversity and ichnodisparity persisted into the Griesbachian, followed by a sharp decline in the Dienerian. This trend contrasted with the body-fossil record. The Griesbachian was characterized by an unusual dominance of shallow-tier trace fossils. The discrepancy between the body-fossil and trace-fossil records was interpreted as a consequence of the flourishing of microbial matgrounds during the Griesbachian, which facilitated the preservation of surface, semi-infaunal, and shallow-tier ichnofossils (Luo et al. 2021).

Bioturbation is a critical ecosystem engineering behavior in marine environments, and trace fossils can provide valuable insights into the impact of the P-Tr mass extinction on bioturbating ecosystem engineers. A dataset was compiled from previously published literature to analyze burrowing ecosystem engineering behaviors across the P-Tr boundary (Cribb and Bottjer 2020). Their findings revealed no significant loss in bioturbation-related ecosystem engineering behaviors following the mass extinction. Instead, complex ecosystem engineering behaviors persisted into the Early Triassic. The continued presence of deep-tier, bioirrigated three-dimensional network burrows suggested that benthic biogeochemical cycling may have been maintained at pre-extinction levels in certain local environments. This stability in biogeochemical processes likely supported ecosystem productivity and played a role in promoting biotic recovery during the Early Triassic (Cribb and Bottjer 2020).

Zhang et al. (2021) compiled a global Permian-Triassic deep-water marine ichnological database and reconstructed the behavioral ecology of trace-making animals in deep marine by using parameters of ecospace utilization and ecosystem engineering. They showed that ichnodiversity, ichnodisparity, ecospace utilization, and ecosystem engineering cubes of deep-water marine infauna did not significantly decline across the P-Tr boundary, indicating that the deep marine ecosystem was under relatively stable conditions during the P-Tr mass extinction. Non-specialized deposit feeding, motile, gallery biodiffusion and compression were the most abundant ecosystem engineers. It was noteworthy that gallery biodiffusors with the highest impacts on sediment modification persisted through the transition of Permian-Triassic. Modes of life in the Early tend to be diversified and dominated by the backfilling conveyors in the shallow tiering. Increased bioirrigation further affected the geochemical cycle of the substrate in the Early Triassic and laid the foundation for the further construction and development of deep-tiering ecospace utilization (Zhang et al. 2021).

In a recent study, the behavioral and ecological diversity of Permian-Triassic trace-making animals was quantitatively analyzed across 26 sections in China using four key parameters: ichnodiversity, ichnodisparity, ecospace utilization, and ecosystem engineering. The findings revealed that infaunal ecological structure had become well established by the early Smithian, approximately three million years (Ma) after the P-Tr mass extinction, which was much earlier than the recovery of most skeletonized organisms in the marine. In addition, the biological bulldozing activities of motile, deposit-feeding infauna likely disrupted fragile ecological habitats on the seafloor, destabilized the substrate, and clogged the filtering structures of non-motile, suspension-feeding animals. These factors may have contributed, at least in part, to the so-called "suppressed" recovery of non-motile, suspension-feeding organisms during the Early Triassic. However, some pioneering ecosystem engineers may have played a crucial role in cultivating the seafloor and occupying vacant ecospace, potentially creating positive feedback effects that facilitated the recovery of nektonic communities. Additionally, trace-fossil evidence suggested a flourishing of soft-bodied (non-skeletal) animals in the late Early Triassic. This supported the hypothesis that soft-bodied bioturbating ecosystem engineers may have triggered evolutionary innovations and radiations to some extent (Feng et al. 2022).

References

Barnosky AD, Matzke N, Tomiya S et al (2011) Has the Earth's sixth mass extinction already arrived? Nature 471(7336):51–57

Baucon A, De Carvalho CN (2016) Stars of the aftermath: asteriacites beds from the Lower Triassic of the Carnic Alps (Werfen Formation, Sauris di Sopra), Italy. PALAIOS 31:161–176

Beatty TW, Zonneveld J-P, Henderson CM (2008) Anomalously diverse Early Triassic ichnofossil assemblages in northwest Pangea: a case for a shallow-marine habitable zone. Geology 36:771–774

Benton MJ, Wu FX (2022) Triassic revolution. Front Earth Sci 10:899541

Benton MJ, Zhang QY, Hu SX, Chen ZQ, Wen W, Liu J, Huang JY, Zhou CY, Xie T, Tong JN, Choo B (2013) Exceptional vertebrate biotas from the Triassic of China, and the expansion of marine ecosystems after the Permo–Triassic mass extinction. Earth Sci Rev 123:199–243

Bond DP, Grasby SE (2017) On the causes of mass extinctions. Palaeogeogr Palaeoclimatol Palaeoecol 478:3–29

Bordy EM, Krummeck WD (2016) Enigmatic continental burrows from the Early Triassic transition of the Katberg and Burgersdorp formations in the main Karoo Basin, South Africa. PALAIOS 31:389–403

Boyle RA, Dahl TW, Dale AW, Shields-Zhou GA, Zhu M, Brasier MD, Canfield DE, Lenton TM (2014) Stabilization of the coupled oxygen and phosphorus cycles by the evolution of bioturbation. Nat Geosci 7:671–676

Brayard A, Escarguel G, Bucher H, Monnet C, Brühwiler T, Goudemand N, Galfetti T, Guex J (2009) Good genes and good luck: ammonoid diversity and the end-Permian mass extinction. Science 325:1118–1121

Brayard A, Krumenacker LJ, Botting JP, Jenks JF, Bylund KG, Fara E, Vennin E, Olivier N, Goudemand N, Saucède T, Charbonnier S, Romano C, Doguzhaeva L, Thuy B, Hautmann M, Stephen DA, Thomazo C, Escarguel G (2017) Unexpected early Triassic marine ecosystem and the rise of the modern evolutionary fauna. Sci Adv 3(2):e1602159

Bromley RG (1996) Trace fossils: biology, taphonomy and application, 2nd edn. Chapman and Hall, London

Buatois LA, Mángano MG (2011) Ichnology: organism-substrate interactions in space and time. Cambridge University Press, Cambridge

Buatois LA, Mángano MG (2013) Ichnodiversity and ichnodisparity: significance and caveats. Lethaia 46:281–292

Buatois LA, Mángano MG (2018) The other biodiversity record: innovations in animal- substrate interactions through geologic time. GSA Today 28:4–10

Buatois LA, Wisshak M, Wilson MA, Mángano MG (2017) Categories of architectural designs in trace fossils: a measure of ichnodisparity. Earth Sci Rev 164:102–181

Buatois LA, Mángano MG, Minter NJ, Zhou K, Wisshak M, Wilson MA, Olea RA (2020) Quantifying ecospace utilization and ecosystem engineering during the early Phanerozoic—the role of bioturbation and bioerosion. Sci Adv 6:eabb0618

Ceballos G, Ehrlich PR, Barnosky AD et al (2015) Accelerated modern human–induced species losses: entering the sixth mass extinction. Sci Adv 1(5):e1400253

Chen ZQ, Benton MJ (2012) The timing and pattern of biotic recovery following the end-Permian mass extinction. Nat Geosci 5:375–383

Chen ZQ, Tong JN, Fraiser ML (2011) Trace fossil evidence for restoration of marine ecosystems following the end-Permian mass extinction in the Lower Yangtze region, South China. Palaeogeogr Palaeoclimatol Palaeoecol 299:449–474

Chen ZQ, Fraiser ML, Bolton C (2012) Early Triassic trace fossils from Gondwana interior sea: implication for ecosystem recovery following the end-Permian mass extinction in south high-latitude region. Gondwana Res 22:238–255

Chen ZQ, Yang H, Luo M, Benton MJ, Kaiho K, Zhao LS, Huang YG, Zhang KX, Fang YH, Jiang HS, Qiu H, Li Y, Tu CY, Shi L, Zhang L, Feng XQ, Chen L (2015) Complete biotic and sedimentary records of the Permian–Triassic transition from Meishan section, South China: ecologically assessing mass extinction and its aftermath. Earth Sci Rev 149:67–107

Chen Z, Chen X, Zhou CM, Yuan XL, Xiao SH (2018) Late Ediacaran trackways produced by bilaterian animals with paired appendages. Sci Adv 4:eaao6691

Cribb AT, Bottjer DJ (2020) Complex marine bioturbation ecosystem engineering behaviors persisted in the wake of the end-Permian mass extinction. Sci Rep 10:203

Cribb AT, Darroch SAF (2024) How to engineer a habitable planet: the rise of marine ecosystem engineers through the Phanerozoic. Palaeontology 67:e12726

Dai X, Davies JH, Yuan Z, Brayard A, Ovtcharova M, Xu GH, Liu XK, Smith CPA, Schweitzer CE, Li MT, Perrot MG, Jiang SY, Miao LY, Cao YR, Yan J, Bai RY, Wang FY, Guo W, Song HY, Song HJ (2023) A Mesozoic fossil lagerstätte from 250.8 million years ago shows a modern-type marine ecosystem. Science 379(6632):567–572

Ding Y, Cao CQ, Zheng QF (2016) Lopingian (Upper Permian) trace fossils from the northern Penglaitan Section, Laibin, Guangxi, South China and their environmental implications. Palaeoworld 25(3):377–387

Ehiro M, Shen SZ (2010) Ammonoid succession across the Wuchiapingian/Changhsingian boundary of the northern Penglaitan Section in the Laibin area, Guangxi, South China. Geol J 45(2–3):162–169

Erwin DH (2008) Macroevolution of ecosystem engineering, niche construction and diversity. Trends Ecol Evol 23:304–310

Erwin DH (2024) Macroevolutionary dynamics of ecosystem-engineering and niche construction. Palaeontology 67:e12718

Erwin DH, Bowring SA, Yugan J (2002) End-Permian mass extinctions: a review. In: Koeberl C, MacLeod KG (eds) Catastrophic events and mass extinctions: impacts and beyond, Geological Society of America special papers. Geological Society of America, Boulder, pp 363–383

Feng XQ, Chen ZQ, Woods AD, Fang YH (2017) A Smithian (Early Triassic) ichnoassemblage from Lichuan, Hubei Province, South China: implications for biotic recovery after the latest Permian mass extinction. Palaeogeogr Palaeoclimatol Palaeoecol 486:123–141

Feng XQ, Chen ZQ, Bottjer DJ, Fraiser ML, Xu YL, Luo M (2018) Additional records of ichnogenus *Rhizocorallium* from the Lower and Middle Triassic, South China: implications for biotic recovery after the end-Permian mass extinction. Geol Soc Am Bull 130:1197–1215

Feng XQ, Chen ZQ, Benton MJ, Wu SQ, Bottjer DJ, Thompson JR (2019a) A diverse trackway-dominated marine ichnoassemblage from the Lower Triassic in the northern Paleotethys: ichnology and implications for biotic recovery. Palaeogeogr Palaeoclimatol Palaeoecol 519:124–140

Feng XQ, Chen ZQ, Bottjer DJ, Wu SQ, Zhao LS, Xu YL, Shi GR, Huang YG, Fang YH, Tu CY (2019b) Unusual shallow marine matground-adapted benthic biofacies from the Lower Triassic of the northern Paleotethys: implications for biotic recovery following the end-Permian mass extinction. Earth Sci Rev 189:194–219

Feng XQ, Chen ZQ, Benton MJ, Su C, Bottjer DJ, Cribb AT, Li Z, Zhao L, Zhu G, Huang Y, Guo Z (2022) Resilience of infaunal ecosystems during the Early Triassic greenhouse Earth. Sci Adv 8:eabo0597

Foster WJ, Danise S, Sedlacek A, Price GD, Hips K, Twitchett RJ (2015) Environmental controls on the post-Permian recovery of benthic, tropical marine ecosystems in western Palaeotethys (Aggtelek Karst, Hungary). Palaeogeogr Palaeoclimatol Palaeoecol 192:374–394

Foster WJ, Danise S, Price GD, Twitchett RJ (2017) Subsequent biotic crises delayed marine recovery following the late Permian mass extinction event in northern Italy. PLoS One 12:0172321e

Fraiser ML, Bottjer DJ (2005) Fossil preservation during the aftermath of the end- Permian mass extinction: taphonomic processes and paleoecological signals. In: Over DJ, Morrow JR, Wignall PB (eds) Understanding late devonian and permian–triassic biotic and climatic events: towards and integrated approach. Elsevier BV, pp 229–311

Fraiser ML, Bottjer DJ (2007) Elevated atmospheric CO_2 and the delayed biotic recovery from the end-Permian mass extinction. Palaeogeogr Palaeoclimatol Palaeoecol 252:164–175

Fraiser ML, Bottjer DJ (2009) Opportunistic behavior of invertebrate marine trace-makers during the Early Triassic aftermath of the end-Permian mass extinction. Aust J Earth Sci 56:841–857

Gougeon RC, Mángano MG, Buatois LA, Narbonne GM, Laing BA (2018) Early Cambrian origin of the shelf sediment mixed layer. Nat Commun 9:1909

Guo WW, Tong JN, Tian L, Chu DL, Bottjer DJ, Shu WC, Ji KX (2019) Secular variations of ichnofossils from the terrestrial Late Permian–Middle Triassic succession at the Shichuanhe section in Shaanxi Province, North China. Glob Planet Chang 181:102978

Hastings A, Byers JE, Crooks JA, Cudddington K, Jones CG, Lambrinos JG, Talley TS, Wilson WG (2007) Ecosystem engineering in space and time. Ecol Lett 10:153–164

Hofmann R, Goudemand N, Wasmer M, Bucher H, Hautmann M (2011) New trace fossil evidence for an early recovery signal in the aftermath of the end-Permian mass extinction. Palaeogeogr Palaeoclimatol Palaeoecol 310:216–226

Hofmann R, Buatois LA, MacNaughton RB, Mángano MG (2015) Loss of the sedimentary mixed layer as a result of the end-Permian extinction. Palaeogeogr Palaeoclimatol Palaeoecol 428:1–11

Hu B, Yang WT, Song HB, Wang M, Zhong MY (2009) Trace fossils and ichnofabrics in the Heshanggou formation of lacustrine deposits, Jiyuan area, Henan Province. Acta Sedimentol Sin 27:573–582. (in Chinese with English abstract)

Hu B, Lu XH, Song HB (2015) Lower Triassic ichnofossils and sedimentary environments in the Dengfeng area, western Henan Province. J Stratigr 39:454–465. (in Chinese with English abstract)

Huang YG, Chen ZQ, Roopnarine PD, Benton MJ, Zhao LS, Feng XQ, Li ZH (2023) The stability and collapse of marine ecosystems during the Permian–Triassic mass extinction. Curr Biol 33(6):1059–1070

Jin YG, Shen SZ, Henderson CM, Wang XD, Wang W, Wang Y, Cao CQ, Shang QH (2006) The Global Stratotype Section and Point (GSSP) for the boundary between the Capitanian and Wuchiapingian stage (Permian). Episodes 29(4):253–262

Jones CG, Lawton JH, Shachak M (1997) Positive and negative effects of organisms as physical ecosystem engineers. Ecology 78:1946–1957

Knaust D (2010) The end-Permian mass extinction and its aftermath on an equatorial carbonate platform: insights from ichnology. Terra Nova 22:195–202

Knaust D (2017) Atlas of trace fossils in well core: appearance, taxonomy and interpretation. Springer, Dordrecht

Knaust D, Bromley RG (2012) Trace fossils as indicators of sedimentary environments. Developments in sedimentology, vol 64, Elsevier, Amsterdam

Knoll AH, Bambach RK, Canfield DE, Grotzinger JP (1996) Comparative Earth history and Late Permian mass extinction. Science 273(5274):452–457

Kristensen E, Penha-Lopes G, Delefosse M, Valdemarsen T, Quintana C, Banta G (2012) What is bioturbation? The need for a precise definition for fauna in aquatic sciences. Mar Ecol Prog Ser 446:285–302

Kristensen E, Røy H, Debrabant K, Valdemarsen T (2018) Carbon oxidation and bioirrigation in sediments along a Skagerrak–Kattegat–Belt Sea depth transect. Mar Ecol Prog Ser 604:33–50

Lai XL, Yang FQ, Hallam A, Wignall PB (1996) The Shangsi section, candidate of the global stratotype section and point of the Permian–Triassic boundary. In: Yin HF (ed) The Palaeozoic–Mesozoic boundary, candidates of the global stratotype section and point of the Permian–Triassic boundary. China University of Geoscience Press, Wuhan, pp 113–124

Lohrer AM, Thrush SF, Gibbs MM (2004) Bioturbators enhance ecosystem function through complex biogeochemical interactions. Nature 431:1092–1095

Lovelace DM, Lovelace SD (2012) Paleoenvironments and paleoecology of a Lower Triassic invertebrate and vertebrate ichnoassemblage from the Red Peak Formation (Chugwater Group), Central Wyoming. PALAIOS 27:636–657

Luo M, George AD, Chen ZQ (2016) Sedimentology and ichnology of two Lower Triassic sections in South China: implications for the biotic recovery following the end-Permian mass extinction. Glob Planet Chang 144:198–212

Luo M, Shi GR, Buatois LA, Chen ZQ (2020) Trace fossil as proxy for biotic recovery after the end-Permian: a critical review. Earth Sci Rev 203:103059

Luo M, Shi GR, Buatois LA, Chen ZQ (2021) Infaunal response during the end-Permian mass extinction. Geol Soc Am Bull 133:91–99

Mángano MG, Buatois LA (2014) Decoupling of body-plan diversification and ecological structuring during the Ediacaran–Cambrian transition: evolutionary and geobiological feedbacks. Proc R Soc B 281:e20140038

Mángano MG, Buatois LA, Minter NJ, Gougeon RC (2024) Bioturbators as ecosystem engineers in space and time. Palaeontology 67:e12732

Marenco KN, Bottjer DJ (2008) The importance of Planolites in the Cambrian substrate revolution. Palaeogeogr Palaeoclimatol Palaeoecol 258:189–199

Mata SC, Bottjer DJ (2011) Origin of Lower Triassic microbialites in mixed carbonate–siliciclastic successions: ichnology, applied stratigraphy, and the end-Permian mass extinction. Palaeogeogr Palaeoclimatol Palaeoecol 300:158–178

Mata SA, Woods AD (2008) Sedimentology and paleoecology of the Lower Member of the Lower Triassic (Smithian-Spathian) Union Wash Formation, east-central California. PALAIOS 23:514–524

McIlroy D, Logan GA (1999) The impact of bioturbation on infaunal ecology and evolution during the Proterozoic- Cambrian transition. PALAIOS 14:58–72

Meysman FJR, Middelburg JJ, Heip CHR (2006) Bioturbation: a fresh look at Darwin's last idea. Trends Ecol Evol 21:688–695

Miller W (2003) Paleobiology of complex trace fossils. Palaeogeogr Palaeoclimatol Palaeoecol 192:3–14

Minter NJ, Buatois LA, Mángano MG, Davies NS, Gibling MR, Macnaughton RB, Labandeira CC (2017) Early bursts of diversification defined the faunal colonization of land. Nat Ecol Evol 1:1–10

Motani R, Jiang DY, Chen GB, Tintori A, Rieppel O, Ji C, Huang JD (2015) A basal ichthyosauriform with a short snout from the Lower Triassic of China. Nature 517:485–488

Payne JL, Summers M, Rego BL, Altiner D, Wei J, Yu M, Lehrmann DJ (2011) Early and Middle Triassic trends in diversity, evenness, and size of foraminifers on a carbonate platform in South China: implications for tempo and mode of biotic recovery from the end-Permian mass extinction. Paleobiology 37:409–425

Petsios E, Bottjer DJ (2016) Quantitative analysis of the ecological dominance of benthic disaster taxa in the aftermath of the end-Permian mass extinction. Paleobiology 42:380–393

Pietsch C, Bottjer DJ (2014) The importance of oxygen for the disparate recovery patterns of the benthic macrofauna in the Early Triassic. Earth Sci Rev 137:65–84

Pruss SB, Bottjer DJ (2004) Early Triassic fossils of the western United States and their implications for prolonged environmental stress from the end-Permian mass extinction. PALAIOS 19:551–564

Qi YA, Hu B, Zhang GC, Gong YM (2007) Ichnofabrics and their environmental interpretation from Middle TriassicYoufangzhuang formation, Jiyuan Region, Western Henan Province. Acta Sedimentol Sin 25:372–379. (in Chinese)

Raup DM, Sepkoski JJ (1982) Mass extinctions in the marine fossil record. Science 215:1501–1503

Ravelo AC, Andreasen DH, Lyle M et al (2004) Regional climate shifts caused by gradual global cooling in the Pliocene epoch. Nature 429(6989):263–267

Rong JY, Huang B (2014) Thirty years study on mass extinction. Sci China Earth Sci 44:377–404

Sepkoski JJ Jr (1984) A kinetic model of Phanerozoic taxonomic diversity, III. Post-Paleozoic families and mass extinctions. Paleobiology 10:246–267

Shen SZ, Zhang H (2017) What causes the Big Five mass extinction? Chin Sci Bull 62(11):1119–1135

Shen SZ, Cao CQ, Zhang H, Bowring SA, Henderson CM, Payne JL, Davydov VI, Chen B, Yuan DX, Zhang YC, Wang W, Zheng QF (2013) High-resolution $\delta^{13}C$ carb chemostratigraphy from latest Guadalupian through earliest Triassic in South China and Iran. Earth Planet Sci Lett 375:156–165

Shi G, Woods AD, Yu MY, Wei HY (2015) Two episodes of evolution of trace fossils during the Early Triassic in the Guiyang area, Guizhou Province, South China. Palaeogeogr Palaeoclimatol Palaeoecol 426:275–284

Shi G, Woods AD, Yu MY, Li XW, Wei HY, Qiao D (2019) Lower Triassic limulid trackways (*Kouphichnium*) from the southwestern margin of the Yangtze carbonate platform: palaeoenvironmental and palaeoecological implications. PALAIOS 34:229–243

Šimo V, Olšavský M (2007) *Diplocraterion parallelum* Torell, 1870, and other trace fossils from the Lower Triassic succession of the Drienok Nappe in the Western Carpathians, Slovakia. Bull Geosci 82:163–173

Song HJ, Wignall PB, Tong JN, Yin HF (2013) Two pulses of extinction during the Permian–Triassic crisis. Nat Geosci 6:52–56

Song HJ, Wignall PB, Dunhill AM (2018) Decoupled taxonomic and ecological recoveries from the Permo-Triassic extinction. Sci Adv 4:eaat5091

Stachacz M, Matysik M (2020) Early Middle Triassic (Anisian) trace fossils, ichnofabrics, and substrate types from the southeastern Germanic Basin (Wellenkalk facies) of Upper Silesia, southern Poland: implications for biotic recovery following the Permian/Triassic mass extinction. Glob Planet Chang 194:103290
Stanley SM (2009) Evidence from ammonoids and conodonts for multiple Early Triassic mass extinctions. Proc Natl Acad Sci USA 106:15264–15267
Tarhan LG, Droser ML, Planavsky NJ, Johnston DT (2015) Protracted development of bioturbation through the early Palaeozoic era. Nat Geosci 8:865–869
Tarhan LG, Zhao M, Planavsky NJ (2021) Bioturbation feedbacks on the phosphorus cycle. Earth Planet Sci Lett 566:116961
Twitchett RJ (1999) Palaeoenvironments and faunal recovery after the end-Permian mass extinction. Palaeogeogr Palaeoclimatol Palaeoecol 154:27–37
Twitchett RJ (2006) The palaeoclimatology, palaeoecology and palaeoenvironmental analysis of mass extinction events. Palaeogeogr Palaeoclimatol Palaeoecol 232:190–213
Twitchett RJ, Barras CG (2004) Trace fossils in the aftermath of mass extinction events. In: McIlroy D (ed) Application of ichnology to palaeoenvironmental and stratigraphic analysis, Geological Society London special publication 228. Geological Society, London, pp 395–415
Twitchett RJ, Wignall PB (1996) Trace fossils and the aftermath of the Permo–Triassic mass extinction: evidence from Northern Italy. Palaeogeogr Palaeoclimatol Palaeoecol 124:137–151
Twitchett RJ, Krystyn L, Baud A, Wheeley JR, Richoz S (2004) Rapid marine recovery after the end-Permian mass extinction event in the absence of marine anoxia. Geology 32:805–808
Uchman A, Hanken NM, Nielsen JK, Grundvag SA, Piasecki S (2016) Depositional environment, ichnological features and oxygenation of Permian to earliest Triassic marine sediments in Central Spitsbergen, Svalbard. Polar Res 35:24782
van de Velde S, Meysman FJR (2016) The influence of bioturbation on iron and Sulphur cycling in marine sediments: a model analysis. Aquat Geochem 22:469–504
van de Velde S, Mills BJW, Meysman FJR, Lenton TM, Poulton SW (2018) Early Palaeozoic Ocean anoxia and global warming driven by the evolution of shallow burrowing. Nat Commun 9:2554
Vermeij GJ (1977) The Mesozoic marine revolution: evidence from snails, predators and grazers. Palaeobiology 3:245–258
Wignall PB, Hallam A, Lai X, Yang F (1995) Palaeoenvironmental changes across the Permian/Triassic boundary at Shangsi (N. Sichuan, China). Hist Biol 10:175–189
Wignall PB, Morante R, Newton R (1998) The Permo–Triassic transition in Spitsbergen: δ^{13}Corg chemostratigraphy, Fe and S geochemistry, facies, fauna and trace fossils. Geol Mag 135:47–62
Xiao SH, Chen Z, Zhou CM, Yuan XL (2019) Surfing in and on microbial mats: oxygen-related behavior of a terminal Ediacaran bilaterian animal. Geology 47:1054–1058
Yang SP, Zhang JP, Yang MF, Zhong G, Yi J, Hua S (2004) Trace fossils of China. Science Press, Beijing
Yuan ZW, Xu GH, Dai X, Wang FY, Liu XK, Jia EH, Miao L, Song HJ (2022) A new perleidid neopterygian fish from the Early Triassic (Dienerian, Induan) of South China, with a reassessment of the relationships of Perleidiformes. PeerJ 10:e13448
Zhang LJ, Buatois LA, Mángano MG, Gong YM, Feng QL, Qi YA, Luo M, Zhang X (2018) Uppermost Permian trace fossils along a shelf to slope transect in South China and their implications for oceanic redox evolution and extinction pattern. Glob Planet Chang 167:74–86
Zhang LJ, Buatois LA, Mángano MG, Qi YA, Tai C (2019) Early Triassic trace fossils from South China marginal-marine settings: implications for biotic recovery following the end-Permian mass extinction. Geol Soc Am Bull 131:235–251
Zhang LJ, Zhang X, Buatois LA, Mángano MG, Shi GR, Gong YM, Qi YA (2020) Periodic fluctuations of marine oxygen content during the latest Permian. Glob Planet Chang 195:103326
Zhang LJ, Dang ZY, Li FY, Yang QQ (2021) Quantifying deep-water marine ecospace utilization and ecosystem engineering from late Permian to Middle Triassic (in Chinese). Chin Sci Bull 66:4295–4306

Zhao X, Tong J, Yao H, Niu Z, Luo M, Huang Y, Song H (2015) Early Triassic trace fossils from the Three Gorges area of South China: implications for the recovery of benthic ecosystems following the Permian–Triassic extinction. Palaeogeogr Palaeoclimatol Palaeoecol 429:100–116

Zhou ZC, Yang H, Li G, Zhu YH, Willems H, Luo H, Cai HW, Xu B, Chen JH, Chen DS (2014) Early Triassic trace fossils from the Xiejiacao Section, Guang'An of Sichuan, and their palaeoecological significance. Acta Palaeontol Sin 53:52–69

Zonneveld J-P, Gingras MK, Beatty TW (2010) Diverse ichnofossil assemblages following the P–T mass extinction, Lower Triassic, Alberta and British Columbia, Canada: evidence for shallow marine refugia on the northwestern coast of Pangaea. PALAIOS 25:368–392

Chapter 2
The Stratigraphical Distributions and Paleoenvironmental Interpretations of the Ichnofossil-Bearing Strata

2.1 Brief Introduction of Geological Background of China During the Permian-Triassic

The Upper Permian to Middle Triassic strata are extensively exposed in both North and South China, encompassing a wide range of sedimentary environments. Lower Triassic successions are particularly well developed across the South China Block (Benton et al. 2013), which was situated within the tropical zone of the eastern Palaeotethys Ocean during the Permian-Triassic transition (Fig. 2.1). During the Early and Middle Triassic, the central part of the South China Block was dominated by the expansive Yangtze Platform (Wang 1985), a massive carbonate platform flanked by distal ramp to shelf-basin settings to the north and south. The southern margin, now referred to as the Nanpanjiang Basin, represents a deep-water basin environment (Feng et al. 1997). In platform settings, the earliest Triassic was characterized by the deposition of massive limestones, which transitioned to thin-bedded lime mudstones during the Olenekian and Middle Triassic. In contrast, basin facies from the Griesbachian interval were marked by alternating black shales and calcareous mudstones, while siliciclastic mudstones and turbidites dominated Olenekian deposits (Lehrmann et al. 2015). Intermediate ramp settings, located between shallow-platform and deep-basin environments, typically feature thick successions of thin-bedded argillaceous limestones and calcareous mudstones (Liu and Xu 1994; Yin et al. 1995, 2001; Zhao et al. 2007, 2013; Ma et al. 2009). Within the Nanpanjiang Basin, several isolated Early–Middle Triassic platforms are prominent, including the Great Bank of Guizhou, which was particularly well-preserved and conspicuous in the field (Lehrmann et al. 1998, 2007). The marine successions spanning the uppermost Permian to Middle Triassic are exceptionally complete across these palaeogeographical settings and are continuously exposed, providing an excellent record of this critical interval in Earth's history (Ji et al. 2011; Jiang et al. 2007, 2011, 2014; Hu et al. 2011; Yan et al. 2013, 2015).

X. Feng et al., *Ichnology of the Permian-Triassic Mass Extinction in China*,
https://doi.org/10.1007/978-981-96-6639-3_2

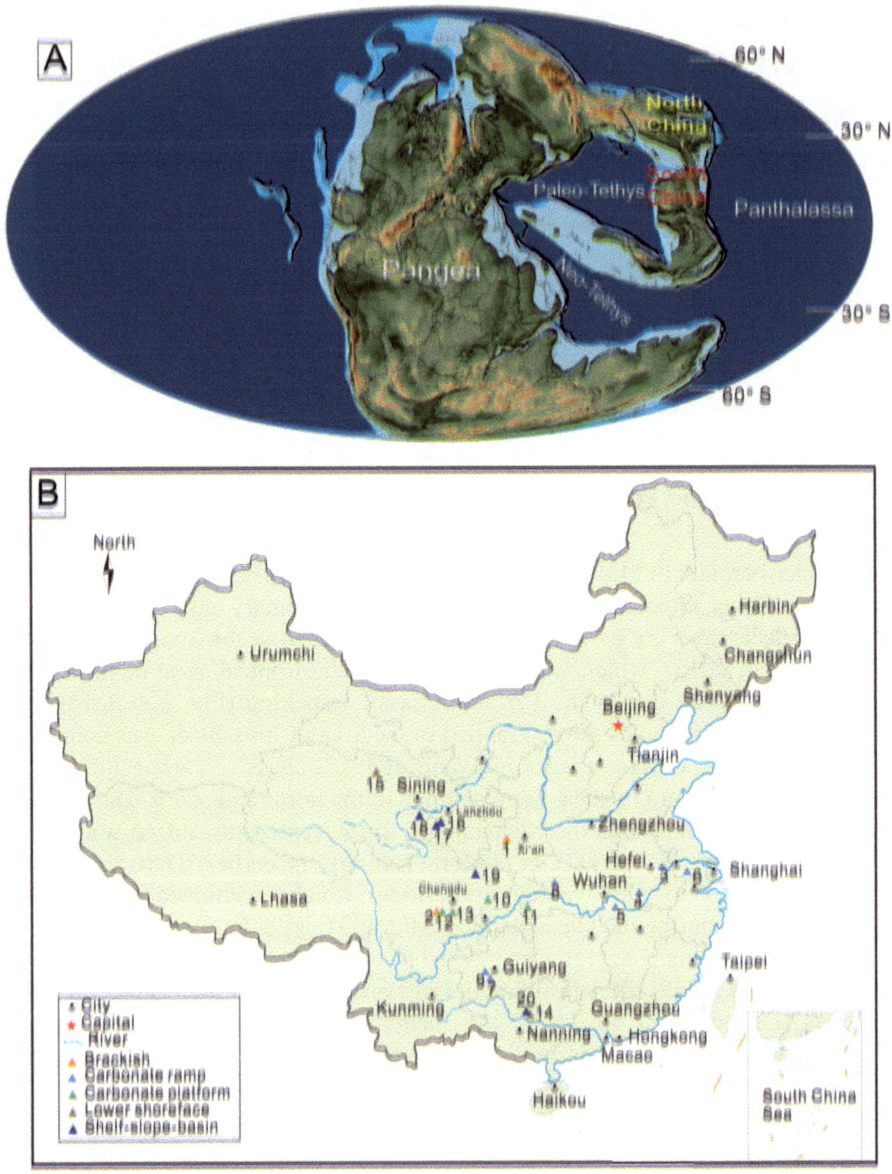

Fig. 2.1 (**a**) Global paleogeographical map during the Permian-Triassic transition showing the locations of North and South China. (**b**) Geographical distribution of the studied sections in China. The number code of the studied sections in B: 1, Houzhougongmiao; 2, Longmendong (lower part); 3, Chaohu; 4, Susong; 5, Daye; 6, Meishan (GSSP); 7, Gaimao; 8, Daxiakou; 9, Dangwu; 10, Xiejiacao; 11, Lichuan; 12, Longmendong (upper part); 13, Shawan; 14, Penglaitan (upper part); 15, Zhihema; 16, Malonggou; 17, Zhayougou; 18, Longwuxia; 19, Shangsi; 20, Penglaitan (lower part). Each colored symbol represents one environmental category. 1–2, brackish; 3–9, carbonate ramp; 10–13, carbonate platform; 14–15, lower shoreface; 16–20, shelf-slope-basin

The extensive Yangtze Platform is commonly divided into three distinct portions—lower, middle, and upper—which correspond geographically to the Lower, Middle, and Upper Yangtze regions, respectively. This subdivision reflects both the spatial distribution and the varying depositional environments across the platform. In the early Triassic, the Lower Yangtze region typically represented shallow-water platform settings in the central and southern parts and extensive ramp facies successions in the northern part, while the Middle Yangtze region encompassed transitional ramp environments. During the Middle-Upper Triassic, both the Lower and Middle Yangtze regions were uplifted due to the suturing of the South China and North China blocks (Wang 1985; Feng et al. 1997), and the Middle Triassic deposits therefore were characterized by shallow marine siliciclastic units (Liu and Xu 1994). In the Upper Yangtze region, the Lower Triassic successions are characterized by widespread ramp facies carbonates in most areas, except for the Nanpanjiang Basin and neighboring areas, in which siliciclastic basin facies and shallow-platform facies muddy limestones are deposited, respectively. During the Middle Triassic, the northern part of the Upper Yangtze region was uplifted to become nonmarine, while the southern part of this region still inherited shallow-platform settings, with some restricted depressions developed in the interiors of the platform (Enos et al. 2006; Benton et al. 2013). This tripartite division provides a useful framework for understanding the stratigraphical and paleoenvironmental variations within the Yangtze Platform during the Early and Middle Triassic.

The North China Craton underwent significant uplift and exposure during the Middle-Late Permian, driven by the amalgamation of the Qaidam, Tarim, and Siberia–Mongolia paleoplates (Wang 1985; Zhai et al. 2000; Zheng et al. 2013; Liu et al. 2015) (Fig. 2.1). As a result, Upper Permian and Lower Triassic successions across much of the North China Craton are predominantly characterized by fluvial and lacustrine facies deposits (Xu et al. 2004; Yu et al. 2005; Chu et al. 2015, 2017; Tu et al. 2016). However, the southern margin of the North China Craton, which faced the northern Palaeotethys Ocean, experienced periodic marine incursions during the earliest Early Triassic. These transgressions led to the formation of embayments in areas such as the northern Weihe and southern Qilian regions (Yin and Lin 1979; Yang et al. 1983; Lai et al. 1995; Liu and He 2000; Xu et al. 2017). These marine-influenced settings contrast with the predominantly terrestrial environments found elsewhere on the craton, highlighting the complex interplay between tectonic activity and sedimentary processes during this period.

In the Ordos Basin of North China, the Permian-Triassic (P-Tr) transition is characterized by terrestrial clastic deposits (Fig. 2.2a). The uppermost Permian is represented by the lower and middle Sunjiagou Formation, while the Lower Triassic comprises the upper Sunjiagou, Liujiagou, and Heshanggou formations (Chu et al. 2015, 2017; Tu et al. 2016). Terrestrial sedimentation during this interval was strongly influenced by significant climatic changes across the P-Tr boundary. The Upper Permian is marked by mixed yellow, greenish, and black siliciclastic deposits, including coal measures, which are overlain by Lower Triassic reddish sandstone and mudstone sequences containing paleosol beds (Xu et al. 2004; Yu et al. 2005; Tu et al. 2016). This shift in sediment color and composition reflects dramatic

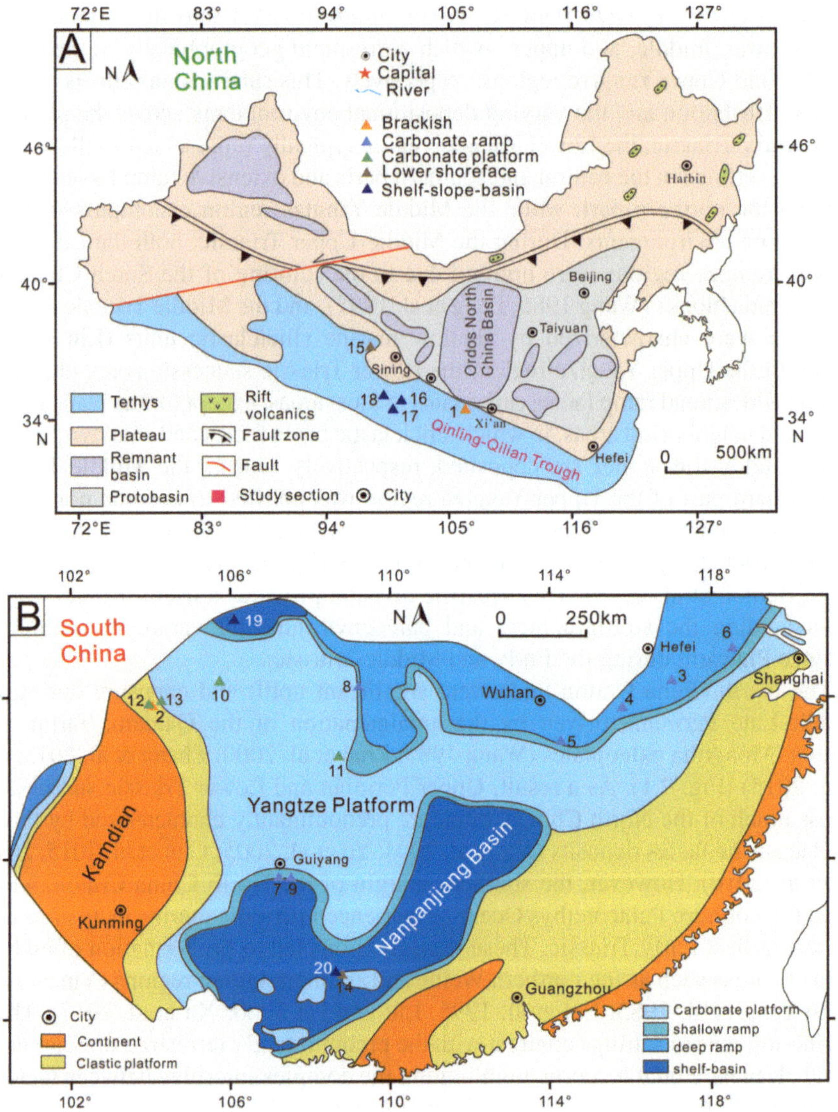

Fig. 2.2 (**a**) Palaeogeographical map of North China showing locations of the studied sections. (**b**) Paleogeographical map of South China showing locations of the studied sections. The number codes are the same as those in Fig. 2.1

environmental and climatic transformations, likely driven by global warming and increased aridity during the Early Triassic. Within the predominantly terrestrial Upper Permian to Lower Triassic successions, marine interbeds have been identified, containing fossils of marine bivalves, brachiopods, and ophiuroids (Yin and Lin 1979; Yang et al. 1983; Liu and He 2000). These marine interbeds are interpreted as evidence of embayments that formed along the southern margin of the North China Craton, influenced by the northern Paleotethys Ocean during the earliest Triassic global transgression (Yin and Lin 1979; Yang et al. 1983). The presence of these marine deposits highlights episodic incursions of seawater into otherwise terrestrial environments, reflecting the dynamic interplay between tectonic activity, sea-level changes, and sedimentary processes during this critical period.

The southern Qilian area was situated on the southern margin of the North China Block during the Early Triassic, occupying a moderate-to-high paleolatitude position along the northern edge of the Palaeotethys Ocean (Ziegler et al. 1998; Zhang et al. 2004, 2016; Pan et al. 2016) (Figs. 2.1 and 2.2a). Geographically, the southern Qilianshan Basin lies on the northeastern edge of the Qinghai-Tibet Plateau. The geological structure of this region is divided into six tectonic uplifts—Tianjun, Datong Mountain, Yangkang, Baixinghada, Shule South Mountain, and Tuolai South Mountain—and five depressions or sags—Tianjun, Xiariha, Muli, Hala Lake, and Shule—based on terminology used for hydrocarbon exploration in the area (Wang et al. 1995; Fu and Zhou 2000). The Lower Triassic succession exposed in the southern Qilian area comprises the Xiahuancang and Jianghe formations (Yang et al. 1983). The Xiahuancang Formation is characterized by alternating sandstone and siltstone deposits, while the Jianghe Formation consists predominantly of massive limestone (Yang et al. 1983; Xu et al. 2017). These formations reflect a transition from clastic-dominated to carbonate-dominated sedimentation, likely influenced by changes in sea level, tectonic activity, and paleoenvironmental conditions during the Early Triassic.

The West Qinling region is situated near the center of the Qinling-Qilian-Kunlun central orogenic belt in China. It is bounded by the Songpan-Garzê terrane to the south, the Qilian terrane to the north, and the Qaidam terrane to the west (Zhang et al. 2001; Dong et al. 2011). The West Qinling has been interpreted as a microcontinental block that originally rifted from the North China Block during the Meso–Neoproterozoic, migrated toward the northern Yangtze Block, and subsequently collided with the North China Block during the Triassic (Meng et al. 2005; Dong et al. 2011). Based on differences in lithostratigraphy, the West Qinling is subdivided into a northern division and a southern division (Li et al. 2014). Well-exposed Permian–Triassic strata are found along the eastern margin of the northern division. This region is characterized by deep-marine deposits spanning the Lopingian (Late Permian) to early Middle Triassic. The Late Permian–early Middle Triassic successions in the northern division are divided into three formations: the Maomaolong Formation, Guomugou Formation, and Gulangdi Formation. These formations consist primarily of deep-marine siliciclastic and calciclastic sediments, which were deposited through a variety of sediment gravity flows and slumps. The deposits exhibit significant spatial and temporal variability, reflecting the dynamic tectonic and sedimentary processes active during this period (Li et al. 2014).

In this book, the studied sections are divided into five environmental categories according to the stratigraphical units: two for siliciclastic systems (brackish, shoreface), two for carbonate systems (ramp, platform), and one for mixed siliciclastic-carbonate systems (shelf-slope-basin) (Figs. 2.2 and 2.3). The brackish environmental

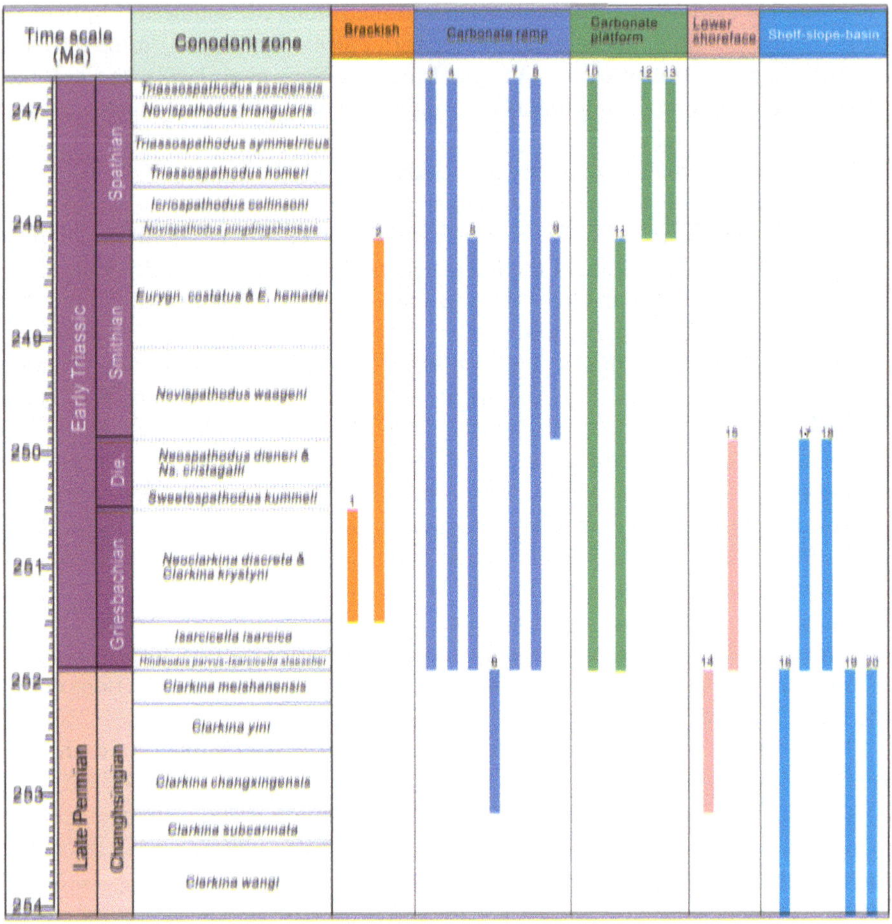

Fig. 2.3 Temporal distribution of the successions from the studied sections across various environmental categories during the P-Tr interval in China. The number codes are the same as those in Fig. 2.1

setting is represented by the Houzhougongmiao and Longmendong sections from South China. The shoreface environmental setting is mainly the Zhihema section from the southern Qilian area of North China and the upper part of the Penglaitan section from South China. The carbonate ramp is composed of the GSSP Meishan, Chaohu, Susong, Daxiakou, Daye Shatian, Dangwu and Gaimao sections from South China. Carbonate platform refers to the Xiejiacao, Shawan, upper Longmendong, and Lichuan sections from South China. The deeper shelf-slope-basin environmental settings comprise Shangsi, Penglaitan sections from South China and Zhayougou, Malongou, Longwuxia sections from the western Qinling area of North China. The trace fossils described in this study are primarily based on field observations and specimens collected over the past decade. Most fossil localities have been visited multiple times (at least three visits per site) to ensure comprehensive specimen collection, detailed photographic documentation, and the excavation of sufficient fossil material. While some data are drawn from earlier publications, all relevant localities have been revisited, and the materials have been reexamined to confirm their taxonomic identification and stratigraphical context. Additionally, a taxonomic reevaluation of certain ichnotaxa has been conducted to ensure accuracy and consistency. In the following sections, the stratigraphical units and geographical background of each study area will be introduced to provide context for the trace fossil assemblages and their paleoenvironmental significance.

2.2 Brackish

2.2.1 Houzhougongmiao Section

The Houzhougongmiao section (HZGM) (GPS: 34°35.033′N, 107°35.833′E) is located in Shibaling Village, Xifang Town, Qishan County, Central Shaanxi Province, Western China. This section records a terrestrial-to-marine siliciclastic succession. The Lower Triassic succession here comprises the upper Sunjiagou, Liujiagou, and Heshanggou formations, which correlate with coeval units across much of North China (Liu and He 2000; Yin and Lin 1979; Tu et al. 2016). However, the HZGM section is distinguished by the presence of marine interbeds near the Permian-Triassic (P-Tr) boundary. The Sunjiagou Formation, approximately 500 m thick, is divided into three parts: the lower part consists of grey, thick-bedded sandstone interbedded with greenish, thin-bedded siltstone. The middle part features reddish sandstone interbedded with greenish mudstone. The upper part is characterized by thin-bedded limestone interbedded with siltstone and lavender, thin-bedded mudstone, with marine interbeds occurring in the uppermost section. Marine invertebrate assemblages have been reported from the greenish and lavender thin-bedded mudstone of the upper Sunjiagou Formation in the HZGM section, as well as from the grey, thin-bedded siltstone of the lower Liujiagou Formation in the nearby Zishiya section (Fig. 2.4). The upper Sunjiagou marine fauna includes bivalves such as *Eumorphotis multiformis*, *Promyalina putiatinensis*, and *Homomya impressa*. Additionally, abundant conchostracans (*Euestheria gutta*) and the *Lundbladispora-Taeniaesporites-Cycadopites* palynological assemblage have been documented from the same beds that yield trace fossils in the Houzhougongmiao section (Liu and He 2000; Chu et al. 2019). The *Lundbladispora-Taeniaesporites-Cycadopites* spore assemblage was widely recognized as the first microfloral zone of the Triassic globally (Tewari et al. 2015; Yu et al. 2015), confirming a Griesbachian age for the marine fossil-bearing strata of the upper Sunjiagou Formation in this section. Abundant trace fossils, including star-shaped *Asteriacites*, hook-shaped *Biformites*, burrows, and arthropod trackways, are preserved in the greenish and lavender thin-bedded mudstone of the upper Sunjiagou Formation.

Fig. 2.4 Field photographs of the Lower Triassic succession from the Houzhougongmiao section. (**a–c**) Thin-bedded siltstone interbedded with mudstone beds, together with a few sandstone beds, makes up the upper Sunjiagou Formation. (**d**) Abundant *Biformites* imprints from the sole surface of siltstone beds from the upper Sunjiagou Formation. (**e–g**) Common Early Triassic bivalve fossil from the siltstone beds of the upper Sunjiagou Formation. **e**, *Neoschizodus laevigatus*; **f**, *Unionites* sp.; **g**, *Neoschizodus* sp.. The coin in D is about 2 cm in diameter

The basal Sunjiagou Formation consists of poorly exposed, grey, thick-bedded sandstones. The middle part, approximately 80 m thick, is dominated by reddish sandstone beds that yield rare body fossils and trace fossils, the latter primarily consisting of small, simple, horizontal *Planolites* burrows (Fig. 2.5). The upper part of the Sunjiagou Formation is composed of thin-bedded siltstone and mudstone, with intercalated limestone and sandstone lenses. The ichnoassemblage examined here is documented from thin-bedded siltstones interbedded with greenish mudstone horizons in the upper Sunjiagou Formation. The studied succession,

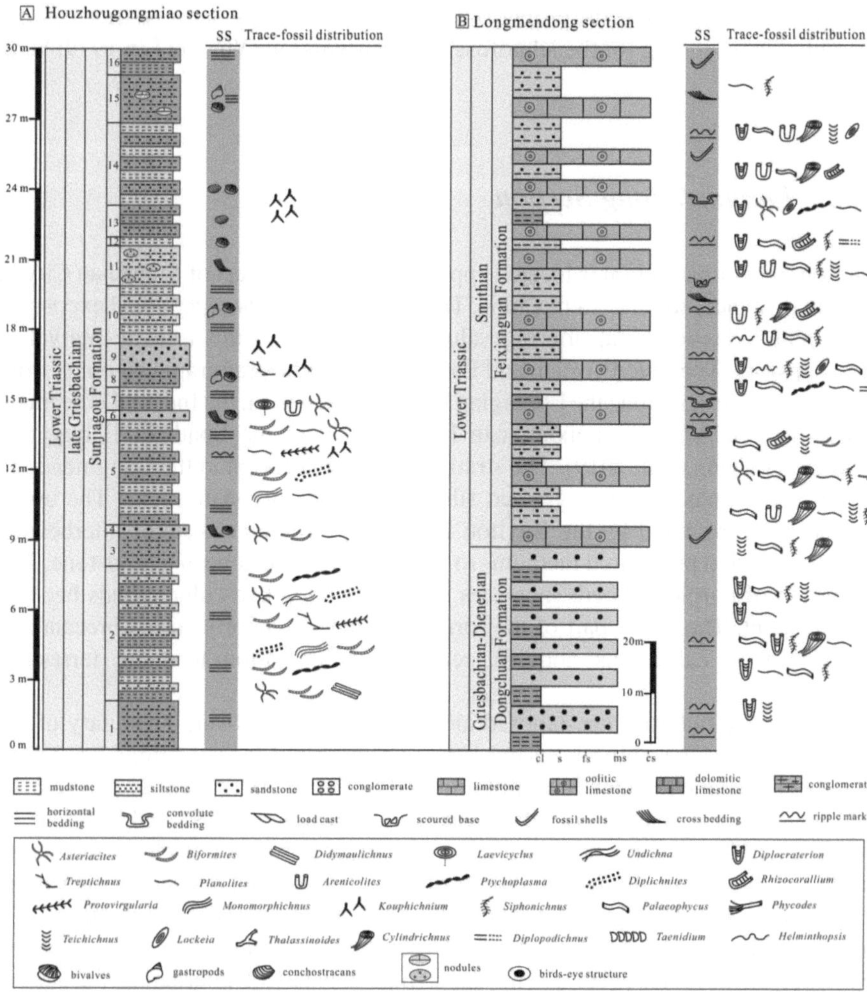

Fig. 2.5 Stratigraphical sections of the Lower Triassic successions exposed at Houzhougongmiao (**a**) and Longmendong (**b**) sections, illustrating stratigraphical distributions of trace fossils, other benthos, and sedimentary structures

approximately 30 m thick, consists of two small cycles of siltstone interbedded with mudstone, capped by thin-bedded, fine-grained sandstone (Beds 4, 9) at the top of each cycle in the lower part. The upper part features greenish and purple, thin-bedded siltstone interbedded with light green mudstone (Fig. 2.5). Faint ripple marks occasionally occur on the upper surfaces of the siltstone beds, while hummocky cross-stratification in the thin-bedded sandstone suggests an intense transgression, indicative of a shoreface depositional environment. Trace fossils are abundant in the two small cycles of the lower part, while bivalve, conchostracan, and gastropod fossils are found in the upper part. Horizontal bedding is common throughout the siltstone beds, further supporting a low-energy, semi-restricted, shallow embayment setting for the ichnofossil-rich succession (Yin and Lin 1979; Yang et al. 1983).

2.2.2 *Longmendong Section*

The Longmendong section is located approximately 5 km west of Emeishan City in Sichuan Province, southwestern China. This section provides exceptional exposures of Permian-Triassic strata, offering valuable insights into this critical interval in Earth's history. The Late Permian and Early Triassic lithostratigraphical units, from oldest to youngest, include the Lopingian Kayitou Formation, the Induan Dongchuan Formation, the Smithian Feixianguan Formation, and the Spathian Jialingjiang Formation (Zhang et al. 2019). The stratigraphical succession in the Longmendong section is dominated by fine-grained siliciclastic and carbonate rocks. The upper Permian is represented by the Kayitou Formation, which consists of interbedded olive, brown, and purple-red medium- to fine-grained sandstone and mudstone. The proportion of purple-red beds increases upward, while olive-colored beds become less abundant. The upper part of the formation includes poorly sorted breccia and paleosols with calcareous nodules, reflecting a transition to more terrestrial conditions.

The Triassic outcrops are exposed along the Longmen River, a tributary of the Qiyinjiang River, spanning a ~5-km-long section. Field investigations were conducted between the Qinying Power Station (GPS: 29°34′50.97″N, 103°24′45.96″E) and the eastern margin of the Huangwan Bridge (GPS: 29°34′58.68″N, 103°25′24.17″E). The lower part of the Dongchuan Formation is characterized by purple-red, fine- to medium-grained sandstone, siltstone, and mudstone, displaying trough cross-stratification, planar cross-stratification, parallel lamination, and current-ripple cross-lamination. Scour structures are locally present, indicating deposition in a fluvial system. The upper part consists of alternating purple-red mudstone and gray-white, thin-bedded, fine- to very fine-grained sandstone (Fig. 2.6). Current-ripple cross-lamination and trace fossils are common. The lower part of the Feixianguan Formation comprises interbedded greyish bioclastic limestone, cross-bedded fine-grained sandstone, and mudstone. The upper part is dominated by gypsum-bearing dolomite intercalated with medium- to coarse-grained sandstone, interpreted as point bar deposits within a meandering river system. Magnetostratigraphical and biostratigraphical data confirmed that the Dongchuan Formation spans the Griesbachian to Dienerian, while the Feixianguan Formation is Smithian in age. Abundant trace fossils have been reported from the upper part of the Dongchuan Formation and the lower part of the Feixianguan Formation (Zhang et al. 2017, 2019). These trace fossils provide critical evidence for understanding the recovery of benthic ecosystems following the Permian-Triassic mass extinction.

Fig. 2.6 Field photographs of the Lower Triassic succession from the Longmendong section. (**a**) Reddish sandstone of the lower Dongchuan Formation. (**b**–**f**) Abundant trace fossils from the sandstone beds of the upper Dongchuan Formation. (Modified from Zhang et al. 2017). The hammer in **a** is about 35 cm in length, and the coin in **b** is about 2 cm in diameter

2.3 Carbonate Ramp

2.3.1 Chaohu Section

The Chaohu section is located in the southeastern suburbs of Chaohu City, Anhui Province, South China. This section features a well-exposed Upper Permian to Lower Triassic marine succession, which is overlain by Middle Triassic evaporitic dolomites. The Chaohu section is particularly notable for its complete record of the Permian-Triassic (P/Tr) boundary and the Induan–Olenekian boundary, as well as its abundant fossil assemblages, including conodonts, ammonoids, bivalves, and brachiopods. Due to its exceptional stratigraphic and paleontological significance, the Chaohu section was nominated as a candidate for the Global Stratotype Section and Point (GSSP) for the Induan–Olenekian boundary (Tong et al. 2003). The Upper Permian Talung Formation is 4–5 m thick and consists of black mudstone and cherty shale. It yields deep-water facies fossils, including conodonts, radiolarians, ammonoids, bivalves, and brachiopods of Changhsingian age. The P/Tr boundary beds are composed of four distinct layers in ascending order: a white claystone layer, a black shale bed, a yellow, medium-bedded marlstone, and another white claystone layer. These beds correlate well with their counterparts at the Meishan section, the GSSP for the P/Tr boundary (Chen et al. 2011). The Lower Triassic Yinkeng Formation is characterized by greenish and black shale interbedded with marlstone. The Helongshan Formation consists of thin-bedded dolomitic limestone intercalated with calcareous mudstone (Fig. 2.7a). The Nanlinghu Formation comprises medium-bedded limestone interbedded with mudstone. This formation yields abundant conodonts assignable to the *Novispathodus pingdingshanensis* and *Triassospathodus homeri* conodont zones, which are key biostratigraphical markers for the Spathian stage in South China and globally. The Chaohu section provides a critical record of marine ecosystems across the P/Tr boundary and into the Early Triassic. Its well-preserved fossil assemblages and stratigraphical continuity make it an invaluable reference for understanding the biotic recovery following the Permian-Triassic mass extinction.

The Yinkeng Formation in the Chaohu section is characterized by distinct lithological and sedimentological features that reflect changes in depositional environments during the Early Triassic. The formation is divided into lower, middle, and upper parts, each exhibiting unique sedimentary patterns and fossil assemblages. The lower part of the Yinkeng Formation consists of alternating shale and marlstone or limestone, organized into small-scale cycles. In these cycles, the lower units (dominated by black shale) become progressively thicker upward, while the upper units (marlstone or limestone) thin upward. This trend results in cycles increasingly dominated by finely laminated black shale, indicating a deepening depositional environment over time (Chen et al. 2011). The lower Yinkeng Formation also yields small, thin-shelled fossils, including the bivalve *Claraia* spp. and the ammonoid *Ophiceras* spp. These fossils form a high-abundance, low-diversity community, typical of post-extinction ecosystems (Chen et al. 2010). The middle part of the Yinkeng Formation is characterized by alternating greenish mudstone and marlstone or limestone. In contrast to the lower part, the upper units (limestone or marlstone) thicken upward, while the lower units (greenish mudstone) thin upward, suggesting a progradational depositional pattern. Hummocky-like cross-stratifications are occasionally observed in the limestone beds, indicating deposition within the storm wavebase action zone. Similar sedimentary structures have been documented in the Yinkeng Formation in the Meishan section, South China, supporting this interpretation (Chen et al. 2007). The presence of storm-related features suggests that the middle Yinkeng Formation was deposited in a proximal ramp setting, within the influence of storm wave action. The upper part of the Yinkeng

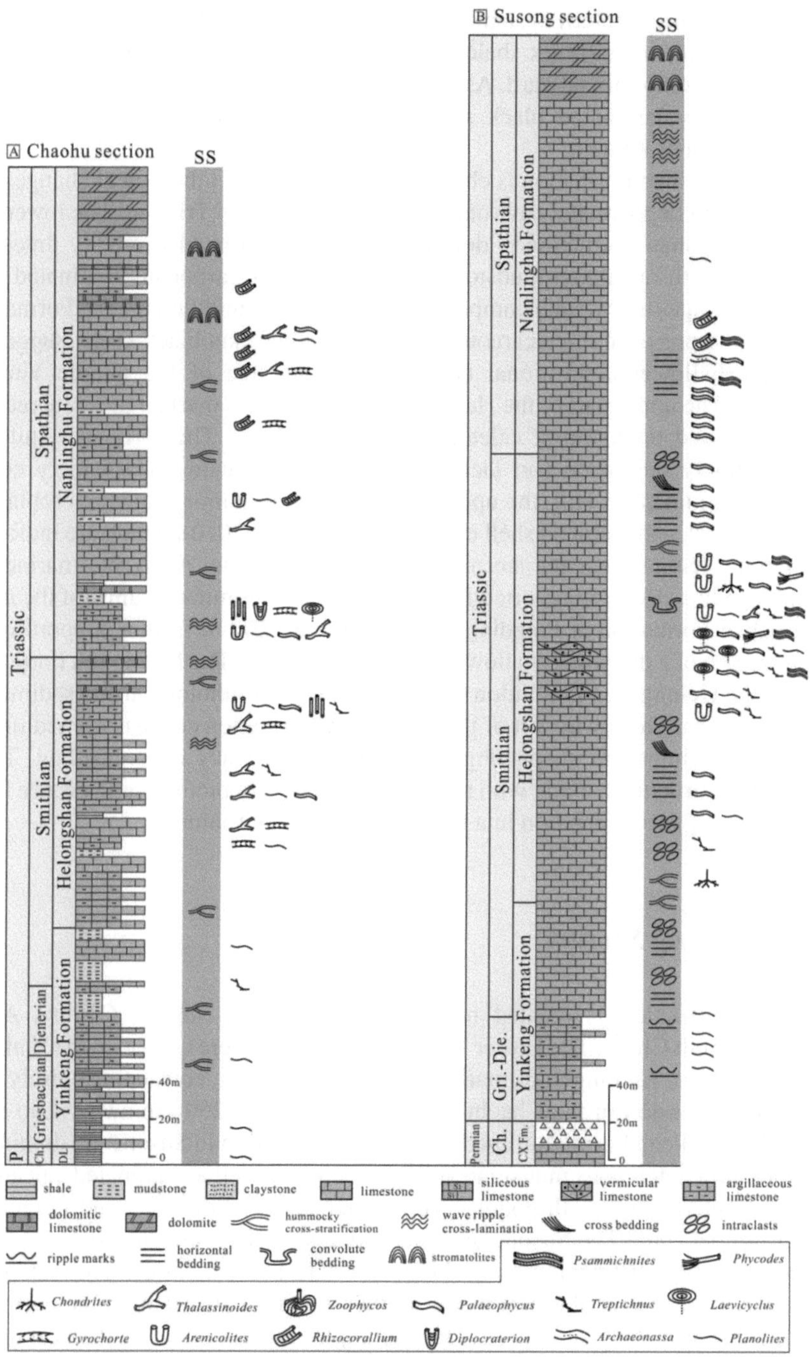

Fig. 2.7 Stratigraphical sections of the Upper Permian to Lower Triassic successions exposed at Chaohu (**a**) and Susong (**b**) sections, illustrating stratigraphical distributions of trace fossils and sedimentary structures

Formation exhibits a return to small-scale cycles dominated by black shale. In these cycles, the lower units (black shale) thicken upward, while the upper units (marlstone or limestone) thin upward. As a result, the cycles become increasingly dominated by finely laminated black shale, indicating a renewed deepening of the depositional environment.

The Helongshan Formation is characterized by distinct lithological changes that reflect shifts in depositional environments during the Early Triassic. The lower part of the Helongshan Formation is dominated by medium-bedded muddy limestone interbedded with calcareous mudstone (Fig. 2.7a). These carbonate-dominated sedimentary packages are broadly comparable to those of the middle Yinkeng Formation. However, the presence of calcareous mudstone, rather than black shale, suggests a relatively shallower depositional environment than that of the middle Yinkeng Formation. The upper part of the Helongshan Formation is marked by an increase in the number and thickness of calcareous mudstone beds. The carbonate units are typically finely laminated and lack cross-bedding, indicating low-energy conditions. At Pingdingshan Hill, the upper Helongshan Formation yields microbialites, microbial mats, and ostracod shell concretions (Jia et al. 2010), which are indicative of a marine habitat near the storm wavebase action zone. A diverse marine fish assemblage has also been reported from the mudstone-dominated units of the upper Helongshan Formation at Pingdingshan (Tong et al. 2004), further supporting the interpretation of a dynamic, shallow marine environment. The Nanlinghu Formation overlies the Helongshan Formation and exhibits distinct lithological and sedimentological features. The lower part of the Nanlinghu Formation consists of medium- to thick-bedded limestone, displaying large ripples and wavy cross-bedding. These sedimentary structures, along with the presence of highly bioturbated massive limestone beds, suggest deposition in a lower shoreface environment.

2.3.2 *Susong Section*

The Susong section is located near Zuoshan Village, Susong County, Anhui Province, South China. The Lower Triassic successions here comprise the Yinkeng, Helongshan, and Nanlinghu formations in ascending order, correlating closely with the same succession in the Chaohu area, approximately 250 km away (Luo et al. 2016). The Yinkeng Formation and Helongshan Formation in Susong are dominated by alternating calcareous mudstone and marlstone, as well as laminated calcareous mudstone interbedded with nodular, vermicular muddy limestone and intraclastic rudstone–floatstone. The Susong area was situated at the western end of a narrow, northeast-southwest-oriented ramp during the late Early Triassic. This ramp was flanked by a carbonate platform to the southeast and a shelf basin to the northwest within the Lower Yangtze Region. The Nanlinghu Formation is the thickest unit in the Lower Triassic succession of the Lower Yangtze region. The lower and middle parts consist of thin-bedded, bioturbated calcareous mudstone with minor sharp-based intraclastic rudstone beds (Fig. 2.7b). The upper part is characterized by peloidal–skeletal packstone–wackestone, interbedded with stromatolitic bindstone, bioclastic floatstone, and calcareous mudstone. The presence of ammonoids such as *Columbites*, *Tirolites*, and *Subcolumbites* indicates a Spathian age for the Nanlinghu Formation (Luo et al. 2016).

Facies in the Susong section have been well-documented based on texture, fabric, sedimentary structures, and fossil content. These facies are grouped into five facies associations, which collectively indicate an overall shallowing-upward succession (Luo et al. 2016). The basal part of the section is dominated by calcareous mudstones and chert, which were deposited in an offshore, low-energy environment. The majority of the section consists of carbonate mudstones with frequent evidence of sediment gravity-flow deposits, including coarser carbonate debris. These features suggest deposition in a carbonate slope setting, where periodic gravity flows transported material from shallower areas into deeper waters. The upper part of the section is characterized by bioclastic limestones and microbialites, with varying degrees of dolomitization. These facies are interpreted as representing increasingly shallow subtidal conditions, where microbial activity and bioclastic accumulation were prominent.

2.3.3 Daye Shatian Section

The Shatian section is located at the Shatian Coal Mine in Bao'an Town, Daye City, Hubei Province, South China. The Early Triassic lithostratigraphical units in this section consist of the Daye Formation and the Jialingjiang Formation. The Daye Formation is subdivided into three members, each with distinct lithological characteristics: Member I is composed of laminated black shale interbedded with grey muddy limestone, indicating deposition in a low-energy, deep-water environment. Member II is dominated by light-grey, medium- to thickly bedded argillaceous limestone, with interbeds of conglomerate and vermicular limestone. These lithologies suggest a transition to more dynamic depositional conditions, possibly influenced by periodic influxes of coarser material. Member III consists primarily of light-grey micritic limestone and argillaceous limestone, with minor amounts of yellow-grey calcareous mudstone (Fig. 2.8). This member reflects a return to relatively stable, shallow marine conditions.

2.3.4 Meishan Section (GSSP)

The Meishan section is located on the south hillside along the road from Meishan Town to Xinhui Village in Changxing County, Zhejiang Province, South China. This section has been ratified as the Global Stratotype Section and Point (GSSP) for the Permian–Triassic boundary (PTB) (Yin et al. 2001), making it a globally significant reference for studying this critical interval in Earth's history. The

Fig. 2.8 Field photographs of the Lower Triassic succession from the Shatian section at Daye City. (**a–d**) Thin-bedded argillaceous limestone interbedded with medium-bedded limestone beds of the Daye Formation. (**e**) Vermicular fabrics in the limestone bed of the Daye Formation. (**f**) Abundant tiny burrows on the bedding surface of argillaceous limestone. The hammer in **c** is about 30 cm in length, and the coin in **d–f** is about 2 cm in diameter

Changxing Formation consists of 247 natural layers, each ranging from 2 to 37 cm in thickness. It is a 41-meter-thick carbonate succession dominated by medium- to thin-bedded limestone. The top two beds of the Changxing Formation, Beds 23–24, preserved important sedimentary and paleontological information from the period immediately preceding the P-Tr mass extinction. The Yinkeng Formation comprises 183 natural layers and is approximately 15 m thick. The lower part is dominated by mudstone and muddy limestone, while the upper part is characterized by thin-bedded limestone. Most of the Yinkeng Formation records the severe biotic extinction and its aftermath, providing critical insights into the ecological consequences of the P-Tr event (Chen et al. 2015). The microstratigraphy and paleoecology of the uppermost Changxing Formation to Yinkeng Formation succession have been extensively studied and summarized (Chen et al. 2015). These beds document the transition from pre-extinction ecosystems to the post-extinction recovery phase. Detailed ichnological studies have been conducted on the P-Tr transition layers at Meishan. Previous works have been re-examined, and updates have been made where necessary (Fig. 2.9a). These studies provide valuable information on the behavior and ecological roles of trace-making organisms during this critical interval.

2.3.5 Gaimao Section

The Gaimao section is located approximately 100 m east of Gaimao Village in the Huaxi District of Guiyang City, Guizhou Province, southwestern China (Fig. 2.10). This section features a well-exposed Lower Triassic succession, which is subdivided into the Shabuwan, Daye, and Anshun Formations (Yan 2013). The Gaimao section records a unique lithological transition between the Permian and Triassic in South China, providing valuable insights into the environmental and biotic changes during this critical interval. The Shabuwan Formation is dominated by gray mudstone, indicating deposition in a low-energy, deep-water environment. The lower part of the Daye Formation is characterized by alternations of calcareous mudstone and thin-bedded argillaceous limestone. The upper part transitions to limestone, reflecting a shift to more carbonate-dominated deposition. The Anshun Formation is composed of massive dolomite with minor thin-bedded limestone (Figs. 2.10, 2.11, and 2.12), indicating a return to shallow marine conditions with significant dolomitization. Several conodont zones have been established within the Shabuwan and Daye Formations, confirming an Early Triassic age for these units (Yang et al. 2012). These biostratigraphic markers provide a robust framework for correlating the Gaimao section with other Early Triassic successions globally.

In particular, the presence of typical Griesbachian conodonts, such as *Isarcicella isarcica* and *I. staeschei*, in the upper Talung Formation indicates that its radiolarian-bearing siliceous mudrocks are of the earliest Triassic age, rather than being restricted to the Permian as previously thought (Yang et al. 2012). The abundance of siliceous plankton in this interval suggests that chert deposition persisted locally, at least in some areas, during the earliest Triassic. This finding makes the Gaimao section an exception to the so-called "Early Triassic Chert Gap," demonstrating that the end of chert deposition does not universally mark the P-Tr mass extinction. The Daye Formation represents deposition in a middle-ramp setting, characterized by alternations of calcareous mudstone and thin-bedded argillaceous limestone in the lower part, transitioning to limestone in the upper part. The Anshun Formation reflects cyclic peritidal settings, dominated by massive dolomite with minor

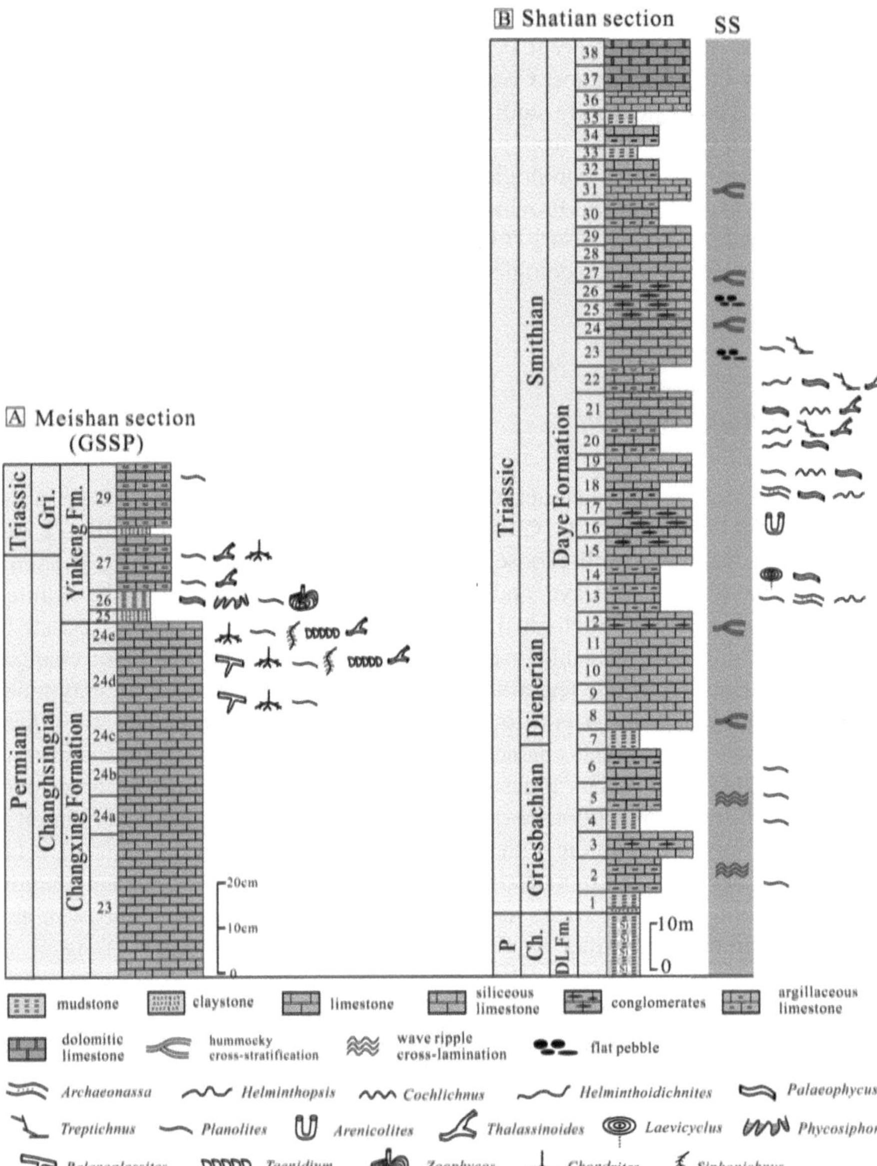

Fig. 2.9 Stratigraphical sections of the Upper Permian to Lower Triassic successions exposed at Meishan (**a**) and Daye Shatian (**b**) sections, illustrating stratigraphical distributions of trace fossils and sedimentary structures

Fig. 2.10 Field photographs of the Permian–Triassic succession from the Gaimao section. (**a**) The location of P-Tr boundary. (**b**) The limestone beds of the upper Daye Formation. (**c**) The mudstone of the lowermost Triassic Shabuwan Formation. (**d**) Thin-bedded argillaceous limestone of the lower Daye Formation. (**e–f**) Thin-bedded argillaceous limestone of the Daye Formation. The bedding surface is irregular. The white cylindrical buildings in **a** are about 30 m high; the pen in **d** and **e** is about 15 cm in length

thin-bedded limestone (Enos et al. 2006; Yang et al. 2012; Yan 2013). The Gaimao section provides critical insights into the persistence of siliceous plankton and chert deposition into the earliest Triassic, challenging the notion of a global Early Triassic Chert Gap. Additionally, the section's well-defined lithostratigraphy and depositional environments offer a valuable record of the transition from middle-ramp to peritidal settings during the Early Triassic.

Fig. 2.11 Field photographs of the Anshun Formation from the Gaimao section. (**a**) Dolomitic limestone of the Anshun Formation, abundant winding burrows were found from the upper bedding surface. (**b**) Dolomitic limestone of the Anshun Formation. (**c**) Flat pebble from the dolomitic limestone bed of the Anshun Formation. (**d**) Ripple marks from the thin-bedded argillaceous dolomite. (**e**) Abundant horizontal burrows from the upper bedding surface of the argillaceous limestone. (**f**) Vertical burrows filling with sparite from the dolomitic limestone of the Anshun Formation. The hammer in **c** is about 35 cm in length, and the pen in **d** and **e** is about 15 cm in length

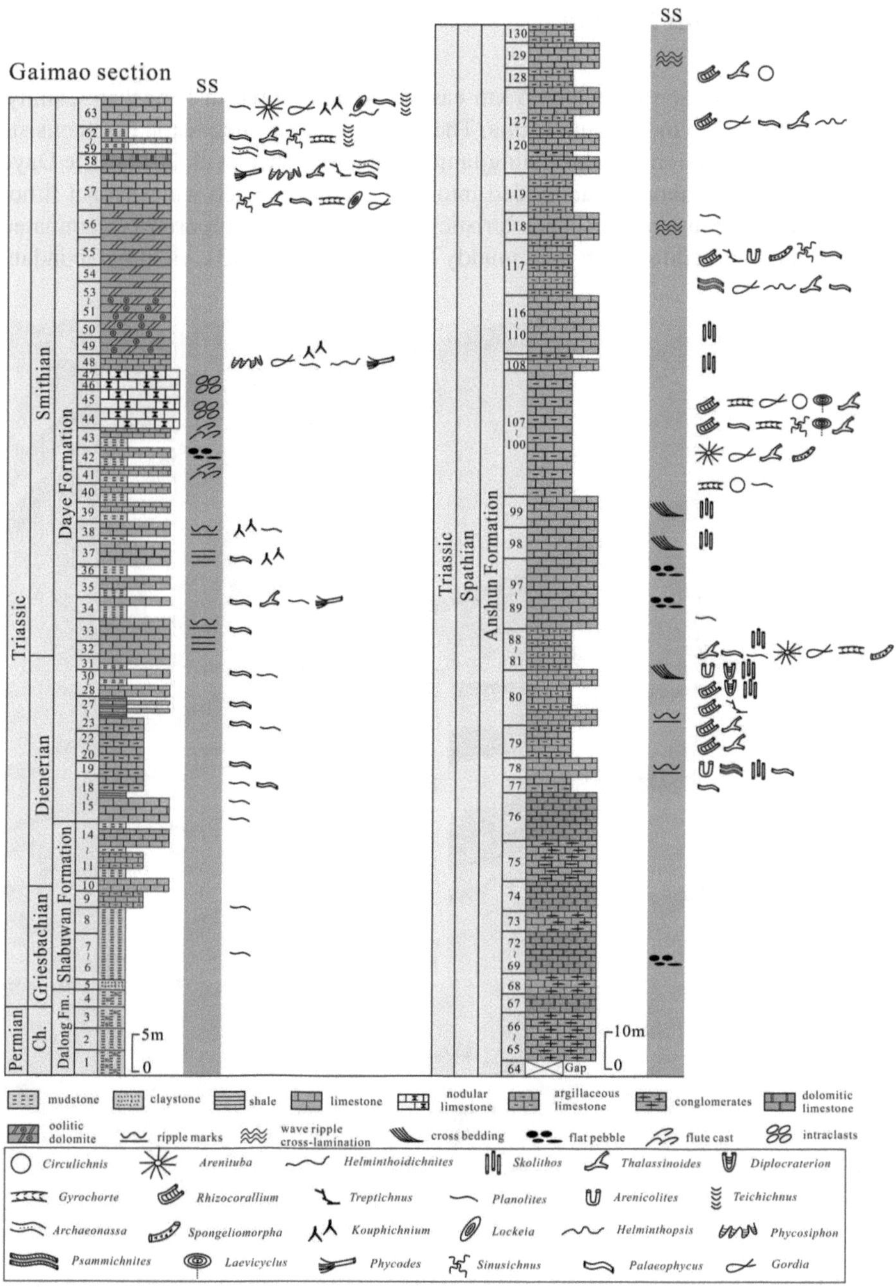

Fig. 2.12 Stratigraphical sections of the Upper Permian to Lower Triassic successions exposed at the Gaimao section, illustrating stratigraphical distributions of trace fossils and sedimentary structures

2.3.6 Daxiakou Section

The Daxiakou section is located 6 km east of Xiakou Town in Xingshan County, western Hubei Province, South China. The Lower Triassic succession here consists of the Daye Formation and the Jialingjiang Formation (Zhao et al. 2015). The Daye Formation at Daxiakou is subdivided into four members, each with distinct lithological and paleontological characteristics: Member I is composed of laminated black shale interbedded with grey muddy limestone (Fig. 2.13), yielding abundant

Fig. 2.13 P-Tr boundary (PTB) and Lower Triassic successions from the Daxiakou section. (**a**) PTB bed exposed at the Daxiakou section; clay and shale of the Talung Formation (at the lower left corner) are exposed below the PTB bed; PTB bed and above strata belong to the Daye Formation. (**b–d**) Thin-bedded limestone of the Daye Formation. (**e**) Thin-bedded argillaceous limestone from the Daye Formation. (**f**) Early Triassic bivalve fossil *Claraia* from the thin-bedded argillaceous limestone of the Daye Formation. The hammer in **a** is about 35 cm in length, the coin in **c** is about 2 cm in diameter, and the steel tape in **e** is about 5 cm in diameter

thin-shelled fossils, including the bivalves *Claraia* and *Eumorphotis*, as well as ammonoids such as *Ophiceras*, *Lytophiceras*, and *Gyronites* (Zhao et al. 2013). Member II is dominated by light-grey, medium- to thickly bedded micritic limestone and vermicular limestone. Member III consists primarily of light-grey micritic limestone and vermicular limestone, with minor amounts of yellow-grey calcareous mudstone. Member IV is characterized by thin- to medium-bedded micritic limestone, oolitic limestone, and vermicular limestone. Trace fossils are abundant and diverse in the micritic limestone and vermicular limestone of the Daye Formation, providing valuable insights into the benthic ecosystems and depositional environments during the Early Triassic.

The Jialingjiang Formation in the Daxiakou section is subdivided into four members, each with distinct lithological and sedimentological features. These members collectively record deposition in a middle ramp setting, as indicated by sedimentary facies and paleoecological analyses (Zhao et al. 2013, 2015). Member I is dominated by light-grey, thin- to medium-bedded muddy dolomite, with minor thin- to medium-bedded micritic limestone. Evaporite solution breccias occasionally occur within the dolomite beds (Fig. 2.14), suggesting periodic exposure and dissolution in a shallow marine environment. Member II is composed of thin- to medium-bedded micritic limestone, vermicular limestone, and calcarenitic limestone. Trace fossils are abundant and diverse in the micritic and vermicular limestone, reflecting active benthic ecosystems. Member III is characterized by greyish-white micritic dolomite, with occasional oolitic beds within the dolostone. Member IV is dominated by light-grey, thickly bedded dolomite with evaporite solution breccias (Fig. 2.15), indicating continued shallow marine conditions with periodic subaerial exposure. Abundant bivalve assemblages and well-defined conodont zones have been established in the Daxiakou section, providing a robust biostratigraphical framework for the Early Triassic. Sedimentary facies and paleoecological analyses suggest that the upper Daye Formation and the Jialingjiang Formation in this locality were deposited in a middle ramp setting, characterized by fluctuating sea levels and periodic evaporitic conditions.

2.3.7 Dangwu Section

The Dangwu section is composed of several limestone slab quarries near Dangwu Town, Guiyang City, Guizhou Province, South China. During the Permian–Triassic transition, the Guiyang area was situated on a ramp setting between the Upper Yangtze Platform to the north and the Nanpanjiang Basin to the south (Enos et al. 2006; Shi et al. 2015, 2019). Continuous sequences of Upper Permian–Lower Triassic strata crop out in the Guiyang area and comprise the Upper Permian Changxing Formation limestone, Permian–Triassic Talung Formation black cherts and cherty shales, lowest Triassic Shabuwan Formation black shale and mudstone, Lower Triassic Daye Formation mudstone, muddy limestone, and dolomitic limestone, and upper Lower Triassic Anshun Formation dolomites (Enos et al. 2006).

Fig. 2.14 Field photographs of the Lower Triassic Daye and Jialingjiang formations at Daxiakou section. (**a–b**) Thin- to medium-bedded limestone interbedded with mudstone layers of the Daye Formation. (**c–f**) Bioturbated limestone beds of the Jialingjiang Formation. The tepee structure can be observed in the central part in **d**

The Daye Formation is the most widely distributed Triassic formation in the Guiyang area and is subdivided into three members. The first member is characterized by alternation of mudstone and thin-bedded limestone; the second member is dominated by thin-bedded, bioturbated muddy limestone; and the third member comprises medium-bedded limestone and dolomitic limestone (Fig. 2.16). The ichnofossil-bearing strata in the Dangwu quarries are horizontal and comprise thin-bedded, highly bioturbated muddy limestones, typical of the second member of the Daye Formation in Guizhou.

At the Dangwu section, the Member II of the Daye Formation is 30 m in thick (Shi et al. 2019). Sampled horizons are argillaceous limestone beds (Fig. 2.16). The newly found conodonts *Novispathodus waageni waageni* and *Guangxidella* cf.

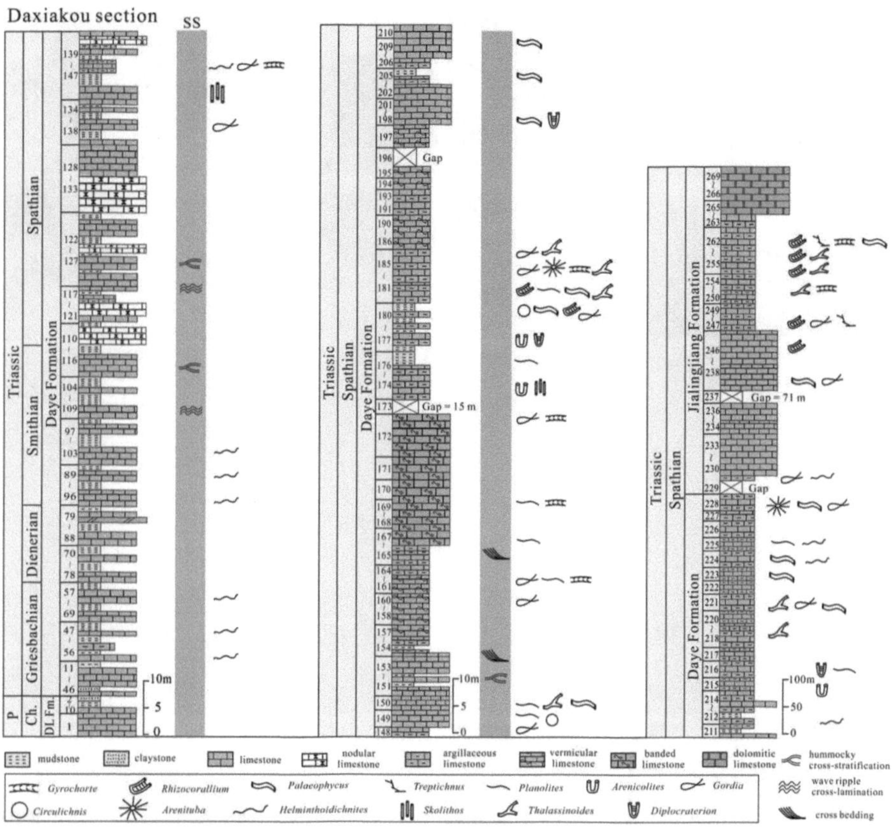

Fig. 2.15 Stratigraphical sections of the Upper Permian to Lower Triassic successions exposed at the Daxiakou section, illustrating stratigraphical distributions of trace fossils and sedimentary structures

bransoni are abundant in the quarry sequence. The former is the zonal species of the *Novispathodus waageni* Zone, marking the beginning of the Smithian (Zhao et al. 2007), and the latter is confined to the lower Smithian in South China (Orchard 2007). The ammonoids *Paranorites* sp. and *Meekoceras* sp. occur in the upper part of the second member and confirm an early-middle Smithian age (Enos et al. 2006). The Dangwu ichnoassemblage therefore is early Smithian in age. This age assignment is reinforced by the 2‰–4‰ ppm of $\delta^{13}C_{carb}$ isotopes (Fig. 2.17). Such positive $\delta^{13}C_{carb}$ values only occur in the uppermost Dienerian to lower Smithian interval throughout the Lower Triassic strata of the shelf basin and distal ramp facies across South China (Song et al. 2013; Tong et al. 2019). Thus, both conodont biostratigraphy and $\delta^{13}C_{carb}$ chemostratigraphy agree on an early Smithian age for the second member of the Daye Formation at Dangwu section (Fig. 2.17). Member II of the Daye Formation has been interpreted to be deposited under a quiet marginal carbonate ramp environmental setting.

Fig. 2.16 Lower Triassic successions from the Dangwu section. (**a**) One of the studied quarries of the Dangwu section. (**b**) Abundant argillaceous limestone slabs excavated from the quarries. (**c–d**) Upper bedding surface containing abundant horizontal burrows. (**e–f**) Polished vertical profile showing relatively high bioturbation degree. The telegraph pole in **a** is about 15 m high

2.4 Carbonate Platform

2.4.1 Xiejiacao Section

The Xiejiacao section is located in Guang'an City, northeastern Sichuan Province, South China (GPS: 30°38′34″N, 107°00′26″E). During the Permian-Triassic (P-Tr) transition, the Xiejiacao area was located on the northern margin of the Yangtze

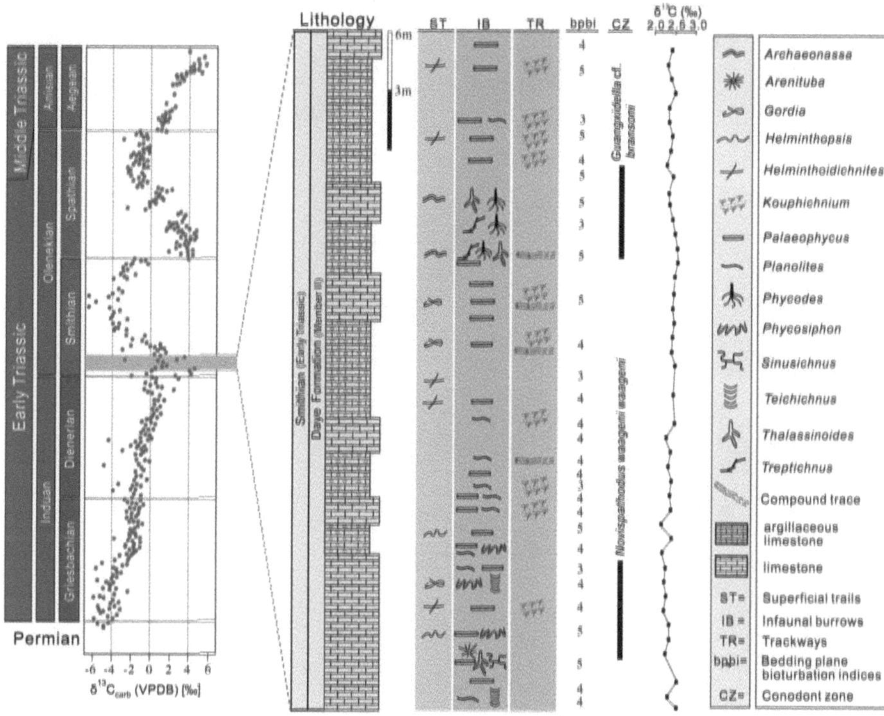

Fig. 2.17 Columnar section of the lower Smithian succession exposed at Dangwu, showing stratigraphical distribution of ichnotaxa, the extent of bedding plane bioturbation (bpbi), distribution of conodonts, and carbon isotope excursions. The left comparative carbon isotope curve through the Early Triassic is from the Chaohu area, South China. (Tong et al. 2007)

Platform (Su et al. 2021). The uppermost Permian Changxing Formation is dominated by dark grey cherty limestone, yielding abundant benthic fossils such as foraminifers, brachiopods, and bivalves (Su et al. 2021). The lowermost Triassic Yelang Formation comprises light grey to yellowish, thick-bedded micritic limestone overlain by alternations of black shale, calcareous mudstone, and marlstone (Fig. 2.18). The micritic limestone in the lower part of the Yelang Formation exhibits thrombolite fabrics, while the upper part displays dendrolite fabrics, both of which are microbialites (Fig. 2.18). The microbialite unit is overlapped by a 1.5-cm-thick, yellowish-brown volcanic ash layer, followed by alternating shale, calcareous mudstone, and marlstone. Abundant conodonts have been extracted from the microbialite unit, with the *Hindeodus parvus* Zone established in the middle part of the thrombolite interval. The first occurrence of *Hindeodus parvus* is a key marker for the base of the Triassic, calibrated to this level (Su et al. 2021). The presence of microbialites (thrombolites and dendrolites) in the basal Yelang Formation suggests a shallow marine environment with conditions favorable for microbial growth, likely in the aftermath of the P-Tr mass extinction. The overlying volcanic ash layer and alternating shale, mudstone, and marlstone indicate continued marine deposition with periodic volcanic input.

Fig. 2.18 Permian-Triassic boundary and Lower Triassic successions from the Xiejiacao section. (**a**) Permian-Triassic boundary (PTB) microbialite (M) and the above volcanic ash layer (between the red dash lines) exposed at Xiejiacao section, and the PTB was placed in the lower part of the microbialite. (**b**) Clotted fabrics of the thrombolite from the lower part of the microbialite. (**c**) Dendrolite fabrics of the thrombolite from the upper part of the microbialite. (**d**) Horizontal burrows of *Planolites* from the argillaceous limestone bed of the lower Yelang Formation. (**e–f**) Siltstone and mudstone beds from the middle part of the Yelang Formation. The hammer in **f** is about 30 cm in length

The Jialingjiang Formation conformably overlies the Yelang Formation and is characterized by distinct lithological changes that reflect its depositional environment. The lower part is composed of thin-bedded micritic limestone and gray argillaceous limestone, indicating a relatively low-energy, shallow marine environment. The middle-upper part is dominated by thick-bedded bioclastic limestone, oolitic limestone, and breccia, suggesting higher-energy conditions typical of a carbonate platform setting (Figs. 2.19 and 2.20). The Jialingjiang Formation is interpreted to have been deposited in a carbonate platform environment. The transition from thin-bedded micritic and argillaceous limestones in the lower part to thick-bedded

Fig. 2.19 Field photographs of the Lower Triassic Jialingjiang Formation at Xiejiacao section. (**a–b**) Thin-bedded limestone from the lower part; bioturbation can be observed in **b**. (**c**) Stromatolite from the middle part of this formation. (**d**) Non-parallel wavy beds from the upper part. (**e**) Flat-pebble conglomerate from the middle part. (**f**) Oolitic limestone from the middle part. The hammer in **a** is about 30 cm in length, and the pen in **e** is about 15 cm in length

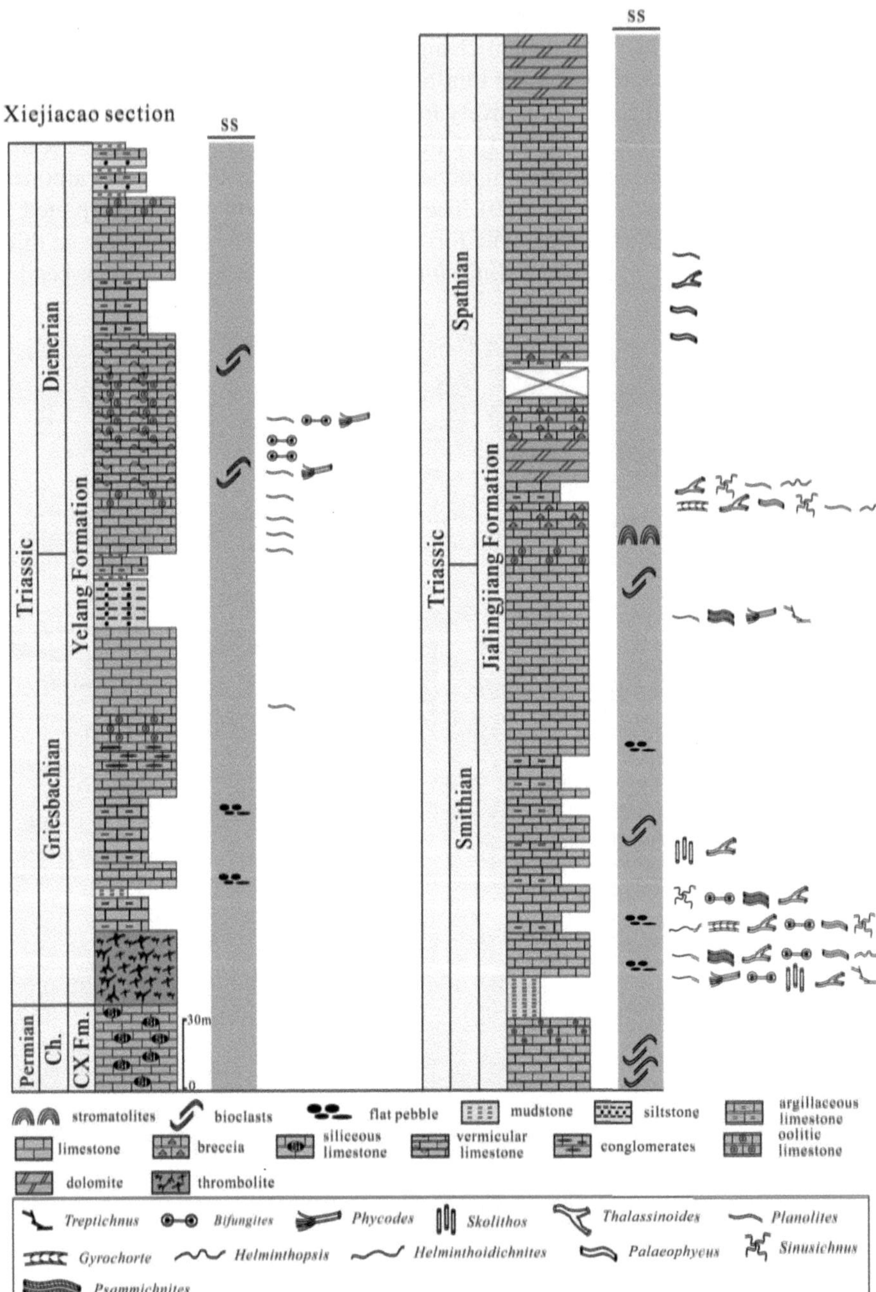

Fig. 2.20 Columnar section of the Upper Permian to Lower Triassic successions exposed at Xiejiacao section, showing stratigraphical distributions of trace fossils and sedimentary structures

bioclastic and oolitic limestones in the middle-upper part reflects a shift from quieter, deeper-water conditions to more dynamic, shallow-water settings with active wave and current action (Su et al. 2021).

2.4.2 Lichuan Section

The Lichuan section (GPS: N30°30.014′, E108°48.209′) is located 22 km north of Lichuan City in western Hubei Province, South China. Continuous outcrops of uppermost Permian to Lower Triassic strata are exposed along roadcuts from Lichuan City to the top of Qiyueshan Mountain. The Changxing Formation is dominated by bioclastic limestone, yielding abundant conodonts assignable to the *Clarkina yini* Zone, which is indicative of the latest Changhsingian in South China (Zhao et al. 2013; Chen et al. 2015). The base of the Daye Formation is characterized by a microbialite that unconformably overlies the bioclastic limestone of the Changxing Formation. The middle part consists of alternating layers of thin-bedded muddy limestone and medium-bedded grainstone to packstone. The upper part features alternations of muddy limestone, bioclastic limestone, and oolite (Fig. 2.21). The Jialingjiang Formation is dominated by thin-bedded, laminated muddy limestone, massive limestone, and dolomitic limestone (Fig. 2.21). The P–Tr boundary is placed within the lower part of the microbialite in the Daye Formation, based on the first occurrence of the conodont *Hindeodus parvus* (Wang and Cao 1981; Wang and Xia 2004; Zhao et al. 2013). This marker is globally recognized as defining the base of the Triassic.

The Daye Formation in the Lichuan section exhibits a clear shallowing-upward depositional pattern, as evidenced by changes in lithology and sedimentary structures. The bioclastic and oolitic limestone units thicken upward in the section, while muddy limestone units become thinner. The thickness of oolitic units increases significantly up-section, culminating in the development of giant ooid banks in the uppermost Daye Formation. The giant ooids, with diameters larger than 2 mm, are indicative of deposition in a high-energy zone above the fair-weather wave base (Mei 2008). The bioclastic and oolitic limestones of the middle-upper Daye Formation represent deposition within a carbonate platform setting. The increasing dominance of high-energy facies (e.g., oolitic and bioclastic limestones) upward in the section reflects a progressive shallowing of the depositional environment, consistent with the development of a carbonate platform.

2.4.3 Shawan Section and Longmendong Section (Upper)

The Shawan section is located 2 km away from Hulu County, Leshan City, Sichuan Province, South China. The section is composed of several neighboring quarries. Abundant carbonate slabs were split open, and trace fossils were observed and

Fig. 2.21 Lower Triassic successions from the Lichuan section. (**a**) The studied quarry with abundant slabs of the Lichuan section. (**b**) Abundant horizontal bilobate burrows from argillaceous limestone bedding surface of the Daye Formation. (**c**) Thin-bedded argillaceous limestone from the upper part of the Daye Formation. (**d**) Vermicular fabrics in the limestone bed of the Daye Formation. (**e**) Successions of the Jialingjiang Formation, white arrow indicates the way up of the succession, (**f**) Argillaceous dolomite of the Jialingjiang Formation. The hammer in **c** is about 35 cm in length

collected. The succession exposed here was dominated by the argillaceous limestone of the Jialingjiang Formation. The Shawan section is about 80 km southern from the Longmendong section. Because the strata are similar to the Jialingjiang Formation of the upper Longmendong section, and they were deposited in a similar environmental setting, thus we described them together (Zhang et al. 2019). The Jialingjiang Formation conformably overlies the Feixianguan Formation and is characterized by distinct lithological changes that reflect its depositional environment. The basal part includes dolomitic shale and micritic dolostone, with rare trace fossils and few bird-eye structures. The lower part is composed of thin-bedded micritic limestone and medium-bedded dolostone (Fig. 2.23). The middle part is dominated by micritic limestone, which contains abundant body fossils and trace fossils. The upper part consists of thick-bedded dolostone and gypsum karst breccia (Fig. 2.22b). The Jialingjiang Formation is interpreted to have been deposited in a

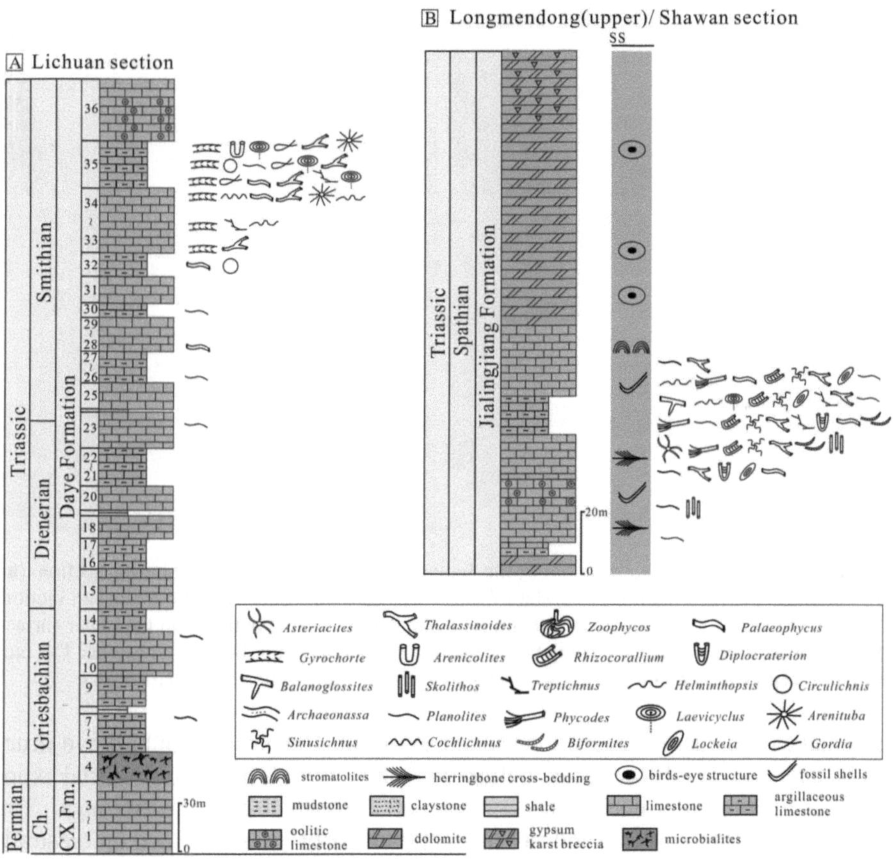

Fig. 2.22 Columnar sections of the Upper Permian to Lower Triassic successions exposed at Lichuan (**a**) and Longmendong (**b**) sections, showing stratigraphical distributions of trace fossils and sedimentary structures

Fig. 2.23 Lower Triassic successions of the Jialingjiang Formation from the Shawan section. (**a**–**b**) Workers are excavating limestone slabs from one of the studied quarries of the Shawan section. The size of slabs is about 50 cm × 100 cm. (**c**) Abundant horizontal burrows on the upper surface of an excavated limestone slab. (**d**–**f**) Polished vertical profile of thin-bedded limestone. The bioturbation degree in **f** and the lower part of **d** are relatively high, while it is rather low in **e**

tidal flat environment flanking a carbonate platform. The presence of bird-eye structures, gypsum karst breccia, and dolomitic facies suggests periodic subaerial exposure and evaporitic conditions, typical of carbonate platform settings. The Jialingjiang Formation is dated as Spathian (late Early Triassic) based on biostratigraphical and lithostratigraphical evidence (Fig. 2.23).

2.5 Lower Shoreface

2.5.1 *Penglaitan Section (Upper)*

The Penglaitan section is located approximately 20 km southwest of Laibin City in Guangxi Province, South China. During the Late Permian to Middle Triassic, this area was situated near the eastern (Heshan) carbonate platform of the Nanpanjiang Basin. The Penglaitan section is internationally significant as it defines the base of the Lopingian Series and preserves the complete Lopingian Series of South China (Jin et al. 2006; Shen et al. 2007, 2019). The uppermost part of the Talung Formation consists of mixed carbonate-clastic turbidites or storm deposits, interbedded with more than 50 layers of massive, immature tuffs and tuffaceous sandstones. Sedimentary structures such as trough cross-bedding followed by plane parallel lamination in sandstone are most consistent with an upper shoreface to foreshore environment. Several thin coal seams with abundant plant fossils and a high proportion of volcanic material in the overlying interval suggest relatively rapid shallowing, likely due to either regional regression or tectonic control. The flora, characterized by the *Gigantonoclea* Assemblage, is found approximately 100 m below the P–T boundary. Ammonoid and conodont zones above the plant-bearing horizons indicate a very late Changhsingian age. The lowest strata of the Luolou Formation are characterized by black shales with centimeter-scale limestone interbeds. These shales contain abundant fossils of the Triassic bivalve *Claraia*, numerous microgastropods, and the conodont *Hindeodus parvus*. This black shale interval is overlain by thin-bedded limestone containing *Isarcicella* species and a few tuffaceous beds. The Penglaitan section records a transition from upper shoreface to foreshore environments in the Talung Formation to deeper-water, black shale deposition in the basal Luolou Formation. The presence of volcanic material and coal seams suggests significant tectonic and volcanic activity during the late Permian, influencing sedimentation patterns.

2.5.2 *Zhihema Section*

The Zhihema section is located approximately 5 km south of Zhihema Town (GPS: 37°27.617′N, 99°04.350′E) in Tianjun County, Qinghai Province, northwestern China. The Zhihema area is part of the southern Qilian Basin, which was situated on the southern margin of the North China Block and the northern margin of the Paleotethys Ocean during the Early Triassic (Ziegler et al. 1998; Zhang et al. 2004, 2016; Pan et al. 2016). The Lower Triassic succession in the Zhihema section consists of the Xiahuancang Formation and the Jianghe Formation. The Xiahuancang Formation is subdivided into two members. Member I comprises a reddish coarse sandstone bed at the base, overlain by medium-grained quartz sandstones and thin-bedded siltstone, yielding abundant bivalves, including *Claraia aurita*, *C.*

yangkangensis, *C. concentrica*, and *Unionites fassaensis*, as well as microbially induced sedimentary structures (MISS) (Xu et al. 2017) (Fig. 2.24f). Large-angle cross-bedding is common in thick-bedded, coarse-grained sandstones of the lower part, indicating high-energy conditions typical of an upper shoreface setting (Feng 2021). The presence of well-sorted, subrounded sandstones further supports this interpretation. The middle and upper parts of Member I are composed of thin- to medium-bedded sandstone interbedded with thin-bedded mudstone or siltstone (Fig. 2.24). Cross-stratification is common in sandstones, with cross-bedding angles of 5–20° and sets 0.1–0.3 m thick (Xu et al. 2017). Horizontal bedding is occasionally present in the siltstone interbeds. Abundant vertical trace fossils, such as *Diplocraterion* and *Skolithos*, along with *Claraia* bivalve shells, are prominent in the sandstones. The lithofacies and ichnoassemblages of Member I indicate a slightly deeper shoreface setting compared to the lower part, with decreasing wave energy upward in the section. This transition reflects a shift from high-energy upper shoreface conditions to a more stable, lower-energy environment.

Member II is characterized by finely laminated siltstone and fine-grained, thin-bedded sandstones, yielding abundant bivalves, including *Eumorphotis multiformis*, *E. inaequicostata*, and *Promyalina putiatinensis* (Feng 2021). Cross-bedding is rare, but hummocky cross-stratification occasionally occurs in sandstones associated with bivalve shell beds. Wrinkle structures are present at the interface between underlying fine sandstones and overlying siltstones. The Jianghe Formation is composed of massive limestone (Yang et al. 1983). The palynological assemblage *Lundbladispora-Taeniaesporites-Cycadopites* has been established in the Xiahuancang Formation across multiple localities in the southern Qilian Basin. This assemblage is widely recognized as the first microfloral zone of the Triassic globally (Tewari et al. 2015; Yu et al. 2015; Ji et al. 2017). The Xiahuancang Formation records a transition from high-energy shoreface settings in Member I to lower-energy, finely laminated deposits in Member II. Oriented shell beds and hummocky cross-stratification in Member II of the Xiahuancang Formation are interpreted as products of storm-wave activity (Dumas and Arnott 2006). The fining-upward succession of Member II, combined with the overlying carbonate strata of the Jianghe Formation, suggests deposition in a lower shoreface to offshore transition setting. This environment was periodically influenced by storm currents, which reworked sediments and created the observed sedimentary structures (Figs. 2.25 and 2.26).

2.6 Shelf-Slope-Basin

2.6.1 Malonggou Section

The Malonggou section is located along the Lanlang Highway, approximately 30 km northwest of Hezuo City in Gansu Province, northwestern China. This section features well-exposed Permian–Triassic strata and is situated on the eastern margin of the northern West Qinling (Zhang et al. 2020b). The West Qinling is

Fig. 2.24 Field photos of the Lower Triassic successions (Member I) of the Xiahuancang Formation from the Zhihema section, south Qilian area. (**a–b**) Reddish, thick-bedded sandstone from the base of the Xiahuancang Formation. (**c**) Sandstone beds from the upper part of Member I. (**d**) Siltstone interbedded with medium-bedded sandstone from the Member I of the Xiahuancang Formation, the upper surface of the sandstone bed contains microbially induced sedimentary structures (MISS) mainly pictograph-like sand cracks/crack-fills (PLSC). (**e**) Vertical burrows of *Diplocraterion parallelum* that penetrated deeply into the coarse sandstone from the lower part of the Member I. (**f**) The common MISS type, pictograph-like sand crack from the lower part of the Member I

Fig. 2.25 Field photos of Member II of the Xiahuancang Formation from the Zhihema section. (**a–b**) Successions of thin-bedded sandstone and siltstone from Member II of the Xiahuancang Formation. (**c**) Ripple marks on the bedding plane of a sandstone, containing abundant dumbbell-shaped burrows of *Diplocraterion* isp. (**d**) Wrinkle structures on the upper surface of sandstone they were preserved together with epifaunal bivalves, and tiny vertical burrows. (**e**) Abundant bivalve fossils of *Eumorphotis* and *Unionites* preserved in the sandstone bed. (**f**) Shell bed made up of bivalve shells in sandstone in vertical section

located near the center of the Qinling-Qilian-Kunlun central orogenic belt of China. It is bounded by the Songpan-Ganzi terrane to the south, the Qilian terrane to the north, and the Qaidam terrane to the west (Zhang et al. 2001; Dong et al. 2011). The West Qinling is interpreted as a microcontinental block that originally rifted from the North China Block during the Meso–Neoproterozoic, migrated toward the northern Yangtze Block, and later collided with the North China Block during the Triassic (Dong et al. 2011). Due to differences in lithostratigraphy, the West Qinling is subdivided into a northern division and a southern division. The northern division primarily consists of deep marine deposits of Lopingian (Late Permian) to early Middle Triassic age (Li et al. 2014). The Lopingian Maomaolong Formation is

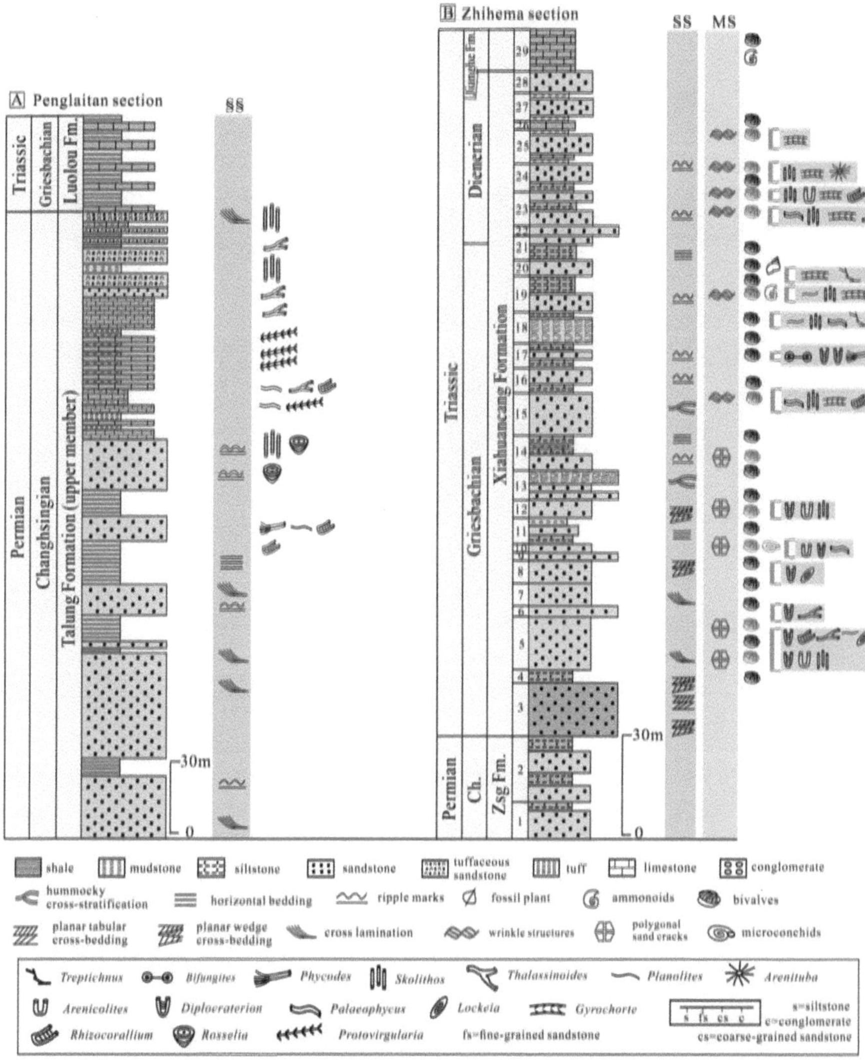

Fig. 2.26 Columnar sections of the Upper Permian to Lower Triassic successions exposed at Penglaitan (**a**) and Zhihema (**b**) sections, showing stratigraphical distributions of trace fossils and sedimentary structures

subdivided into two units: the lower unit is composed of gray- to dark-gray, medium- to thick-bedded, medium- to fine-grained sandstones and massive brecciated carbonates, intercalated with thin-bedded siltstones, slates, and conglomerates, containing abundant Pennsylvanian to Guadalupian brachiopods, corals, and fusulinids in the brecciated limestones (Fig. 2.27). Syn-depositional folds (soft-sediment deformation structures) intercalated with undeformed layers, as well as well-developed Bouma sequences, indicate deposition from turbidity flows. The upper unit is characterized by gray-dark, thin-bedded silty slates intercalated with gray,

Fig. 2.27 Field photos of the Late Permian Maomaolong Formation from the Malonggou section, western Qinling. (**a**) Thick-bedded conglomeratic sandstone in the lower part. (**b**) Successions of the section. (**c**) Silty slate yields abundant trails and burrows. (**d**) Slate interbedded with sandstone beds from the upper part of the Maomaolong Formation. (**e–f**) Trace fossils preserved on the upper surface of the sandstone. The person in **a** is about 1.6 m in height, the telegraph pole in **b** is about 30 m high, and the pen in **d** and **f** is about 12 cm in length

medium-bedded fine-grained sandstones and calcareous siltstones, suggesting lower-energy conditions. Based on lithology and sedimentary structures, the Maomaolong Formation is interpreted as having been deposited in a proximal slope environment, influenced by submarine rockfalls and turbidity flows. The presence of syn-depositional folds and Bouma sequences supports this interpretation.

2.6.2 Zhayougou Section

The Zhayougou section is located approximately 15 km west of Hezuo City in Gansu Province, northwestern China. Similar to the Malonggou section, the Zhayougou section is situated on the eastern margin of the northern West Qinling. The Lower Triassic succession in the Zhayougou section is represented by the Shangaling Group, which conformably overlies the Lopingian Maomaolong Formation and underlies the Middle Triassic Gulangdi Formation. The Shangaling Group primarily consists of shale, siltstones, slates, and a few sandstone beds. Abundant Early Triassic fossils, such as *Claraia* cf. *stachei* and *Claraia* cf. *griesbachi*, occurred in the slates, providing biostratigraphical evidence for their age. A gabbro intrusion in the lower part of the Shangaling Group has been dated using U-Pb methods to 250.1 ± 2.2 Ma, confirming an Early Triassic age for this group (Wang et al. 2010). The lithology of the Shangaling Group, dominated by fine-grained sediments such as shale, siltstone, and slate, suggests deposition in a low-energy, deep-water environment.

The Shangaling Group is interpreted as having been deposited in a base-of-slope apron environment characterized by fine-grained sediments such as shale, siltstone, and slate. This setting reflects a low-energy, deep-water depositional system. The overlying Gulangdi Formation consists of thick-bedded sandstones, conglomerates, and pebbly sandstones, which are interpreted as deposits of a base-of-slope fan environment (Fig. 2.28). The lithofacies and spatial-temporal changes in paleo-current data from these strata suggest the opening of a back-arc basin system during the Lopingian (Late Permian) to early Middle Triassic in the northern West Qinling (Li et al. 2014). The Permian-Triassic strata in the northern West Qinling are predominantly composed of deep-marine mixed siliciclastic and calciclastic sediments deposited by turbidity currents and debris flows (Fig. 2.29).

2.6.3 Longwuxia Section

The Longwuxia section is located at a road cut along the Longwu River, approximately 20 km south of Jianzha County in Qinghai Province, northwestern China. This section features complete exposures of Late Permian to Middle Triassic successions. The Upper Permian (Changhsingian) is dominated by siliceous limestone, representing a deep marine environment. The Lower Triassic Guomugou Formation is divided into three members: the lower member is composed of coarse-grained sandstone, pebbly sandstone, and conglomerate. The middle member is dominated by fine- to medium-grained sandstone interbedded with argillaceous limestone, siliceous limestone, and several volcanic rocks (Fig. 2.30). Trace fossils are abundant in the argillaceous limestone and sandstone beds (Fig. 2.30). The upper member consists of quartz sandstone and slate. Syn-depositional folds (soft-sediment deformation structures), convolute bedding, and Bouma sequences are well-developed in

Fig. 2.28 Field photos of the Lower Triassic Shangaling Group from the Zhayougou section, western Qinling. (**a**) Thick-bedded sandstone in the lower part. (**b**) Silty slate. (**c**) Slate interbedded with sandstone. (**d**) The silty slate contains gravel. (**e**) Ripple marks on the upper surface of the sandstone. (**f**) Flute cast on the sole of sandstone. The person in **a** is about 1.7 m in height, the hammer in **c** and **f** is 30 cm in length, the coin in **d** and **e** is 2 cm in diameter

this formation, indicating deposition in deep-sea turbidite facies. The base of the Jiangligou Formation is defined by the disappearance of coarse-grained feldspar quartz sandstone and the appearance of sandstone (Fig. 2.31). The Jiangligou Formation is divided into two members: the lower member is composed of gray, medium-bedded sandy limestone interbedded with thin-bedded mudstone, feldspar quartz sandstone, or calcareous slate. The upper member consists of sandy limestone, medium- to thick-bedded limestone, and banded limestone. The contact with

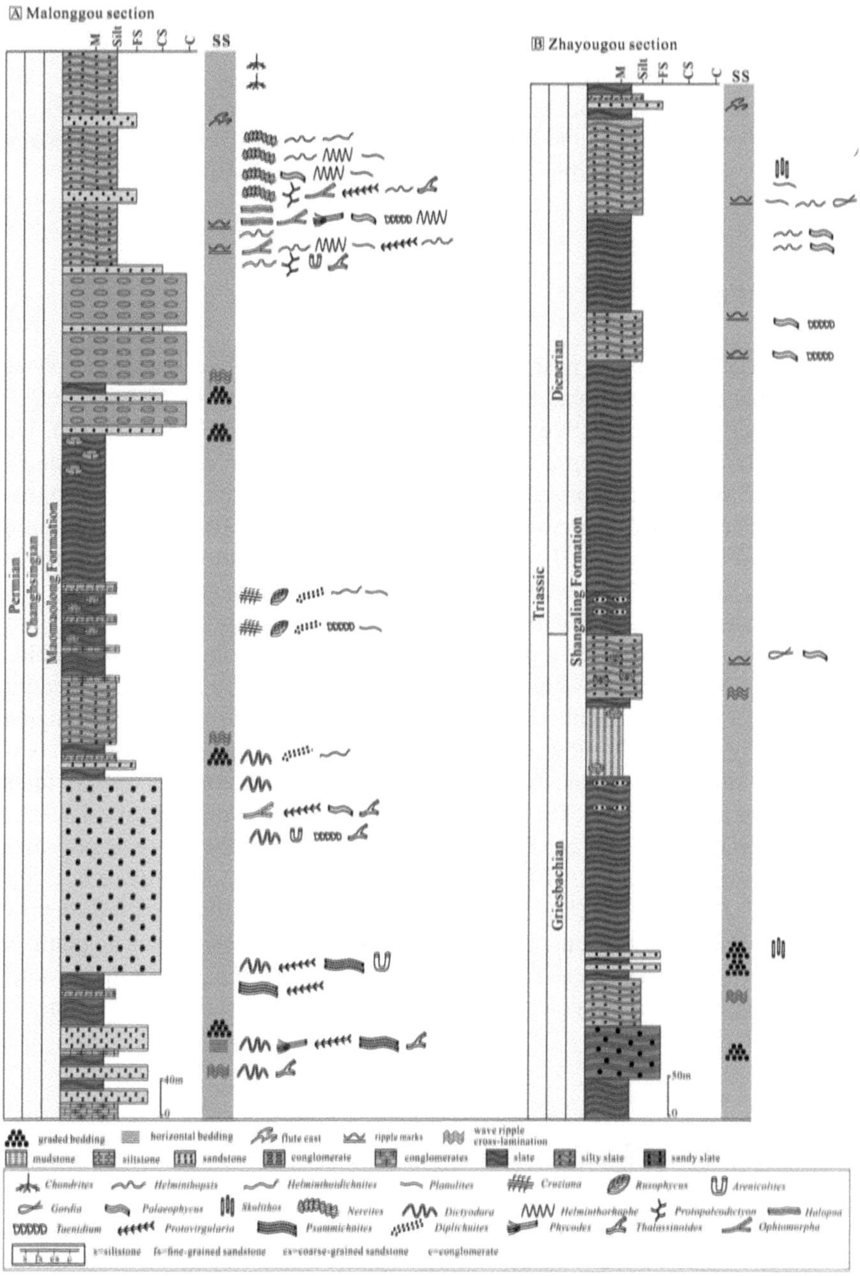

Fig. 2.29 Columnar sections of the Late Permian to Lower Triassic successions exposed at Malonggou (**a**) and Zhayougou (**b**) sections, showing stratigraphical distributions of trace fossils and sedimentary structures

the overlying Gulangdi Formation is marked by the disappearance of limestone and the appearance of conglomerate or calcareous breccia. The Guomugou Formation records deposition in a deep-sea turbidite environment, as evidenced by syn-depositional folds, convolute bedding, and Bouma sequences. The Jiangligou Formation reflects a transition to shallower marine conditions, with sandy limestone and banded limestone indicating a more stable, carbonate-dominated setting.

2.6.4 *Shangsi Section*

The Shangsi section, located near Guanyuan in Sichuan Province, once was considered a candidate for the Global Stratotype Section and Point (GSSP) of the P-Tr Boundary (PTB) (Lai et al. 1996; Jiang et al. 2011; Zhang et al. 2018, 2020a). The section is well-exposed along a creek in the Shangjianggou Valley near Shangsi. The Shangsi section comprises the Talung Formation and the Feixianguan Formation. The studied interval includes the uppermost 10 m of the Talung Formation and the lowermost 1 m of the Feixianguan Formation, which spans the P–T boundary (Fig. 2.32). The Talung Formation (Lopingian) is dominated by siliceous limestone intercalated with a few thin-bedded mudstone layers. A 25 cm-thick limestone bed near the top contains abundant conodonts and ammonoids, confirming a Changhsingian age. This limestone is overlain by three claystone/ash beds (totaling 15 cm in thickness), which correspond to the base of the Feixianguan Formation (Fig. 2.32). The lowest part of the Feixianguan Formation consists of yellow to yellowish-green calcareous mudstone and argillaceous limestone, interbedded with a few ash beds. The exact location of the PTB at the Shangsi section remains uncertain. The first occurrence of the index species *Hindeodus parvus* is relatively high in the section. U-Pb dating of ash layers provides precise age constraints at 252.28 ± 0.13 Ma (Shen et al. 2011). The Shangsi section records deposition in a deep shelf environment on the margin of an intracratonic carbonate platform during the Permian and earliest Triassic. The presence of ash beds and fine-grained sediments suggests periodic volcanic activity and low-energy conditions during the P–T transition (Fig. 2.33).

2.6.5 *Penglaitan Section (Lower)*

The Penglaitan section is located approximately 20 km southwest of Laibin City in Guangxi Province, South China. During the Late Permian to Middle Triassic, this area was situated near the eastern (Heshan) carbonate platform of the Nanpanjiang Basin (Shen et al. 2007, 2019). The Penglaitan section, located on the southern bank of the Hongshui River, is the Global Stratotype Section and Point (GSSP) for the Guadalupian–Lopingian boundary (Jin et al. 2006). The Permian-Triassic record at Penglaitan is consistent with tectonic convergence and magmatic arc development

Fig. 2.30 Field photos of the Lower Triassic Guomugou Formation of the Longwuxia section, western Qinling. (**a**) The nearly perpendicular strata of the section. (**b**) The dark siliceous limestone (left) intruded by andesite (red A). (**c**) Abundant flute cast preserved on the sole of a sandstone bed. (**d**) The intruded andesite. (**e–f**) Sole bedding of a sandstone bed with ripple marks. Abundant trace fossils are observed on the surface. (**f**) A close-up view of an area in **e** shows the ripple marks and trace fossils. The person in **a** is about 1.9 m high, and the hammer in **c** and **f** is 35 cm in length

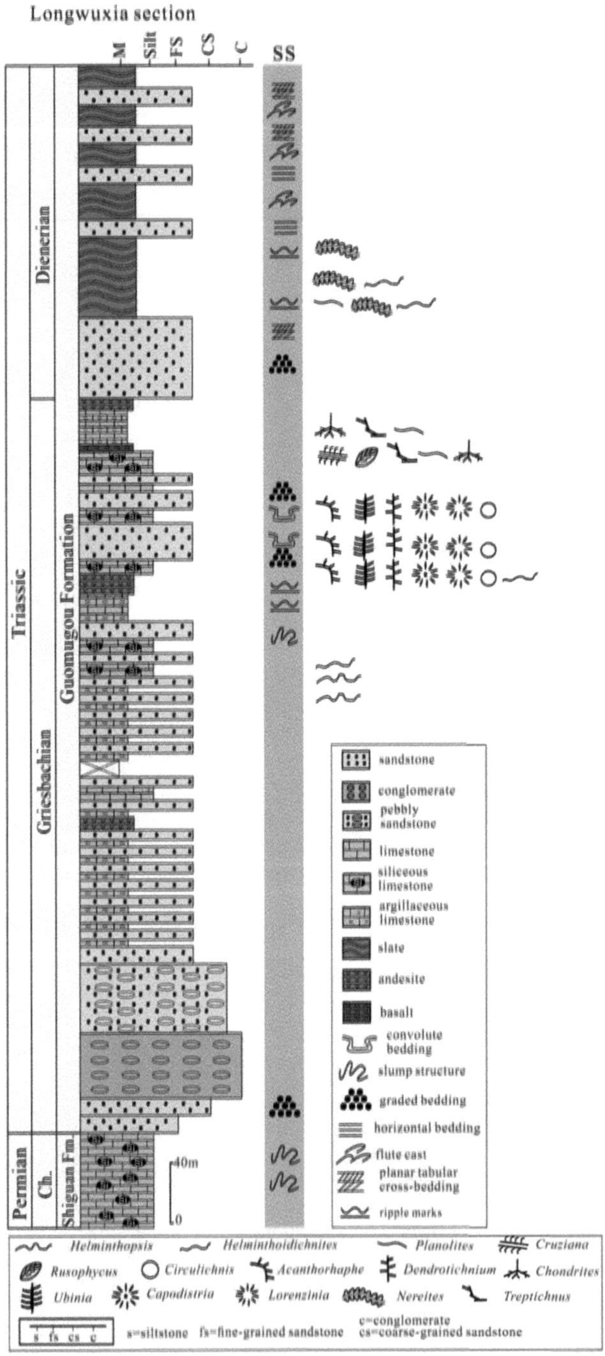

Fig. 2.31 Columnar section of the Late Permian to Lower Triassic successions exposed at Longwuxia section, showing stratigraphical distributions of trace fossils and sedimentary structures

Fig. 2.32 Permian-Triassic boundary and Lower Triassic successions of the Shangsi section. (**a**) Limestone of the Late Permian Talung Formation. (**b**) Permian-Triassic boundary horizons and the lowermost Triassic horizons exposed at the Shangsi section, and the PTB was placed in the mudstone of the upper part. (**c**) Close-up of the PTB mudstone of the Feixianguan Formation. (**d**) The mudstone interbedded with argillaceous limestone of the lower Feixianguan Formation. (**e**) Thin-bedded limestone of the lower Feixianguan Formation. (**f**) Early Triassic ammonoids from the argillaceous limestone of the Feixianguan Formation

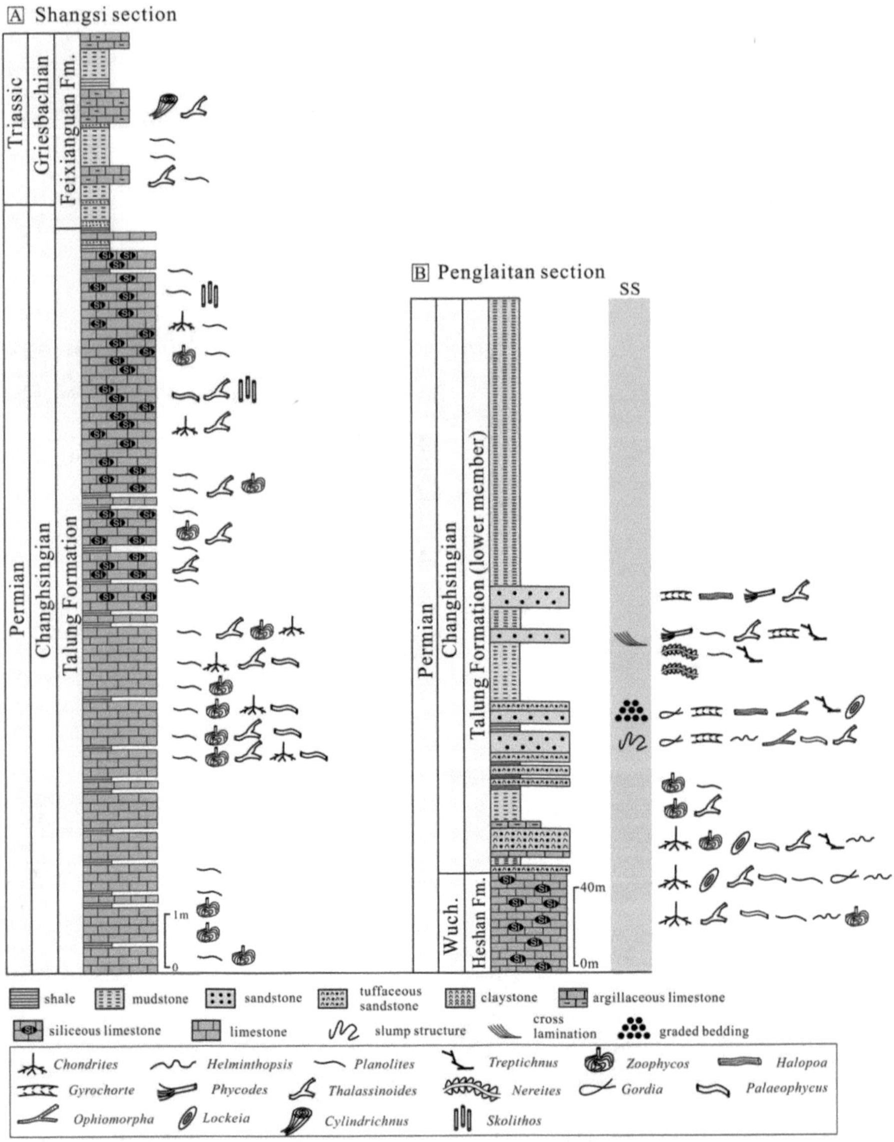

Fig. 2.33 Columnar sections of the Late Permian to Lower Triassic successions exposed at Shangsi (**a**) and Penglaitan (**b**) sections, showing stratigraphical distributions of trace fossils and sedimentary structures

to the south. This tectonic activity influenced sedimentation patterns, resulting in thick deposits with abundant volcanic and siliciclastic input. The Lopingian Series at Penglaitan consists of the following units in ascending order: uppermost Maokou Formation, which represents the top of the Guadalupian Series; the Heshan Formation overlies the Maokou Formation and is characterized by carbonate-dominated deposits. The Talung Formation is remarkably thick and subdivided into two members: Ergou Member is composed of silicified carbonate rocks, containing diverse ammonoids belonging to the *Penglaites* Zone; the Penglaitan Member is extremely thick (523.6 m), likely due to high rates of tectonic subsidence, local syn-sedimentary faulting, and abundant input of siliciclastic and pyroclastic materials from nearby arc-related volcanic eruptions (Fig. 2.34). The base of this member is marked by a resistant bed of coarse tuffaceous sandstone. Late Changhsingian deposits exhibit a variety of sedimentary structures, including

Fig. 2.34 Late Permian of the Penglaitan section. (**a**) Thick-bedded tuffaceous sandstone of the Lower part of the Talung Formation. (**b**) Thin-bedded fine sandstone interbedded with mudstone of the lower member of the Talung Formation. (**c**) Cross-bedding in a sandstone bed. (**d**) Ripple marks on the upper surface of a sandstone bed. (**e**) Spherical pyrite particles. (**f**) Late Permian ammonoids

graded bedding, flute casts, load casts, dish structures, and slump-originated folded layers, indicating deposition in a slope to submarine fan delta setting (Shen et al. 2007, 2019).

References

Benton MJ, Zhang QY, Hu SX, Chen ZQ, Wen W, Liu J, Huang JY, Zhou CY, Xie T, Tong JN, Choo B (2013) Exceptional vertebrate biotas from the Triassic of China, and the expansion of marine ecosystems after the Permo-Triassic mass extinction. Earth Sci Rev 125:199–243

Chen ZQ, Tong J, Kaiho K, Kawahata H (2007) Onset of biotic and environmental recovery from the end-Permian mass extinction within 1–2 million years: a case study of the Lower Triassic of the Meishan section, South China. Palaeogeogr Palaeoclimatol Palaeoecol 252:176–187

Chen ZQ, Tong J, Liao Z, Chen J (2010) Structural changes of marine communities over the Permian–Triassic transition: ecologically assessing the end-Permian mass extinction and its aftermath. Glob Planet Chang 73:123–140

Chen ZQ, Tong JN, Fraiser ML (2011) Trace fossil evidence for restoration of marine ecosystems following the end-Permian mass extinction in the Lower Yangtze region, South China. Palaeogeogr Palaeoclimatol Palaeoecol 299:449–474

Chen ZQ, Yang H, Luo M, Benton MJ, Kaiho K, Zhao LS, Huang YG, Zhang KX, Fang YH, Jiang HS, Qiu H, Li Y, Tu CY, Shi L, Zhang L, Feng XQ, Chen L (2015) Complete biotic and sedimentary records of the Permian–Triassic transition from Meishan section, South China: ecologically assessing mass extinction and its aftermath. Earth Sci Rev 149:67–107

Chu DL, Tong JN, Song HJ, Benton MJ, Bottjer DJ, Song HY, Tian L (2015) Early Triassic wrinkle structures on land: stressed environments and oases for life. Sci Rep 5:10109e

Chu DL, Tong JN, Bottjer DJ, Song HJ, Song HY, Benton MJ, Tian L, Guo WW (2017) Microbial mats in the terrestrial Lower Triassic of North China and implications for the Permian–Triassic mass extinction. Palaeogeogr Palaeoclimatol Palaeoecol 474:214–231

Chu DL, Tong JN, Benton MJ, Yu JX, Huang YF (2019) Mixed continental-marine biotas following the Permian-Triassic mass extinction in South and North China. Palaeogeogr Palaeoclimatol Palaeoecol 519:95–107

Dong Y, Zhang G, Neubauer F, Liu X, Genser J, Hauzenberger C (2011) Tectonic evolution of the Qinling orogen, China: review and synthesis. J Asian Earth Sci 41:213–237

Dumas S, Arnott RWC (2006) Origin of hummocky and swaley cross-stratification–the controlling influence of unidirectional current and aggradation rate. Geology 34:1073–1076

Enos P, Lehrmann DJ, Wei JY, Yu YY, Xiao JF, Chaikin DH, Minzoni M, Berry AK, Montgomery P (2006) Triassic evolution of the Yangtze Platform in Guizhou Province, People's Republic of China. Geological Society of America special paper 417, 105p, The Geological Society of America, Boulder, USA.

Feng XQ (2021) Reassessing Early Triassic wrinkle structures from moderate-high latitudes: an updated interpretation of metazoan colonization in matground ecosystems after the Permian-Triassic mass extinction. Glob Planet Chang 205:103590

Feng Z, Bao Z, Liu S (1997) Lithofacies palaeogeography of Early and Middle Triassic of South China. Petroleum Industry Press, Beijing, 222 pp

Fu J, Zhou L (2000) Triassic stratigraphic provinces of the southern Qilian basin and their petrogeological features. Northwest Geol 21(2):64–72

Hu SX, Zhang QY, Chen ZQ, Zhou CY, Lü T, Xie T, Wen W, Huang JY, Benton MJ (2011) The Luoping biota: exceptional preservation, and new evidence on the Triassic recovery from end-Permian mass extinction. Proc R Soc B 278:2274–2282

Ji W, Tong J, Zhao L, Zhou S, Chen J (2011) Lower–Middle Triassic conodont biostratigraphy of the Qingyan section, Guizhou Province, Southwest China. Palaeogeogr Palaeoclimatol Palaeoecol 308:213–223

Ji LX, Liu F, Luo W, Ouyang S (2017) Lower Triassic spores-pollen assemblages of Qinghai Province and their stratigraphic significance. J Stratigr 41:450–466. (in Chinese with English abstract)

Jia ZH, Yang FM, Hong TQ, Zhang LW (2010) The Smithian microstromatolites in Triassic in Chaohu, East China. J Earth Sci 21:137–140

Jiang HS, Lai XL, Luo GM, Aldridge R, Zhang K, Wignall P (2007) Restudy of conodont zonation and evolution across the P/T boundary at Meishan section, Changxing, Zhejiang, China. Glob Planet Chang 55:39–55

Jiang HS, Lai XL, Yan CB, Aldridge RJ, Wignall P, Sun YD (2011) Revised conodont zonation and conodont evolution across the Permian-Triassic boundary at the Shangsi section, Guangyuan, Sichuan, South China. Glob Planet Chang 77:103–115

Jiang HS, Lai XL, Sun YD, Wignall P, Liu JB, Yan CB (2014) Permian-Triassic conodonts from Da jiang (Guizhou, South China) and their implication for the age of microbialite deposition in the aftermath of the end-Permian mass extinction. J Earth Sci 25:413–430

Jin YG, Shen SZ, Henderson CM, Wang XD, Wang W, Wang Y, Cao CQ, Shang QH (2006) The Global Stratotype Section and Point (GSSP) for the boundary between the Capitanian and Wuchiapingian stage (Permian). Episodes 29:253–262

Lai XL, Yin HF, Yang FQ (1995) Reconstruction of the Qinling Triassic Paleo-ocean. J Earth Sci 20:648–656

Lai XL, Yang FQ, Hallam A, Wignall PB (1996) The Shangsi section, candidate of the global stratotype section and point of the Permian–Triassic boundary. In: Yin HF (ed) The Paleozoic–Mesozoic boundary candidates of Global Stratotype Section and Point of the Permian–Triassic Boundary. China University of Geosciences Press, Wuhan, pp 113–124

Lehrmann DJ, Wei JY, Enos P (1998) Controls on facies architecture of a large Triassic carbonate plat form: The Great Bank of Guizhou, Nanpanjiang Basin, South China. J Sediment Res 68:311–326

Lehrmann DJ, Payne JL, Pei D, Enos P, Ellwood B, Orchard MJ, Zhang J, Wei J (2007) Record of the end-Permian extinction and Triassic biotic recovery in the Chongzuo–Pingguo platform, southern Nanpanjiang Basin, Guangxi, South China. Palaeogeogr Palaeoclimatol Palaeoecol 252:200–217

Lehrmann DJ, Chaikin DH, Enos P, Minzoni M, Payne JL, Yu MY, Goers A, Wood T, Richter P, Kelley BM, Li XW, Qin YJ, Liu LY, Lu G (2015) Patterns of basin fill in Triassic turbidites of the Nanpanjiang Basin: implications for regional tectonics and impacts on carbonate-platform evolution. Basin Res 27:587–612

Li L, Meng Q-R, Pullen A, Garzione C, Wu G, Wang Y, Ma S, Duan L (2014) Late Permian–early Middle Triassic back-arc basin development in West Qinling, China. J Asian Earth Sci 87:116–129

Liu SW, He ZJ (2000) Marine conchostracans from the 'Sunjiagou Formation' of Qishan, Shaanxi. Acta Palaeontol Sin 39:230–236

Liu BJ, Xu XS (eds) (1994) Paleogeography and lithofacies Atlas of South China (Sinian–Triassic): the series of paleogeography and lithofacies of South China. Science Press, Beijing, 192p

Liu YQ, Kuang HW, Peng N, Xu H, Zhang P, Wang NS, An W (2015) Mesozoic basins and associated palaeogeographic evolution in North China. J Palaeogeogr 4:189–202

Luo M, George AD, Chen ZQ (2016) Sedimentology and ichnology of two Lower Triassic sections in South China: implications for the biotic recovery following the end-Permian mass extinction. Glob Planet Chang 144:198–212

Ma YS, Chen HD, Wang GL (eds) (2009) Atlas of tectonics and sequence lithofacies palaeogeography in Southern China (Sinian–Neogene). Science Press, Beijing, 301 p

Mei XM (2008) Significance of the unusual giant oolite and its morphologic diversity in Phanerozoic, a case of the Daye Formation of the Lower Triassic, Lichuan, Hubei Province. Mod Geosci 22:683–698

Meng Q-R, Wang E, Hu J (2005) Mesozoic sedimentary evolution of the Northwest Sichuan Basin: implication for continued clockwise rotation of the South China Block. Bull Geol Soc Am 117:396–410

Orchard MJ (2007) Conodont diversity and evolution through the latest Permian and early Triassic upheavals. Palaeogeogr Palaeoclimatol Palaeoecol 252:93–117

Pan GT, Lu SN, Xiao QH, Zhang KX, Yi FG, Hao GJ, Luo MS, Ren F, Yuan SH (2016) Division of tectonic stage and tectonic evolution in China. Front Earth Sci 6:1–23

Shen SZ, Wang Y, Henderson CM, Cao CQ, Wang W (2007) Biostratigraphy and lithofacies of the Permian system in the Laibin-Heshan area of Guangxi, South China. Palaeoworld 16:120–139

Shen SZ, Crowley JL, Wang Y, Bowring SA, Erwin DH, Sadler PM, Cao CQ, Rothman DH, Henderson CM, Ramezani J (2011) Calibrating the end-Permian mass extinction. Science 334(6061):1367–1372

Shen SZ, Ramezani J, Chen J, Cao CQ, Erwin DH, Zhang H, Xiang L, Schoepfer SD, Henderson CM, Zheng QF, Bowring SA, Wang Y, Li XH, Wang XD, Yuan DX, Zhang YC, Mu L, Wang J, Wu YS (2019) A sudden end-Permian mass extinction in South China. Geol Soc Am Bull 131:205–223

Shi G., Woods AD, Yu MY, Wei HY, (2015) Two episodes of evolution of trace fossils during the Early Triassic in the Guiyang area, Guizhou Province, South China. Palaeogeogr Palaeoclimatol Palaeoecol 426:275–284

Shi G, Woods AD, Yu MY, Li XW, Wei HY, Qiao D (2019) Lower Triassic limulid trackways (Kouphichnium) from the southwestern margin of the Yangtze carbonate platform: Palaeoenvironmental and palaeoecological implications. Palaios 34:229–243

Song HY, Tong JN, Algeo TJ, Horacek M, Qiu HO, Song HJ, Tian L, Chen ZQ (2013) Large vertical $\delta^{13}C$ gradients in Early Triassic seas of the South China craton: implications for oceanographic changes related to Siberian Traps volcanism. Glob Planet Chang 105:7–20

Su CM, Chen ZQ, Wang X, Wu SQ, Guo Z (2021) Biotic and palaeoecological variations in the Permian-Triassic boundary microbialite (Xiejiacao, South China): implication for a two-phase ecological crisis in microbialite ecosystems. Glob Planet Chang 207:103679

Tewari R, Ram-Awatar, Pandita SK, McLoughlin S, Agnihotri D, Pillai SSK, Singh V, Kumar K, Bhat GD (2015) The Permian–Triassic palynological transition in the Guryul Ravine section, Kashmir, India: implications for Tethyan-Gondwanan correlations. Earth Sci Rev 149:53–66

Tong J, Zakharov YD, Orchard MJ, Yin H, Hansen HJ (2003) A candidate of the Induan–Olenekian boundary stratotype in the Tethyan region. Sci China Earth Sci 46:1182–1200

Tong J, Zakharov YD, Wu SB (2004) Early Triassic ammonoid succession in Chaohu, Anhui Province. Acta Palaeontol Sin 43:192–204

Tong J, Zuo J, Chen ZQ (2007) Early Triassic carbon isotope excursions from South China: Proxies for devastation and restoration of marine ecosystems following the end-Permian mass extinction. Geol J 42:371–389

Tong JN, Chu DL, Liang L, Shu WC, Song HJ, Song T, Song HY, Wu YY (2019) Triassic integrative stratigraphy and timescale of China. Sci China Earth Sci 62:189–222

Tu CY, Chen ZQ, Retallack GJ, Huang YG, Fang YH (2016) Proliferation of MISS related microbial mats following the end-Permian mass extinction in terrestrial ecosystems: evidence from the Lower Triassic of the Yiyang area, Henan Province, North China. Sediment Geol 333:50–69

Wang HZ (ed) (1985) Atlas of the palaeogeography of China. Cartographic Publishing House, Beijing, 281 p

Wang ZH, Cao YY (1981) Conodonts of the Early Triassic from the Lichuan area, Hubei Province. Acta Palaeontol Sin 20:363–378

Wang GQ, Xia WC (2004) Conodont zonation across the Permian–Triassic boundary at the Xiakou section, Yichang city, Hubei province and its correlation with the global stratotype section and point of the PTB. Can J Earth Sci 41:323–330

Wang Y, Liu X, Fu D, Wei G (1995) Permian–Triassic sedimentary-tectonic evolution of the south Qilian area. Reg Geol China 14(3):245–251

Wang HQ, Zhu YH, Lin QX, Li YL, Wang K (2010) LA-ICP-MS zircon U-Pb dating of the gabbro from Longwu Gorge ophiolite, Jianzha-Tongren Area, Qinghai, China and its geological significance. Geol Bull China 29:86–92

Xu H, Zhao Z, Lu F, Yang Y, Tang Z, Sun G, Xu Y (2004) Tectonic evolution of the Nanhuabei area and analysis about its petroleum potential. Geotecton Metallog 28:450–463

Xu YL, Chen ZQ, Feng XQ, Wu SQ, Tu CY (2017) Proliferation of MISS-related microbial mats following the end-Permian mass extinction in northern margins of the Paleo-Tethys Ocean: evidence from southern Qilianshan region, western China. Palaeogeogr Palaeoclimatol Palaeoecol 474:198–213

Yan CB (2013) Conodont biostratigraphy of the Nan pan jiang Basin during the Early and Middle Triassic. Ph.D. thesis. China University of Geo sciences, China University of Geosciences Press, Wuhan, 99p

Yan CB, Wang LN, Jiang H, Wignall PB, Sun YD, Chen YL, Lai XL (2013) Uppermost Permian to Early Triassic conodonts at Bianyang section, Guizhou Province, South China. PALAIOS 28:509–522

Yan CB, Jiang HS, Lai XL, Sun YD, Yang B, Wang LN (2015) The relationship between the "Green-Bean Rock" layers and conodont Chiosella timorensis and implications on defining the Early- Middle Triassic boundary in the Nanpanjiang Basin, South China. J Earth Sci 26:236–245

Yang ZY, Yin HF, Xu GR, Wu SB (1983) The Triassic of the South Qilian Mountain. Geological Publishing House, Beijing, 259 pp

Yang B, Lai XL, Wignall PB, Jiang HS, Yan CB, Sun YD (2012) A newly discovered earliest Triassic chert at Gaimao section, Guizhou, southwestern China. Palaeogeogr Palaeoclimatol Palaeoecol 344–345:69–77

Yin HF, Lin HM (1979) Marine Triassic faunas and the geologic time from Shihchienfeng Group in the northern Weihe River Basin. J Stratigr 3:233–241

Yin H, Ding M, Zhang K, Tong J, Yang F, Lai X (1995) Dongwuan–Indosinian (Late Permian–Middle Triassic) ecostratigraphy of the Yangtze Region and its margins. Science Press, Beijing, 338p

Yin H, Zhang K, Tong J, Yang Z, Wu S (2001) The global stratotype section and point (GSSP) of the Permian- Triassic boundary. Episodes 24:102–114

Yu HZ, Lu FL, Guo QX, Lu WZ, Wu JY, Han SH (2005) Proto sediment basin types and tectonic evolution in the southern edge of North China Plate. Pet Geol Explor 27:111–117

Yu JX, Broutin J, Chen ZQ, Shi X, Li H, Chu DL, Huang QS (2015) Vegetation changeover across the Permian–Triassic Boundary in Southwest China: extinction, survival, recovery and paleoclimate: a critical review. Earth Sci Rev 149:203–224

Zhai MG, Bian AG, Zhao TP (2000) The amalgamation of the supercontinent of North China craton at the end of Neoarchean, and its break up during the Late Paleoproterozoic and Mesoproterozoic. Sci China Earth Sci 43:219–232

Zhang GW, Zhang BR, Yuan XC, Xiao QH (2001) Qinling orogenic belt and continental geodynamics. Beijing Science Press, Beijing

Zhang GW, Guo AL, Yao AP (2004) Western Qinling–Songpan tectonic node in China's continental tectonics. Front Earth Sci 11:23–32

Zhang KX, He WH, Xu YD, Luo MS, Song BW, Kou XH, Zhang ZY, Xiao QH, Pan GT (2016) Palaeogeographic distribution and tectonic evolution of OPS in China. Front Earth Sci 6:24–30

Zhang LJ, Buatois LA, Gabriela Mángano M, Qi YA, Zhang X, Sun S, Tai C (2017) Early Triassic estuarine depauperate *Cruziana* Ichnofacies from the Sichuan area of South China and its implications for the biotic recovery in brackish-water settings after the end-Permian mass extinction. Palaeogeogr Palaeoclimatol Palaeoecol 485:351–360

Zhang LJ, Buatois LA, Mángano MG, Gong YM, Feng QL, Qi YA, Luo M, Zhang X (2018) Uppermost Permian trace fossils along a shelf to slope transect in South China and their implications for oceanic redox evolution and extinction pattern. Glob Planet Chang 167:74–86

Zhang LJ, Buatois LA, Mángano MG, Qi YA, Tai C (2019) Early Triassic trace fossils from South China marginal-marine settings: implications for biotic recovery following the end-Permian mass extinction. Geol Soc Am Bull 131:235–251

Zhang LJ, Zhang X, Buatois LA, Mángano MG, Shi GR, Gong YM, Qi YA (2020a) Periodic fluctuations of marine oxygen content during the latest Permian. Glob Planet Chang 195:103326

Zhang LJ, Fan RY, Dang ZY et al (2020b) The youngest known *Dictyodora* from the Late Permian (Lopingian) deep sea in West Qinling, Central China. Palaeogeogr Palaeoclimatol Palaeoecol 558:109948

Zhao LS, Orchard MJ, Tong JN, Sun Z, Zuo JX, Zhang S, Yun A (2007) Lower Triassic conodont sequence in Chaohu, Anhui Province, China, and its global correlation. Palaeogeogr Palaeoclimatol Palaeoecol 252:24–38

Zhao LS, Chen YL, Chen ZQ, Cao L (2013) Uppermost Permian to Lower Triassic conodont zonation from Three Gorges area, South China. PALAIOS 28:523–540

Zhao X, Tong J, Yao H, Niu Z, Luo M, Huang Y, Song H (2015) Early Triassic trace fossils from the Three Gorges area of South China: implications for the recovery of benthic ecosystems following the Permian Triassic extinction. Palaeogeogr Palaeoclimatol Palaeoecol 429:100–116

Zheng YF, Xiao WJ, Zhao GC (2013) Introduction to tectonics of China. Gondwana Res 23:1189–1206

Ziegler AM, Gibbs MT, Hulver ML (1998) A mini-atlas of oceanic water masses in the Permian period. Proc R Soc Vic 110:323–344

Chapter 3
Systematic Ichnology and Fossil Illustrations

3.1 General Summary

Trace fossils are documented from twenty sections of the five environmental categories (brackish, carbonate ramp, carbonate platform, siliciclastic shoreface, and shelf-slope-basin). Trace fossils from these sections are predominantly derived from field observations and specimens collected over the past decade. Most fossil sites have been revisited multiple times to gather specimens and capture images, enhancing the likelihood of obtaining adequate fossil materials. While some data have been referenced from prior studies, the corresponding localities have been revisited, and the associated materials have been reassessed. Additionally, taxonomic evaluations of certain ichnotaxa have been performed to refine their classification.

The brackish environmental setting is represented by the Houzhougongmiao and Longmendong sections. The ichnoassemblage examined in the Houzhougongmiao section is preserved within the thin-bedded siltstones and greenish mudstones of the upper Sunjiagou Formation, which is Griesbachian in age. The Griesbachian ichnoassemblage includes 11 ichnogenera: *Arenicolites*, *Asteriacites*, *Biformites*, *Didymaulichnus*, *Diplichnites*, *Kouphichnium*, *Laevicyclus*, *Monomorphichnus*, *Planolites*, *Protovirgularia*, and *Treptichnus*. Of these, *Asteriacites*, *Biformites*, and trackways dominate the ichnocoenosis. Trace fossils from the Lower Triassic Dangchuan and Feixianguan formations of the Longmendong section were well examined (Zhang et al. 2017, 2019). No trace fossils were identified in the lower and middle parts of the Dongchuan Formation. Six ichnogenera, *Cylindrichnus*, *Diplocraterion*, *Palaeophycus*, *Planolites*, *Siphonichnus*, and *Teichichnus*, occurred

Provide the reader with a great deal of high-quality photographs, figures, and accompanying text for the identification and description of trace fossils of the Late Permian to Early Triassic from China

X. Feng et al., *Ichnology of the Permian-Triassic Mass Extinction in China*,
https://doi.org/10.1007/978-981-96-6639-3_3

in the uppermost part of the Dongchuan Formation. The ichnofauna in the lower part of the Feixianguan Formation is diverse and abundant, including 14 ichnogenera, *Arenicolites*, *Asteriacites*, *Biformites*, *Cylindrichnus*, *Diplocraterion*, *Diplopodichnus*, *Helminthopsis*, *Lockeia*, *Palaeophycus*, *Planolites*, *Ptychoplasma*, *Rhizocorallium*, *Siphonichnus*, and *Teichichnus*. No trace fossils occur in the upper part of the Feixianguan Formation.

The carbonate ramp is composed of the GSSP Meishan, Chaohu, Susong, Daye Shatian, Daxiakou, Dangwu, and Gaimao sections from South China. The Meishan site serves as the Global Stratotype Section and Point (GSSP) for the Permian-Triassic Boundary (PTB), with the PTB occurring at the base of Bed 27c, characterized by the first appearance datum (FAD) of the conodont *Hindeodus parvus*. Trace fossils across the PTB in the Meishan section have been well studied (Zhao and Tong 2010; Zheng et al. 2013; Chen et al. 2015; Zhang et al. 2018). The Late Permian ichnofauna includes 7 ichnogenera: *Balanoglossites*, *Taenidium*, *Planolites*, *Thlassinoides*, *Skolithos*, *Chondrites*, *Phycosiphon* and *Zoophycos*. Few trace fossils have been found from other sections of the carbonate ramp during the Late Permian. In the Chaohu section, the Early Triassic ichnofauna includes 9 ichnogenera: *Arenicolites*, *Diplocraterion*, *Gyrochorte*, *Laevicyclus*, *Palaeophycus*, *Planolites*, *Rhizocorallium*, *Thalassinoides*, *Treptichnus*. In Susong section, the Early Triassic ichnofauna includes 11 ichnogenera according to a systematic investigation (Luo et al. 2016), namely, *Archaeonassa*, *Arenicolites*, *Chondrites*, *Laevicyclus*, *Palaeophycus*, *Phycodes*, *Planolites*, *Psammichnites*, *Thalassinoides*, *Treptichnus*, *Rhizocorallium*. The Lower Triassic in the Shatian section contains 9 ichnogenera: *Archaeonassa*, *Arenicolites*, *Cochlichnus*, *Helminthoidichnites*, *Palaeophycus*, *Planolites*, *Thalassinoides*, *Treptichnus*, and *Rhizocorallium*. In the Daxiakou section, *Arenicolites*, *Arenituba*, *Circulichnis*, *Gordia*, *Gyrochorte*, *Helminthoidichnites*, *Palaeophycus*, *Planolites*, *Thalassinoides*, *Treptichnus*, and *Rhizocorallium* are documented from the Lower Triassic strata of the Daye and Jialingjiang formations. In the Dangwu section, imprints and trackways of *Kouphichnium* are the most distinctive features when compared to other Early Triassic ichnoassemblages. Burrows from the Dangwu section are dominated by *Palaeophycus*, *Planolites*, *Phycosiphon*, *Phycodes*, *Thalassinoides*, and *Sinusichnus*, with rare occurrences of *Teichichnus*, *Helminthopsis* and *Arenituba*. The simple horizontal trails *Helminthoidichnites*, *Archaeonassa*, and *Gordia* are subordinate elements. The Early Triassic ichnotaxa in the Gaimao section are dominated by *Archaeonassa*, *Arenituba*, *Arenicolites*, *Circulichnis*, *Diplocraterion*, *Gordia*, *Gyrochorte*, *Helminthoidichnites*, *Kouphichnium*, *Laevicyclus*, *Lockeia*, *Palaeophycus*, *Phycodes*, *Planolites*, *Psammichnites*, *Sinusichnus*, *Skolithos*, *Teichichnus*, *Thalassinoides*, *Treptichnus*, and *Rhizocorallium*.

The carbonate platform comprises Xiejiacao, Shawan, upper Longmendong, and Lichuan sections from South China. The late Permian trace fossils in carbonate platform environmental settings are rare. The Early Triassic trace fossils of the Xiejiacao section are documented from the Yelang and Jialingjiang formations. 10 ichnogenera are collected: *Gyrochorte*, *Helminthoidichnites*, *Palaeophycus*, *Phycodes*, *Planolites*, *Psammichnites*, *Sinusichnus*, *Skolithos*, *Thalassinoides*,

Treptichnus. The Jialingjiang Formation of the Shawan and upper part of the Longmendong sections yield 15 ichnogenera, namely *Balanoglossites*, *Biformites*, *Diplocraterion*, *Helminthopsis*, *Helminthoidichnites*, *Laevicyclus*, *Lockeia*, *Palaeophycus*, *Planolites*, *Phycodes*, *Rhizocorallium*, *Sinusichnus*, *Skolithos*, *Thalassinoides*, and *Treptichnus*. The Smithian (Early Triassic) ichnoassemblage from Member II of the Daye Formation of the Lichuan section contains 12 ichnogenera: *Arenicolites*, *Arenituba*, *Circulichnis*, *Cochlichnus*, *Gordia*, *Gyrochorte*, *Helminthopsis*, *Laevicyclus*, *Palaeophycus*, *Planolites*, *Thalassinoides*, and *Treptichnus*.

The siliciclastic shoreface environmental setting contains the Zhihema section from the southern Qilian area of North China and the upper part of the Penglaitan section from South China. Trace fossils from the Zhihema section are documented from the Early Triassic Xiahuancang Formation. A total of 14 ichnogenera: *Arenicolites*, *Arenituba*, *Cochlichnus*, *Diplocraterion*, *Gyrochorte*, *Lockeia*, *Palaeophycus*, *Phycodes*, *Planolites*, *Rhizocorallium*, *Rosselia*, *Skolithos*, *Taenidium*, and *Thalassinoides* are identified. Trace fossils from the upper part of the Penglaitan section are documented from the upper Talung and lowermost Luolou formations, which are Changhsingian (late Permian) in age. Seven ichnogenera: *Phycodes*, *Planolites*, *Protovirgularia*, *Rhizocorallium*, *Rosselia*, *Skolithos*, and *Thalassinoides* are identified.

Trace fossils within the Zhihema section are identified from the Lower Triassic Xiahuancang Formation, A total of 14 distinct ichnogenera have been recognized, including *Arenicolites*, *Arenituba*, *Cochlichnus*, *Diplocraterion*, *Gyrochorte*, *Lockeia*, *Palaeophycus*, *Phycodes*, *Planolites*, *Rhizocorallium*, *Rosselia*, *Skolithos*, *Taenidium*, and *Thalassinoides*. In the upper part of the Penglaitan section, trace fossils are documented within the upper Talung and lowermost Luolou formations, which are Changhsingian (late Permian) in age. Seven ichnogenera: *Phycodes*, *Planolites*, *Protovirgularia*, *Rhizocorallium*, *Rosselia*, *Skolithos*, and *Thalassinoides* are identified.

The deeper shelf-slope-basin environmental settings contain Shangsi, lower Penglaitan sections from South China, and Zhayougou, Malongou, and the Longwuxia sections from the western Qinling area of North China. The Changhsingian ichnoassemblage from the Shangsi section contains five ichnogenera: *Chondrites*, *Palaeophycus*, *Planolites*, *Thalassinoides*, and *Zoophycos* (Zhang et al. 2018, 2020), and the Early Triassic ichnoassemblage contains two ichnogenera: *Cylindrichnus* and *Thalassinoides* (Godbold et al. 2017). The Changhsingian ichnoassemblage from the lower Penglaitan section includes 12 ichnogenera: *Chondrites*, *Gordia*, *Gyrochorte*, *Helminthopsis*, *Lockeia*, *Nereites*, *Palaeophycus*, *Phycodes*, *Planolites*, *Thalassinoides*, *Treptichnus*, and *Zoophycos*. The Changhsingian ichnoassemblage from the Malonggou section includes 14 ichnogenera: *Chondrites*, *Cruziana*, *Dictyodora*, *Diplichnites*, *Halopoa*, *Helminthoidichnites*, *Helminthopsis*, *Nereites*, *Palaeophycus*, *Phycodes*, *Planolites*, *Protovirgularia*, *Taenidium*, and *Thalassinoides*. The Early Triassic ichnoassemblage from the Zhayougou section includes 4 ichnogenera: *Helminthopsis*, *Nereites*, *Palaeophycus*, and *Planolites*. The Early Triassic ichnoassemblage from the

Longwuxia section includes 11 ichnogenera: *Acanthorhaphe*, *Capodistria*, *Chondrites*, *Circulichnis*, *Dendrotichnium*, *Helminthoidichnites*, *Helminthopsis*, *Lorenzinia*, *Nereites*, *Planolites*, and *Treptichnus*.

3.2 Systematic Ichnology

Arenicolites Salter 1857
Arenicolites isp.
Figure 3.1

Description Simple vertical U-shaped tubes without spreiten, preserved as full reliefs in sandstones, mudstones, or argillaceous limestones. Limbs are closely spaced, circular to sub-circular in cross section, and parallel to each other. The distance between two limbs is about 8 mm and can extend ~80 mm. The depth of the burrows is 9–11 cm; the diameter of the tubes is 0.8–1 cm. On bedding planes, traces are shown as pairs of holes. The width of the burrow and the diameter of the limb remain uniform throughout the entire height of the burrow. The burrow fills are distinct and differ in composition from the surrounding host sediment.

Remarks More than ten ichnospecies of *Arenicolites* are described in the literature; however, many of them are poorly studied and their validity needs further identification. Some other U-shaped burrows such as *Tisoa* and *Diplocraterion* are often confused with *Arenicolites*. The type ichnospecies *Arenicolites carbonarius* (but in Knaust (2019) is referred to as *A. carbonaria*) and is a branched form (Pollard 1999). *A. sparsus* stands out as the most distinctive ichnospecies, characterized by a simple, regular U-shaped burrow with vertical limbs and without a thick inner lining (Rindsberg and Kopaska-Merkel 2005). *A. longistriatus* is subhorizontal and longitudinally striate (Rindsberg and Kopaska-Merkel 2005). *A. curvatus* is characterized by inclined U-shaped burrows, ranging from vertical to nearly horizontal, that penetrate a thick sandstone bed. The cross-sectional profiles of the inclined tubes exhibit elliptical, whereas sub-vertical tubes appear with a circular cross-section. The margins of the burrows range from unlined to densely lined and have been regarded as synonymous with *Tisoa habichi* (Knaust 2019). *A. tenuis* was initially erected on incomplete specimens (Kulkarni et al. 2008); according to its characteristics, it has been re-classified as *Tisoa siphonalis* (Knaust 2019). *A. helixus* is established using a single field specimen, which conflicts with Recommendation 73A of the ICZN (2018) (Rajkonwar et al. 2013). Furthermore, the original description of the elongated U-shaped burrow primarily reflects the division of contrasting burrow fill (in relation to the host rock) as a response to processes like compaction, shearing, or weathering, and thus is also now regarded as *Tisoa siphonalis* (Knaust 2019). The orientation of the limbs is variable in the deep U-shaped burrow of *A. variabilis*, which also shows some morphological affinity with *Tisoa* (Knaust 2019).

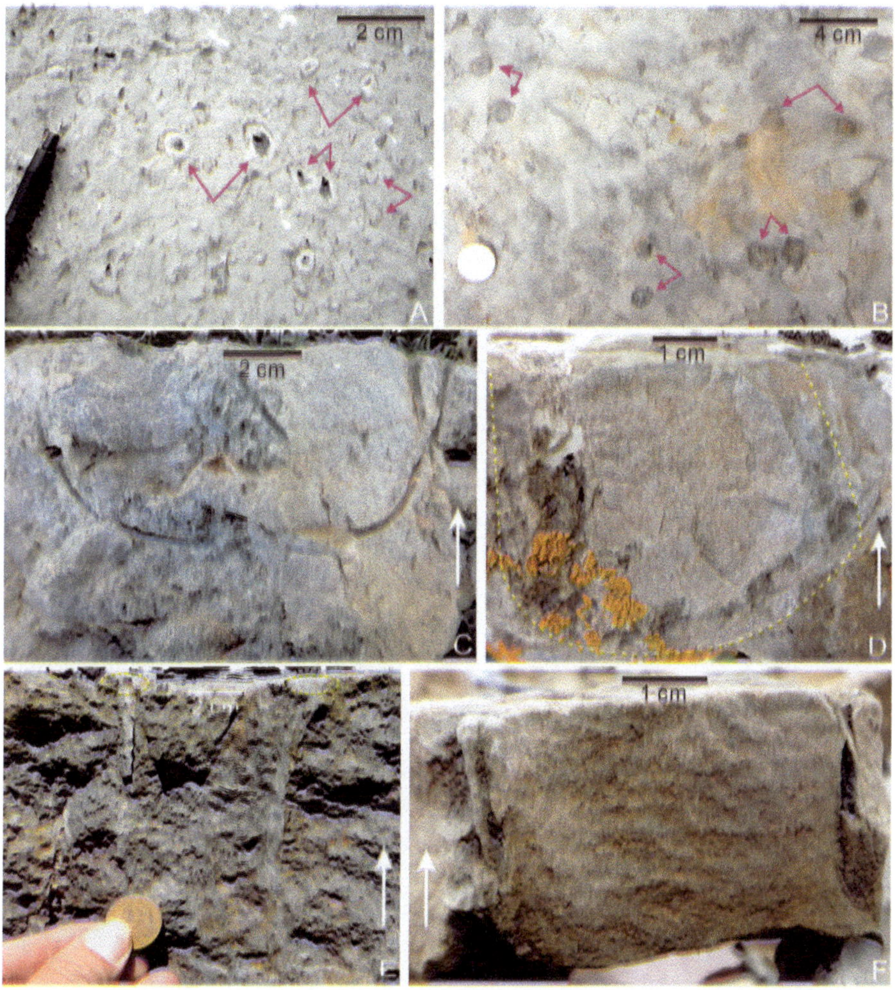

Fig. 3.1 *Arenicolites* in outcrop. (**a–b**), *Arenicolites* isp., preserved as pairs of holes or shafts (arrows) on the upper surface of argillaceous limestone. Lower Triassic Anshun Formation, Gaimao section, Guizhou Province, South China. (**c–f**), *Arenicolites* isp., preserved as vertical U-shaped tubes in sandstone beds. Lower Triassic Xiahuancang Formation, Zhihema section, Qinghai Province, northwestern China

U-shaped burrows are probably created by varieties of worm-like organisms, which could be filter feeders, sediment feeders or both. The polychaetes, a diverse group of primarily marine worms, are known for creating various burrows, such as U-shaped ones. These burrows are coated with mucous and are used for extended periods. Amphipod crustaceans are a frequent creator of modern *Arenicolites*-like burrows within marine environments (Bradshaw 2010; Baucon et al. 2014). Additionally, insects might create comparable traces in terrestrial settings (Rindsberg and Kopaska-Merkel 2005).

Archaeonassa Fenton and Fenton 1937
Archaeonassa fossulata Fenton and Fenton 1937
Figure 3.2

Description The trails are characterized as horizontal, unbranched, straight or curved with irregular levees on both sides. The cross-section exhibits a V-shape, and the diameter ranges from 5 to 18 mm. The height of levees is 2–8 mm. The length of a whole trail is about 20 cm. These trails are saved as negative epirelief. This ichnogenus is only documented from the argillaceous limestone of the Daye Formation from the Dangwu section.

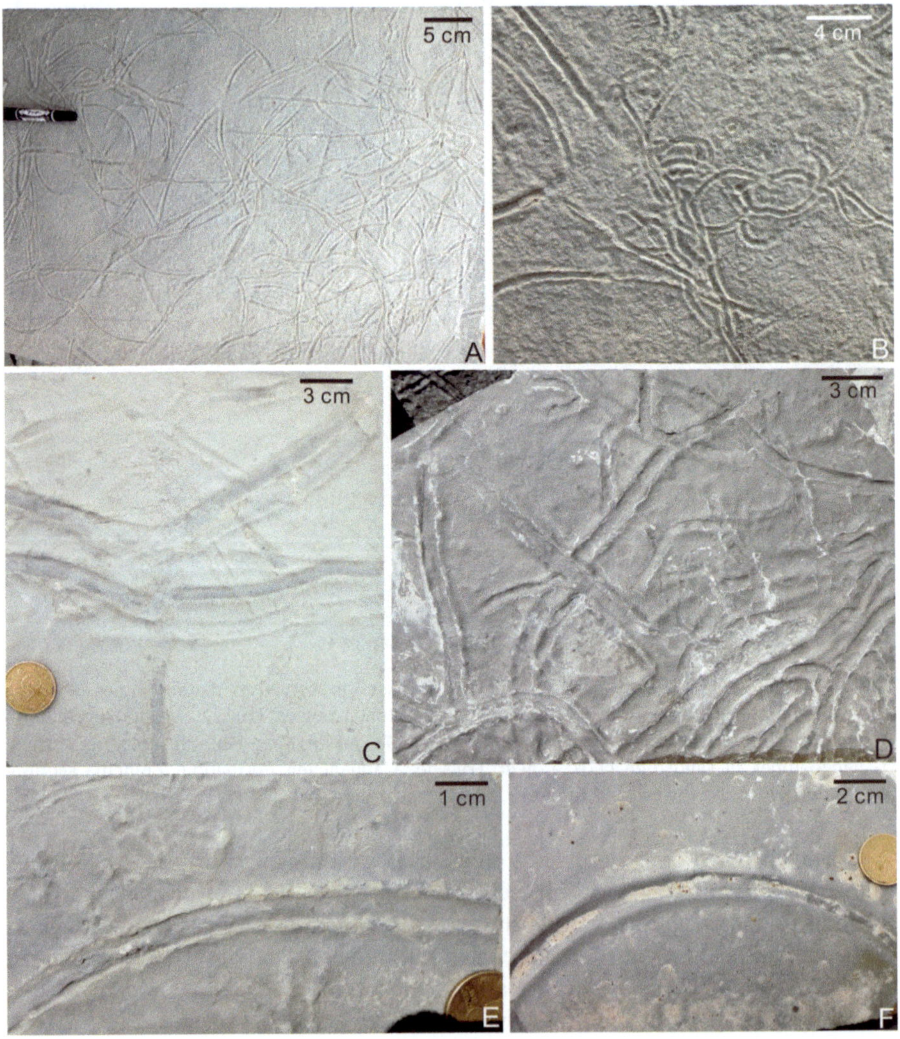

Fig. 3.2 *Archaeonassa* in outcrop. (**a–f**), *A. fossulata* preserved on the upper surface of argillaceous limestone. Lower Triassic Daye Formation, Dangwu section, Guizhou Province, South China

Remarks This ichnogenus has been examined and reviewed by Buckman (1994) and Yochelson and Fedonkin (1997). These trace fossils described in this study, characterized by features such as a median groove bordered by rounded ridges, are classified under *A. fossulata*. *Archaeonassa* is very similar to the so-called Paleozoic "*Scolicia*-group" traces such as *Psammichnites* and *Aulichnites*. However, the formation of *Archaeonassa* results from the surface activity of the producer, creating a simple negative epirelief trail. In contrast, *Scolicia*-group traces establish a complex backfill pattern, which is always formed by endogenic subsurface movement (Uchman 2005). Jensen (2003) describes *Archaeonassa* as having a flat, wide central area, and the feature is also present in our specimen. *A. fossulata* features two parallel convex levees, often with transverse or oblique ornamentation. *Archaeonassa* is predominantly formed in shallow marine environments (Jensen 2003; Uchman and Martyshyn 2020) as well as in continental environments, especially within fluvial floodplains (Buatois and Mángano 2002). *Archaeonassa* is considered a crawling trail of gastropods (Fenton and Fenton 1937; Buckman 1994; Stanley and Feldmann 1998) or crustaceans (Yochelson and Fedonkin 1997).

Arenituba Chamberlain 1971
Arenituba verso Chamberlain 1971
Figure 3.3

Description Irregularly arranged tunnels that are straight or have a slight curve, emanating from a central gallery, either discrete or grouped together, and varying in shape from straight or curved, with surfaces ranging from smooth to finely annulate, partly preserved as full relief or concave epireliefs. The distal parts of the horizontal tunnels turn upwards. Branching of the horizontal tunnels is rarely observed. If present, branches are arranged in flat bunches. The central gallery probably represents a cast of a vertical shaft. Tunnels are not distributed evenly around the shaft. Both vertical shaft and horizontal burrows are filled with sandy material that is lithologically similar to the host sediment. The filling is structureless. The central shaft, characterized by indistinct borders, measures approximately 4 mm in diameter. The radial tunnels have a width ranging from 3 to 4 mm. A total of six to ten well-defined tunnels are present. The length of these tunnels varies and is significantly influenced by selective preservation. The longest existing ray measures 26 mm in length. The overall width of a burrow system ranges between 25 and 50 mm.

Remarks Chamberlain (1971) initially classified *Arenituba* under the name *Micatuba*; however, since *Micatuba* was already designated for a genus of agglutinated foraminifer, Stanley and Pickerill (1995) suggested renaming it to *Arenituba*. *Arenituba* is recognized as a monospecific ichnotaxon, *A. verso* (Chamberlain 1971) is recognized as a monospecific ichnospecies. *Arenituba* is likely formed by a worm-like organism that maintains a permanent residence and continuously creates radial tubes to search for food while utilizing sand grains bound by mucus for defense (Chamberlain 1971; Stanley and Pickerill 1995). *Arenituba* is recognizable

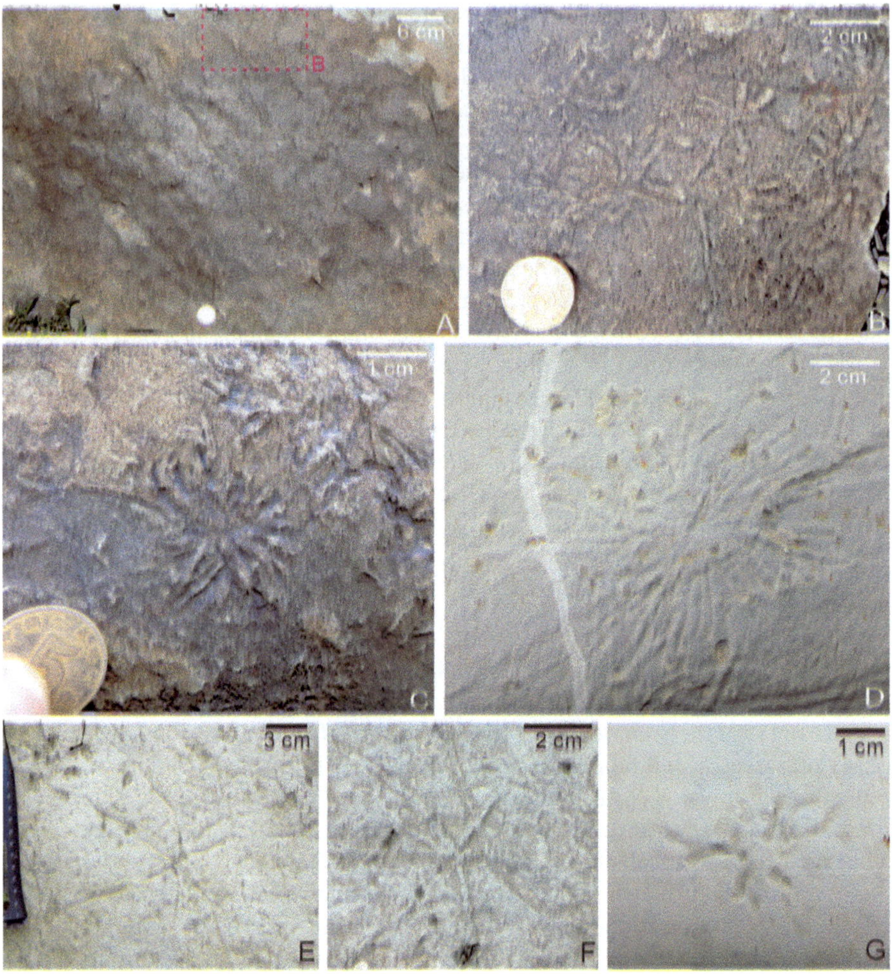

Fig. 3.3 *Arenituba* in outcrop. (**a–c**), *A. verso* preserved on the sole bedding surface of sandstone, **b** and **c** are from the same bedding surface with **a**, and B is a close-up view of the dashed block in **a**. Lower Triassic Xiahuancang Formation, Zhihema section, Qinghai Province, northwestern China. (**d, g**), *A. verso* preserved on the upper surface of argillaceous limestone, preserved as full reliefs. Lower Triassic Daye Formation, Dangwu section, Guizhou Province, southwestern China. (**e–f**), *A. verso* in the argillaceous limestone, **e** is preserved as concave epirelief, and **f** is convex hyporelief. Lower Triassic Anshun Formation, Gaimao section, Guizhou Province, southwestern China

from other comparable radiate ichnogenera due to its simple horizontal tunnels and a central vertical shaft. *Arenituba*, together with other similar radial to rosette trace fossils and their evolution through the Phanerozoic, have been well documented (Muñoz et al. 2019). The initial appearance of *Arenituba* is recorded in Upper Ordovician offshore deposits (Stanley and Pickerill 1995). Subsequent findings have been observed in both shallow and deep marine environments throughout the

Phanerozoic (Chamberlain 1971; Miller 1986, 1993; Lockley et al. 1987; Głuszek 1998; Uchman 1998; Petter and Steel 2006; Mikuláš et al. 2013; Wesolowski et al. 2018).

Asteriacites von Schlotheim 1820
Asteriacites lumbricalis von Schlotheim 1820
Figure 3.4

Description Star-shaped traces, characterized by a central disc-shaped area with a pentagonal outline, feature five arms with a large length-to-width ratio emanating from the disc's corners. The arms, as recorded from the Houzhougongmiao section (HZGM), measure 10–20 mm in length and 2 mm in width, with the central disc approximately 5 mm in size. The arms exhibit a length-to-width ratio of 5–10. These fossils were only found in the upper Sunjiagou Formation of the HZGM section, central Shaanxi Province, Western China. They were preserved as convex hyporeliefs on sandstone beds and were distinguished by a pentagonal central disc and slender arms with a high length-to-width ratio.

Remarks The ichnogenus *Asteriacites* comprises eight recognized ichnospecies. Knaust and Neumann (2016) established a novel classification framework.

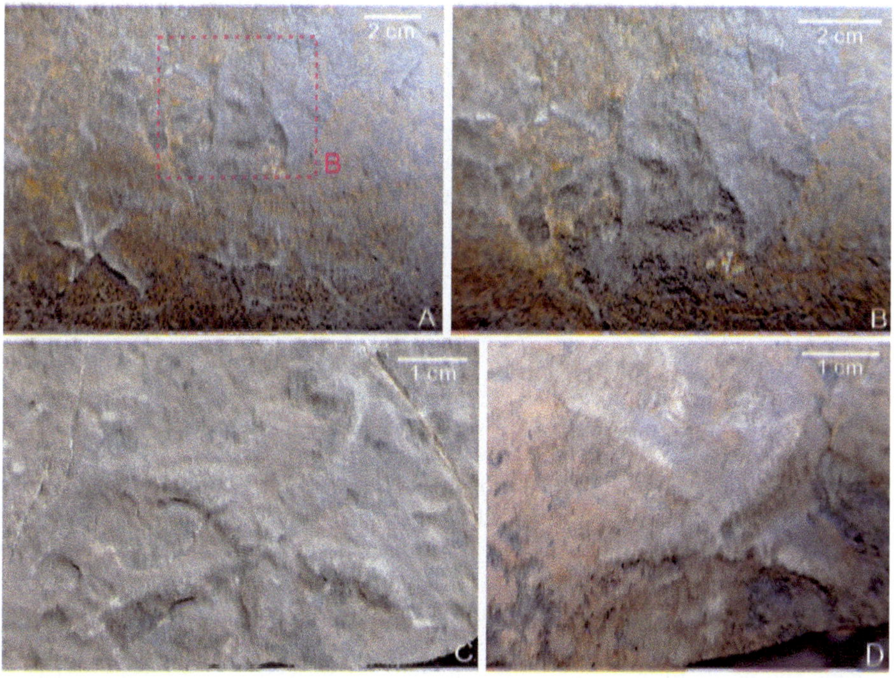

Fig. 3.4 *Asteriacites* in outcrop. (**a**–**d**), *A. lumbricalis* preserved as convex hyporeliefs in sandstone beds, a star-shaped central disc is visible in **c**. Lower Triassic Sunjiagou Formation, Houzhougongmiao section, Shaanxi Province, northwestern China

A. lumbricalis von Schlotheim 1820, *A. stelliformis* (Miller and Dyer 1878) Osgood 1970, and *A. quinquefolius* (Quenstedt 1876) Seilacher 1953, represent three valid ichnospecies distinguished by the reducing length/width ratios of their arm imprints. *A. lumbricalis* is the type ichnospecies, which is distinguished by a pentagonal central disc and narrow arms characterized by a significant length-to-width proportion (Knaust and Neumann 2016). The arms of *A. lumbricalis* from the HZGM section range in length from 10 to 20 mm and are 2 mm wide, while the central disc measures approximately 5 mm. Consequently, the specimens from HZGM are identified as *A. lumbricalis*. Moreover, the production of *A. lumbricalis* is solely attributed to ophiuroids due to the unique imprint of a central disc, which is not present in asteroids (Mikuláš 1990; Mángano et al. 1999; Wilson and Rigby 2000; Bell 2004; Schatz et al. 2013; Baucon and De Carvalho 2016; Knaust and Neumann 2016; Belaústegui et al. 2017). *A. lumbricalis* is characterized by burrowing or resting behavior with both the disc and arms buried together. The ophiuroid buries itself in sediment by moving its central disc in swinging motions, while maintaining its arms spread and at rest (Häntzschel 1975; Bell 2004; Schatz et al. 2013; Baucon and De Carvalho 2016; Morton et al. 2023). Unburied ophiuroid arms can move sideways, resulting in repetitive traces characterized by branching arms (Mángano et al. 1999).

Acanthorhaphe Książkiewicz 1970
Acanthoraphe delicatula Książkiewicz 1977
Figure 3.5

Description Hypichnial, arch-shaped burrows with short appendages usually on the convex side. These appendages branch off at right angles from the main burrow, spaced between 4 and 8 mm apart. The main burrow is about 50–80 mm in length and 2 mm in width; the short appendages are about 5–10 mm in length and 1.5–2 mm in width and are consistent in length and width, more or less among individuals. The traces are smooth, without any transverse or longitudinal striae. Curvity of the main burrow is relatively high and consistent among individuals. Burrow fill is indistinct with the host rock. Burrows were preserved as convex hyporelief on the siltstone beds from the Guomugou Formation of the Longwuxia section.

Remarks *Acanthorhaphe* owned three ichnospecies, the type ichnospecies *Acanthorhaphe incerta*, and another two ichnospecies *A. delicatula*, and *A. longispinosa*. Książkiewicz (1970) erected *A. incerta* and *A.* isp, and later he redescribed the material and erected the former *A.* isp as *A. delicatula* (Książkiewicz 1977). On the same basis as the Książkiewicz specimen, Seilacher (1977) proposed *A. pectinata*, which has been regarded as a junior synonym of *A. delicatula* (Uchman 1998). Kozur et al. (1996) proposed *A. longispinosa* from the deep-water turbidites in Lercara Formation of Early Permian from western Sicily (Italy), but did not illustrate and discuss this new ichnospecies; thus the validity of *A. longispinosa* is problematic and needs further discussion. The type ichnospecies *A. incerta* consists of multiple meanders that are interrupted to form a sequence of narrow, semielliptical traces. These traces feature branches of varying lengths on their convex sides. Thus,

Fig. 3.5 *Acanthorhaphe* in outcrop. (**a–f**), *A. delicatula* preserved as convex hyporeliefs in sandstone beds (arrows). Lower Triassic Guomugou Formation, Longwuxia section, Qinghai Province, northwestern China

our material from the western Qinling area is not consistent with this diagnosis. In contrast, *A. delicatula* is characterized by a single bend of semi-elliptical or semicircular form, and its short appendages emanate from the main burrow at various angles. The resulting trace is smooth. We therefore treated our Early Triassic specimens as *A. delicatula*. *Acanthorhaphe* is a typical graphoglyptids, which is predominantly identified in deep-sea flysch deposits. Generally, *Acanthorhaphe* has been rarely reported from around the world since its erection from the Miocene flysch deposits of the Polish Carpathians. Kozur et al. (1996) reported an early

Permian case. Here we reported this ichnogenus from the deep marine strata following the P-Tr mass extinction.

Balanoglossites Mägdefrau 1932
Balanoglossites triadicus Mägdefrau 1932
Figures 3.6 and 3.7

Description Irregular, three-dimensional complex burrow system, with Y-shaped, U-shaped, or J-shaped burrows in vertical profile, often with multiple openings, some of which are funnel-shaped. The cross-section is often nearly circular or elliptical with slight indentations. In vertical directions, tunnels branched irregularly in different directions, often with short blind lateral branches in the shape of acorns, bulbs, or spears. The diameter of the burrow is not uniform, with a large range of variation, mostly 5–40 mm. The maximum depth of the burrow can reach 20 cm, and there are often irregular leaf-shaped, tiny, and short lateral branches at the edge of the burrow. The burrow fillings were yellow siltstone of the upper layer, which was significantly different from the host rock of dark gray mudstone limestone.

Remarks Mägdefrau (1932) originally described and named *Balanoglossites* in the Middle Triassic limestone of the German basin. *Balanoglossites* was recognized as cavity limestone by early sedimentologists and stratigraphers from the German Muschelkalk (Middle Triassic) long before its formal establishment (Wagner 1897) and vermicular limestone (Zenker 1836; von Eck 1872). This unique ichnofabric has been identified in various regions worldwide but is linked to differing ichnotaxonomic classifications or described using distinct terminologies. Examples include dolomitic mottling (Beales 1953), nodular chalk/limestone (Fürsich 1973; Kennedy and Garrison 1975; Abed and Schneider 1980), rubbly limestone (Jones et al. 1979), pseudochickenwire texture (Pedley 1992), swiss cheese texture (Beach 1995), pseudobreccias (Horbury and Qing 2004), biogenic leopard patch structure (Qi et al. 2014), and spongy–like burrows (Todaro et al. 2016). Mägdefrau (1932) established the ichnogenus *Balanoglossites* in the Middle Triassic limestone formations from the Jena region of Germany and described *Balanoglossites* as "Thick borings with irregular branches and several openings." According to whether the openings of borings are funnel-shaped, two ichnospecies, *B. teradicus* and *B. urystomus*, are divided. *B. teradicus* is the type ichnospecies and does not have funnel-shaped openings, while *B. urystomus* has funnel-shaped openings. After Kazmierczak and Pszczółkowski (1969) questioned Mägdefrau's basis for dividing the two ichnospecies, they found the burrows' openings of samples from the Holy Defrau Mountains of Poland were incomplete due to the erosion of the river surface, and the morphology of burrows' openings changed with late erosion and other effects. Therefore, it is unreasonable to divide the two ichnospecies only according to the morphology of the opening. Based on a re-examination of the type specimens and supplementary materials from the type locality, Knaust (2008) re-evaluated *Balanoglossites*, identifying it as a complex trace fossil that incorporates both burrowing and boring elements. Morphological analysis led to the categorization of

Fig. 3.6 *Balanoglossites* in outcrop. (**a–g**), *B. triadicus* preserved in limestone, **d** is bedding plane view, and others are in vertical profile view. Lower Triassic Yongningzhen Formation, Yongningzhen section, Guizhou Province, southwestern China

Fig. 3.7 *Balanoglossites* in outcrop. (**a–f**), *B. triadicus* preserved in limestone, all are in vertical profile view. Lower Triassic Yongningzhen Formation, Yongningzhen section, Guizhou Province, southwestern China

Balanoglossites into two ichnospecies: *B. triadicus*, primarily featuring deep U- or Y-shaped tunnels, and *B. ramosus*, characterized by irregularly branching galleries. Our fossil material was mainly collected from the Early Triassic Yongningzhen

Formation of the Yongningzhen section. Based on the characteristics of the whole burrow and lateral branches, we attributed the material to *B. triadicus*.

An analysis of existing literature combined with new insights from 87 stratigraphical units indicates that *Balanoglossites* is a frequent component of predominantly monospecific ichnofabrics in carbonate rocks. It exhibits a broad geographic range in shallow marine settings, spanning from the Ediacaran through the Phanerozoic (Knaust 2021). *Balanoglossites* shares surface similarities with other ichnotaxa, particularly those thought to represent crustacean burrows like *Thalassinoides*, *Psilonichnus* or *Pholeus*, which can complicate identification. Nonetheless, distinctive features (e.g., irregularity, U- and Y-shaped components, blind side branches) are key to accurate classification (Knaust 2008). *Thalassinoides* burrow systems are distinguished by Y-shaped and/or T-shaped tunnel junctions with swellings (Häntzschel 1975; Schlirf 2000). Such a regular branching pattern is uncommon in *Balanoglossites*, even though the burrow diameter can vary significantly. *Balanoglossites* can be differentiated from *Psilonichnus* through its highly variable burrow diameters, elliptical burrow cross-sections, and multiple side branches (Knaust 2008). The tracemakers of *Balanoglossites* were originally thought to be Enteropneusta or Polychaete (Mägdefrau 1932). Kazmierczak and Pszczółkowski (1969) compared the *Balanoglossites* with burrows of modern Enteropneusta, finding that the morphology of the two is similar, and thought the tracemakers were Enteropneusta. Knaust (2008, 2013, 2021) thought that polychaetes are the most likely tracemakers of *Balanoglossites*.

Biformites Linck 1949
Biformites zhadaensis (Yang and Song 1985)
Figures 3.8 and 3.9

Description Narrow, hook-shaped, or sinuous imprints characterize this ichnogenus, which has only been found in siltstone beds of the Lower Triassic Sunjiagou Formation at the Houzhougongmiao section, located on the southern margin of the Ordos Basin. Hook-shaped imprints are particularly abundant on thin-bedded siltstone bedding planes, especially in horizons adjacent to the *Asteriacites lumbricalis*-bearing sandstone beds. A few specimens have also been found on the same sandstone beds as *A. lumbricalis*. Series of hook-shaped imprints are commonly observed in the examined beds, though isolated individuals are occasionally present. The imprints are always parallel to the bedding plane and exhibit extremely shallow penetration. Most imprints are preserved as convex hyporeliefs, though concave epireliefs also occur occasionally. The combination of clustered arcuate or hook-shaped imprints with slightly tapering terminations, along with bisymmetrically arranged transverse protuberances, distinguishes these ichnofossils from most burrow-form traces and makes them resemble casts of ophiuroid arms.

Remarks A new ichnogenus and ichnospecies, *Zhadaichnus zhadaensis* was established based on material from Triassic turbidites in Tibet (Yang and Song 1985). Later, Bell (2004) described similar traces as *Ophioichnus aysenensis* Bell 2004,

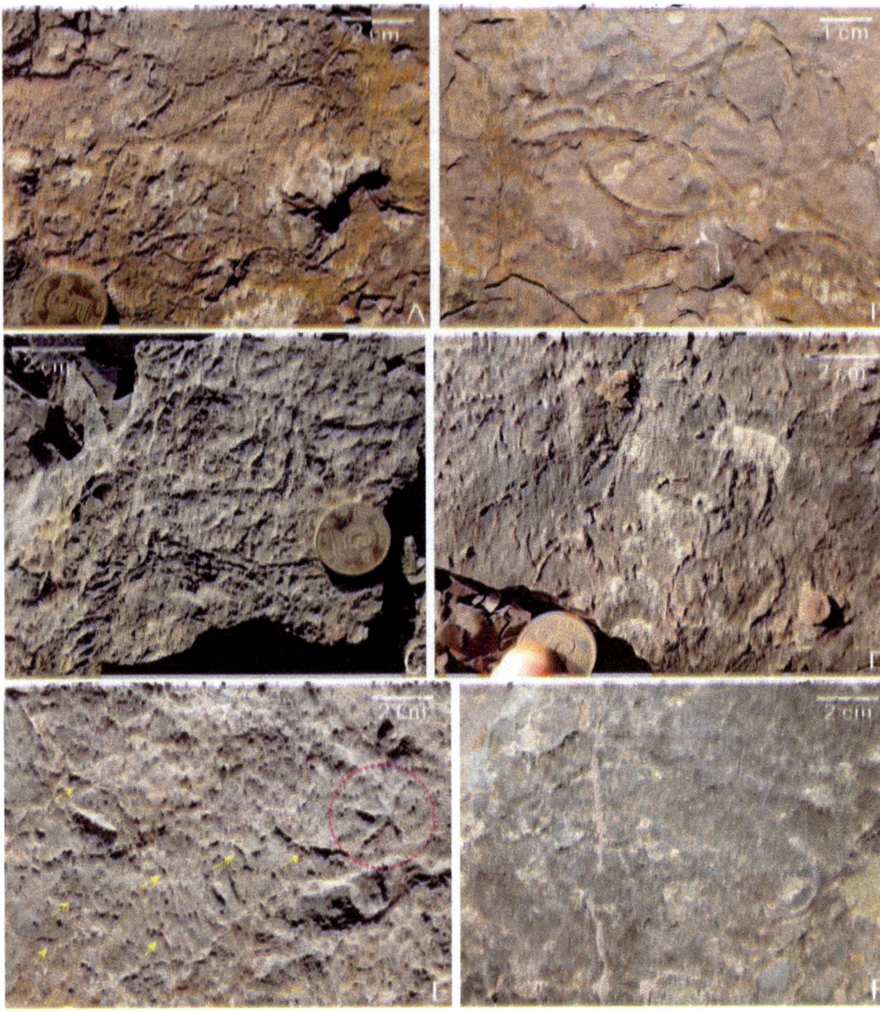

Fig. 3.8 *Biformites* in outcrop. (**a–f**), *B. zhadaensis* from thin-bedded siltstone and mudstone beds, a *Asteriacites lumbricalis* (red dashed block) was preserved together with abundant *B. zhadaensis* (yellow arrows) in **e**. All specimens were preserved as convex hyporeliefs. Lower Triassic Sunjiagou Formation, Houzhougongmiao section, Shaanxi Province, northwestern China

which shares an identical overall morphology with *Z. zhadaensis*. Consequently, Knaust et al. (2012) regarded *O. aysenensis* Bell 2004 as a junior synonym of *Z. zhadaensis*. More recently, Knaust and Neumann (2016) reclassified *Z. zhadaensis* under *Biformites*, based on the distinctive imprints of their central discs. As a result, Knaust and Neumann (2016) proposed a revised classification of *Biformites*, which includes *B. insolitus* (Linck 1949) and *B. zhadaensis* (Yang and Song 1985). The newly obtained specimens are assignable to *Biformites zhadaensis* due to the presence of delicate, ideally bisymmetrically arranged transverse (elongate or

Fig. 3.9 *Biformites* in outcrop. (**a–f**), *B. zhadaensis* from thin-bedded siltstone and mudstone beds, all specimens were preserved as convex hyporeliefs. Lower Triassic Sunjiagou Formation, Houzhougongmiao section, Shaanxi Province, northwestern China

polygonal) protuberances, which are characteristic of imprints left by the arms of walking ophiuroids (Bell 2004; Schlirf 2012; Knaust and Neumann 2016; Morton et al. 2023). *Biformites* is a trace fossil produced by ophiuroids and asteroids during locomotion on the sediment surface (Belaústegui et al. 2017; Morton et al. 2023). Secondary branching phenomenon which is unique to this group of fossils had been reported in the literature. In contrast to other similar ichnofossil genera, *Biformites* displays a specific branching pattern that is a result of the animal's movement and overlapping of its earlier traces (Morton et al. 2023). The Sunjiagou Formation specimens are exclusive to the study area and have not been found in coeval or younger Early Triassic successions elsewhere globally. Both *A. lumbricalis* and *B. zhadaensis* are assumed to be made by ophiuroids, whose body and trace fossils were frequent in the lowest Triassic successions.

Capodistria Vialov 1968
Capodistria vettersi Vialov 1968
Figure 3.10

Description Radiating, short ridges extending from a central area. A distinct knob was positioned approximately at the central point. The ridges vary in length and might display asymmetric arrangement. The central area has a diameter of 0.5 cm, and the ridges vary in length from less than 0.8 to 2 cm. The knob in the center area is about 0.3 cm in diameter. The trace displays less than 10 ridges, with the overall diameter ranging from 25 to 45 mm. The ridges are cylindrical in cross-section and vary in length and spacing, not forming a perfectly uniform circle. The separate

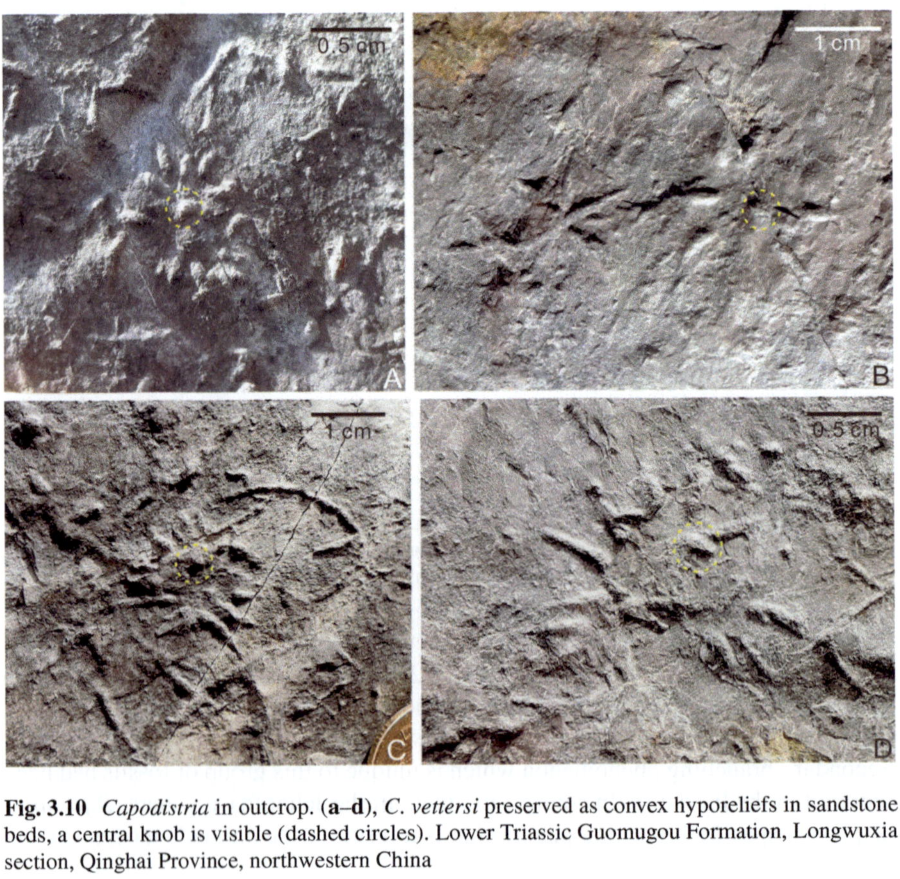

Fig. 3.10 *Capodistria* in outcrop. (**a–d**), *C. vettersi* preserved as convex hyporeliefs in sandstone beds, a central knob is visible (dashed circles). Lower Triassic Guomugou Formation, Longwuxia section, Qinghai Province, northwestern China

ridges have no surface ornamentation, preserved in convex hyporelief. Burrow fill is indistinct with host rock indicating passive filling. These traces were only found on the sole beds of the siltstone layers of the Lower Triassic Guomugou Formation from the Longwuxia section. This ichnotaxon was preserved on the same beds with other graphoglyptids such as *Dendrotichnium*, *Acanthoraphe*, *Lorenzinia*, and *Ubinia*.

Remarks *Capodistria* looks very similar to *Lorenzinia* according to the radiation pattern of the ridges and the size of the separate ridges and the whole trace. The only difference may be the distinct knob located in the center. Therefore, Uchman et al. (1998) proposed that the separation of *Capodistria* from *Lorenzinia* at the ichnogeneric level may be questioned. In some cases, the distinction between irregular *Sublorenzinia* (=*Lorenzinia*) and *Capodistria* was difficult. Thus, the relationship between this ichnogenus and *Lorenzinia*, as well as several other similar ichnotaxa, needs reexamination. However, the lack of well-preserved specimens prevented our further comparison. On the contemporary deep-sea floor, similar grooves radiating from a central hole were photographed. They were produced by bonnelid echiurians (Ohta 1984) and may represent behavioral counterparts of *Capodistria* (Uchman 1998). *Capodistria* occurs in flysch deposits from the Lower Devonian to the Holocene (Buatois and Mángano 2018).

Chondrites von Sternberg 1833
Chondrites targionii (Brongniart 1828).
Figures 3.11 and 3.12

Description Tunnels winding or arcuate, with sharp-angled branches. Some vertical or oblique shafts are noticeable on the bedding surface. Second-order branches are common, third-order branches are rare, and radiating branches are also rare. Single tunnels, 1–3 mm wide, filled with darker sediments than host rock. Specimens preserved, showing only a few branching tunnels, and an entire root-like tunnel system not observed. These burrows are mainly preserved as full relief within fine-grained beds. They are characterized by flattened, straight, and rarely branched tunnels, often accompanied by numerous rounded to ovoid forms, with widths of 1–2 mm. The burrows are filled with material that is darker than the surrounding host rock. This ichnogenus was found in the shale and muddy siltstone beds of the Maomaolong Formation from the Malonggou section, Western Qinling area.

Remarks *Chondrites* tunnel systems are complex, typically featuring a single or few master shafts that extend from the surface and branch at acute angles to form a dendritic or root-like structure (Uchman 1999). Initially, 170 ichnospecies were assigned to *Chondrites* due to their broad morphological range. However, after revising this ichnogenus, Fu (1991) identified only four valid ichnospecies: *C. intricatus*, *C. targionii*, *C. patulus*, and *C. recurvus*. Our specimens despite being incompletely preserved, exhibit burrow morphology and structures characteristic of *Chondrites* ichnospecies, particularly resembling *C. targionii* due to their relatively

Fig. 3.11 *Chondrites* in outcrop. (**a–f**), *C. targionii* preserved in siliceous rocks, all preserved on the upper bedding surface, except for **b**, which is vertical profile view of **a**. Changhsingian (late Permian) Shaiwa Formation, Sidazhai section, Guizhou Province, southwestern China

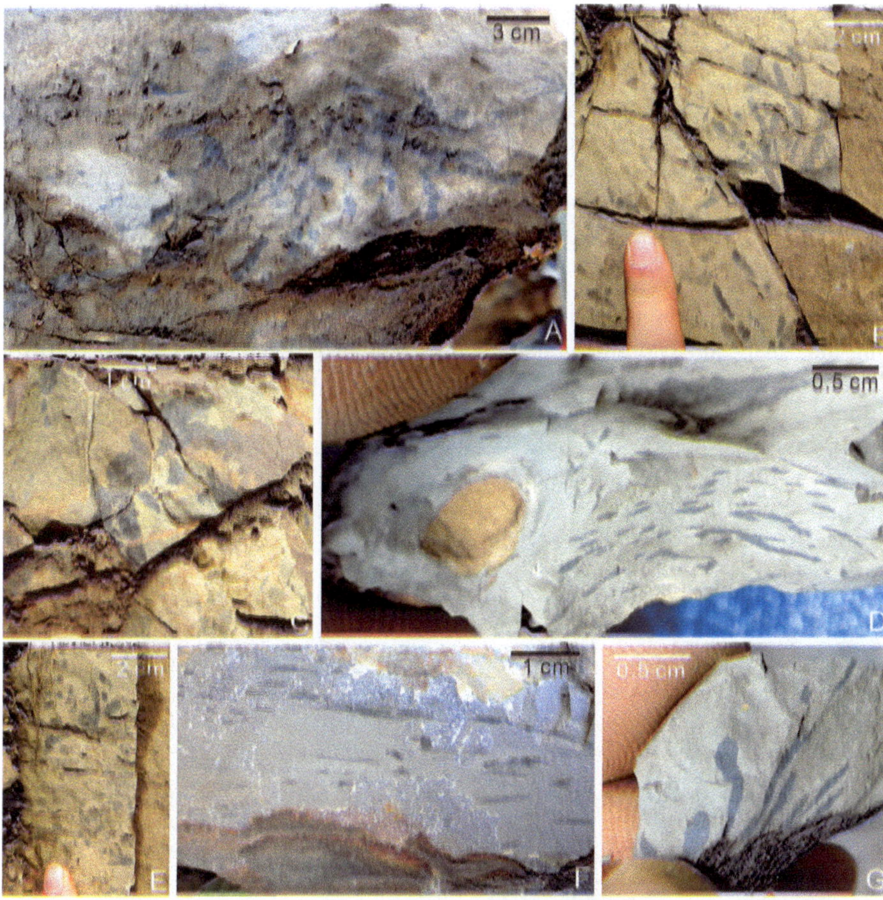

Fig. 3.12 *Chondrites* in outcrop. (**a–f**), *C. targionii* preserved in mudstone, **a–c** and **e** are from upper bedding plane view, while **d** and **f–g** are vertical profile or side view. All specimens are from the Changhsingian (late Permian). **a** is from the Maomaolong Formation, Malonggou section, Gansu Province, northwestern China. The remaining are from the Shaiwa Formation, Sidazhai section, Guizhou Province, southwestern China

larger burrows and predominant second-order branches. Previously, *Phymatoderma Brongniart* was misidentified as "giant *Chondrites*" or "pellet-filled *Chondrites*" but is distinguished by larger tunnels and fecal pellet infills (Miller 2011; Izumi 2012). The trace-maker of *Chondrites burrows* was interpreted to be able to live under dysaerobic conditions as a chemosymbiotic organism (Fu 1991) or subsurface deposit-feeder (Osgood 1970). These burrows are commonly found in marginal marine to deep-sea settings and serve as indicators of dysoxic conditions (Bromley and Ekdale 1984). *Chondrites* are widespread trace fossils, ranging from the Middle Ordovician to the Holocene (Wetzel 2008; Buatois and Mángano 2018). *Chondrites* is a debated ichnogenus in terms of its ethology and trace-makers. Baucon et al. (2020) demonstrated that, depending on the producer, *Chondrites* can represent a fodichnion (produced by annelids or sipunculoids), a chemichnion (produced by chemosymbiotic bivalves), or an agrichnion (produced by bivalves cultivating and ingesting bacteria). This ichnogenus is often associated with low-oxygen or dysoxic conditions (Bromley and Ekdale 1984; Fu 1991; Mekki et al. 2023), though its presence alone is not definitive proof of anoxic seawater.

Circulichnis Vialov 1971
Circulichnis montanus Vialov 1971.
Figure 3.13

Description Hypichnial, horizontal, smooth, with a circular to slightly elliptical, ring-like shape on bedding planes; some were incompletely preserved. Burrow diameters are about 2 mm, whole circle or ellipse diameters are 20–40 mm. Burrow fillings are identical to the host rock. Burrow is smooth, except for some specimens displaying local nodes and slight vertical undulations. The trace fossil was preserved in full relief in muddy limestone of the Daye Formation from the Lichuan section, South China.

Fig. 3.13 *Circulichnis* in outcrop. (**a–d**), *C. montanus* preserved on the upper surface of argillaceous limestone. **a** and **b** are from the Lower Triassic Anshun Formation, Gaimao section, Guizhou Province, southwestern China. **c** and **d** are from the Lower Triassic Daye Formation, Lichuan section, Hubei Province, South China

Remarks Vialov (1971) defined *Circulichnis* as a "ring trace, almost circular (or elliptical), developed in one cylinder." Keighley and Pickerill (1997) suggested changing the original name *Circulichnis* to *Circulichnus* due to its incorrect ending. They thus officially established *Circulichnus* Keighley and Pickerill 1997, as a younger objective synonym of *Circulichnis* Vialov 1971. However, Uchman and Rattazzi (2019) argued that because *Circulichnus* is a younger objective synonym of *Circulichnis*, according to the International Code of Zoological Nomenclature, *Circulichnus* is an invalid name and should not be recommended for further use, and *Circulichnis* is the valid name. Besides the type ichnospecies *C. montanus*, Uchman and Rattazzi (2019) amended several other ichnospecies of this ichnogenus and excluded them as *Circulichnis*. They also proposed another new ichnospecies *Circulichnis ligusticus*.

Circulichnis ligusticus Uchman and Rattazzi 2019.
Figure 3.14

Description A horizontal, cylindrical ring with an irregular course, accompanied by a short, lateral cylindrical tunnel extending outside the ring. The cylinder of the ring is more or less uniform in width, ranging from 1 to 2 mm. The tunnel is mostly smooth, though some specimens may exhibit local nodes and slight vertical undulations. The ring is about 20–40 mm wide. They were preserved as convex hyporelief in thin turbiditic beds of fine-grained sandstone beds of the Guomugou Formation, Longwuxia section, northwestern China. They co-occurred with many other deep marine graphoglyptids such as *Acanthoraphe*, and *Dendrotichnium* in the Guomugou Formation; the burrow sizes of these burrows were very similar.

Remarks *Circulichnis ligusticus* distinguishes it from *C. montanus* in its winding and irregular course. They show very close morphometric parameters with other graphoglyptids from the same beds of deep-sea turbiditic sediments. *Circulichnis* was initially interpreted as a crawling trace produced by a worm that, after completing its circular path, either moved upward from the seafloor or burrowed into the sediment. However, other researchers have proposed alternative interpretations, suggesting that *Circulichnis* could represent a fodinichnion produced by annelids or an unspecialized grazing trail (Buatois et al. 1998a, 2006; Mángano et al. 1997). Uchman and Rattazzi (2019) proposed another ethological model of *Circulichnis* and interpreted it as an exploration of the trace maker of *Helminthoidichnites*, *Gordia*, or *Helminthopsis*. In their model, the trace-makers of *Helminthoidichnites*, *Gordia* and *Helminthopsis* were likely exploring environments at deeper or shallower sediment levels for feeding, typically along bedding interfaces. The most efficient way to create such an exploratory burrow is to drive a shaft through less nutritionally attractive sediment and form a ring-shaped burrow in a more nutrient-rich horizon. Back to our material in the Guomugou Formation, they co-occurred with graphoglyptids such as *Acanthoraphe* and *Dendrotichnium*, and their burrow sizes are very similar. The main burrows of some *Acanthoraphe* are also highly bent to an oval or cylindrical shape. This means that the trace-makers of these graphoglyptids may also be the same as *C. ligusticus*, but with different behaviors.

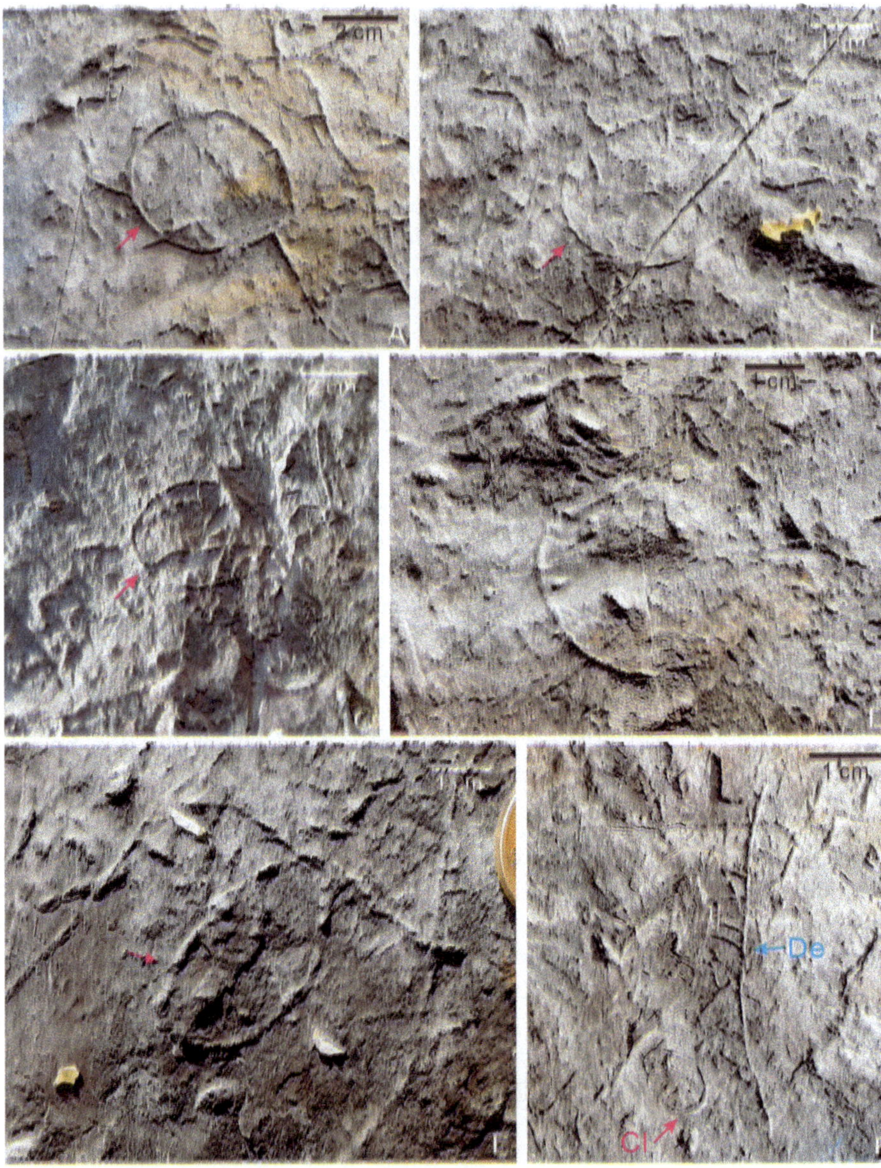

Fig. 3.14 *Circulichnis* in outcrop. (**a–f**), *C. ligusticus* preserved on the sole bedding plane of sandstone (red arrows), other deep marine graphoglyptids, such as *Dendrotichnium* (*De* in **f**) can be observed on the same bed. Lower Triassic Guomugou Formation, Longwuxia section, Qinghai Province, northwestern China

Cochlichnus Hitchcock 1858
Cochlichnus anguineus Hitchcock 1858
Figure 3.15

Description Small, sinuous, or meandering horizontal trail with a smooth surface resembling a sine curve. These trace fossils are found on the upper bedding surfaces of muddy limestone in the Daye Formation in the Lichuan section, South China. The burrows are 2–4 mm in diameter and maintain a consistent width throughout their entire observed length. The wavelength is approximately 15 mm, with wave

Fig. 3.15 *Cochlichnus* in outcrop. (**a–f**), *C. anguineus* preserved in argillaceous limestone and limestone beds. **a**, and **e–f** are from the Lower Triassic Daye Formation, Lichuan section, Hubei Province, South China; **b–c** are from the Lower Triassic Jialingjiang Formation, Shawan section, Sichuan Province, southwestern China; and **d** is from the Lower Triassic Daye Formation, Shatian section, Hubei Province, South China

amplitudes ranging from 2 to 3 mm. No overlapping among specimens has been observed. They are preserved as positive hyporeliefs.

Remarks *Cochlichnus* differs from *Belorhaphe* in having smooth, rounded curves instead of angular ones and differs from *Helminthopsis* in its regular sinusoidal curves, while *Helminthopsis* outlines resemble river meanders rather than sinusoidal waves (Buatois et al. 1997). *C. anguineus* differs from other ichnospecies of *Cochlichnus* by the absence of annulations, as well as its smaller size. Therefore, our materials are assigned to this ichnospecies. *Cochlichnus* is interpreted as grazing, feeding, and locomotion, or a combination of these ethological categories, depending on environmental conditions and potential producers (Buatois et al. 1997). Potential trace makers include worms, nematodes (Metz 1998), annelids (Hitchcock 1858), cyclostomates (Głuszek 1995), or insect larvae. *Cochlichnus* is known from the entire Phanerozoic (Buatois et al. 1997) and was generally distributed across a wide range of environments, including both non-marine and marine settings (Buatois et al. 1997; Keighley and Pickerill 1997; Smith et al. 2011; Buatois and Mángano 2018).

Cruziana d'Orbigny 1842
Cruziana isp.
Figure 3.16

Description Positive hyporeliefs; horizontal, elongate, and bilobate scratch marks with median grooves covered by herringbone-shaped or transverse striations or imbricated sediment pads. The scratch marks are oriented obliquely to the long axis of the trace. The trackway measures 15 cm in length and 4 cm in width; the median groove is 6 mm deep. The transverse striations are relatively robust and arranged asymmetrically on both sides along the median groove. This ichnotaxon is collected from the Upper Changhsingian Maomaolong Formation of the Malonggou section, western Qinling, North China.

Remarks *Cruziana* comprises more than 30 ichnospecies, which are distinguished based on morphological features resulting from the appendage morphology of the tracemaker. In addition, the difference and validity surrounding the classification of *Cruziana* and *Rusophycus* had once been extensively debated, which reflected the complexities of ichnology and the challenges inherent in categorizing trace fossils. Now, it has been widely accepted that the "length/width" ratio (2:1) is an important criterion distinguishing *Rusophycus* from *Cruziana*. For instance, traces having "length/width" ratio > 2:1 are assignable to *Cruziana*, otherwise assignable to *Rusophycus* (Keighley and Pickerill 1996). Although this ichnogenus may represent complex behaviors, *Cruziana* is typically interpreted as representing a combination of locomotion and feeding behaviors (Seilacher 2007), with possible producers of trilobites or trilobitomorphs or other arthropods (Seilacher 1992; Keighley and Pickerill 1996; Hofmann et al. 2012). The newly obtained specimens possess elon-

Fig. 3.16 *Cruziana* in outcrop. (**a–b**), *Cruziana* isp., preserved as convex hyporeliefs on the sole bedding plane of sandstone. Changhsingian Maomaolong Formation, Malonggou section, Gansu Province, northwestern China

gate, bilobate trackways separated by a median groove, typical of *Cruziana* characteristics, but differ from *C. problematica* by possessing more robust and asymmetrical transverse striations and a wider and shorter outline. However, insufficient material prevents a precise identification of *Cruziana* ichnospecies. *Cruziana* ranges from the early Cambrian to the Holocene (Seilacher 2007; Buatois and Mángano 2018) and has been documented in a wide range of paleoenvironments, from shallow marine settings to freshwater habitats. *Cruziana* was once thought to represent the locomotory traces of trilobites or trilobite-like arthropods and was considered characteristic of the Paleozoic. However, large *Cruziana* were collected from marine strata of the Middle Triassic Liard Formation from western Canada. They were interpreted as being produced by post-extinction stomatopod or decapod crustaceans (Zonneveld et al. 2002).

Dendrotichnium Farrés 1967
Dendrotichnium seilacheri Kozur et al. 1996
Figure 3.17

Description Hypichnial burrows display a tree-like structure with limited branches. The trace is characterized by a central burrow that is straight to slightly curved, measuring 2–4 mm in width and 90–150 mm in length. Lateral branches are short, widely spaced, mostly perpendicular to the central main burrow, and the branches

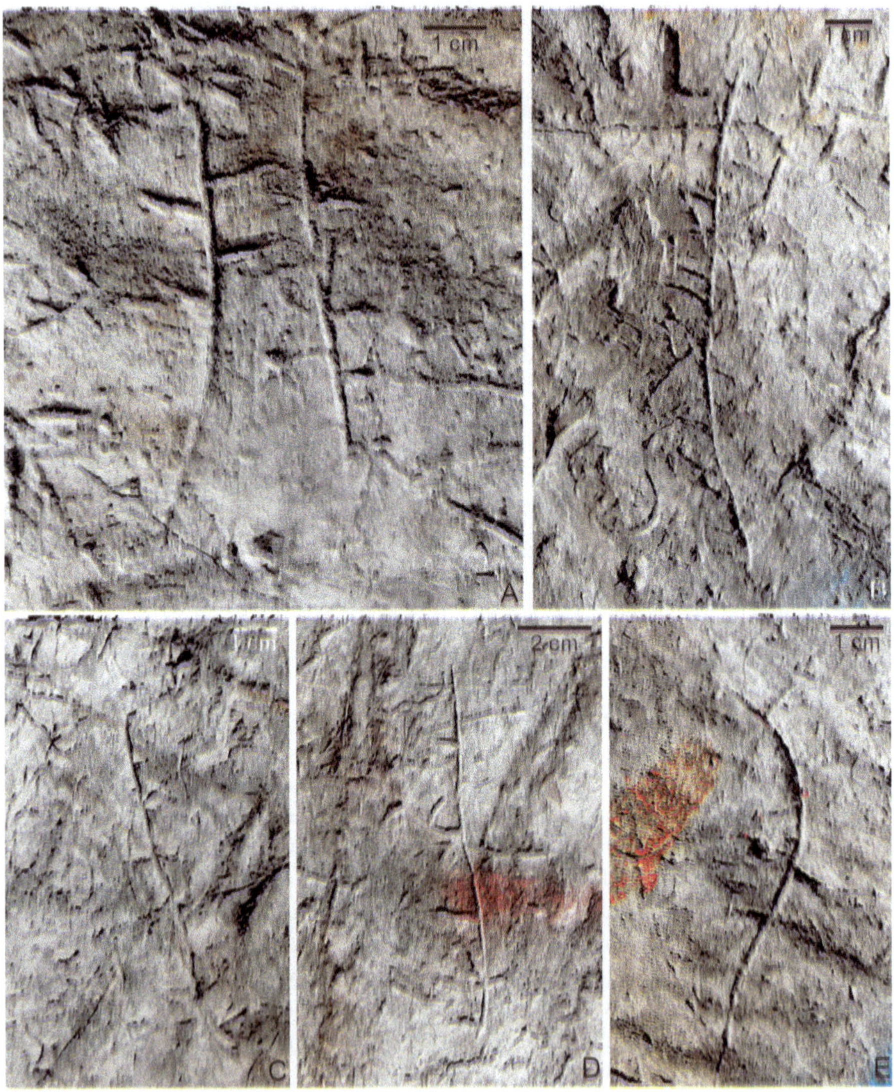

Fig. 3.17 *Dendrotichnium* in outcrop. (**a–e**), *D. seilacheri* preserved as convex hyporeliefs on the sole bedding plane of sandstone beds. Lower Triassic Guomugou Formation, Longwuxia section, Qinghai Province, northwestern China

are 10–20 mm long and 2 mm wide. One side displays branches spaced between 10 and 30 mm and nearly perpendicular, while the other side includes a single oblique branch. They were preserved as positive hyporeliefs on the soles of the sandstone beds of the Guomugou Formation, Longwuxia section. They were also preserved together with other similar graphoglyptids traces such as *Ubinia* and *Acanthoraphe*. The burrow sizes of these ichnogenera are also similar to each other.

Remarks *Dendrotichnium* has been a relatively rare ichnogenera since its erection. There are still debates about *Dendrotichnium*, *Dendrorhaphe*, and *Ubinia* (Kozur et al. 1996; Wetzel and Uchman 1997; Jensen 1999). There are four valid ichnospecies of *Dendrotichnium*, which are primarily distinguished by the distribution of their short branches. In the type ichnospecies, *Dendrotichnium llarenai*, the branches on both sides of the central main burrow are closely spaced, with intervals ranging from 3 to 10 mm. In *D. haentzscheli*, the branches on either side of the central burrow are irregularly spaced, and most branches are inclined relative to the axial burrow. The branches in *D. haentzscheli* also reach a greater maximum length compared to those in *D. seilacheri*. Seilacher (1977) established the new monospecific ichnogenus *Dendrorhaphe* for *D. haentzscheli*, proposing that the three specimens on the holotype slab of *D. haentzscheli* represent components of a radial system. However, his decision to separate *Dendrotichnium haentzscheli* was not positively accepted. In other words, if the only one ichnospecies *Dendrorhaphe haentzscheli* of ichnogenus *Dendrorhaphe* (suggested by Seilacher) is attributed to *Dendrotichnium haentzscheli*, which is earlier erected, then the validity of the name of *Dendrorhaphe* need to be reconsideration (Kozur et al. 1996). The last ichnospecies *D. alternans* has a zig-zag-shaped central main burrow and regularly alternating branches on both sides and thus is the easiest to distinguish among ichnospeices of *Dendrotichnium*. However, *D. alternans* has also been disputed with *Ubinia* and *Chondrites patulus*. Wetzel and Uchman (1997) considered *Dendrotichnium* to be a subjective junior synonym of *Ubinia* (Grossgeim 1961). They suggested reclassifying all ichnospecies of *Dendrotichnium* to the ichnogenus *Ubinia*. However, this decision is vague due to the lack of discussion between these two ichnogenera in their paper. Indeed, we have observed the co-occurrence of *Dendrotichnium*, *Acanthoraphe* and *Ubinia*, as well as the distribution of their branches, and found there are differences among their branches, according to the specimens from the Longwuxia section. They are all preserved on the same bed, and all have similar burrow sizes. We thus suggest they may be produced by the same organisms. The discussion on these ichnogenera will be published anywhere, but obviously beyond the scope of this book.

Dictyodora Weiss 1884
Dictyodora zimmermanni Hundt 1913
Figure 3.18

Description Smoothly winding burrow with relatively long wavelengths and high amplitudes, amplitudes are up to 20 cm. Burrows are about 12 mm wide and 20–150 mm long. Burrow fillings are darker than host rock; some specimens dis-

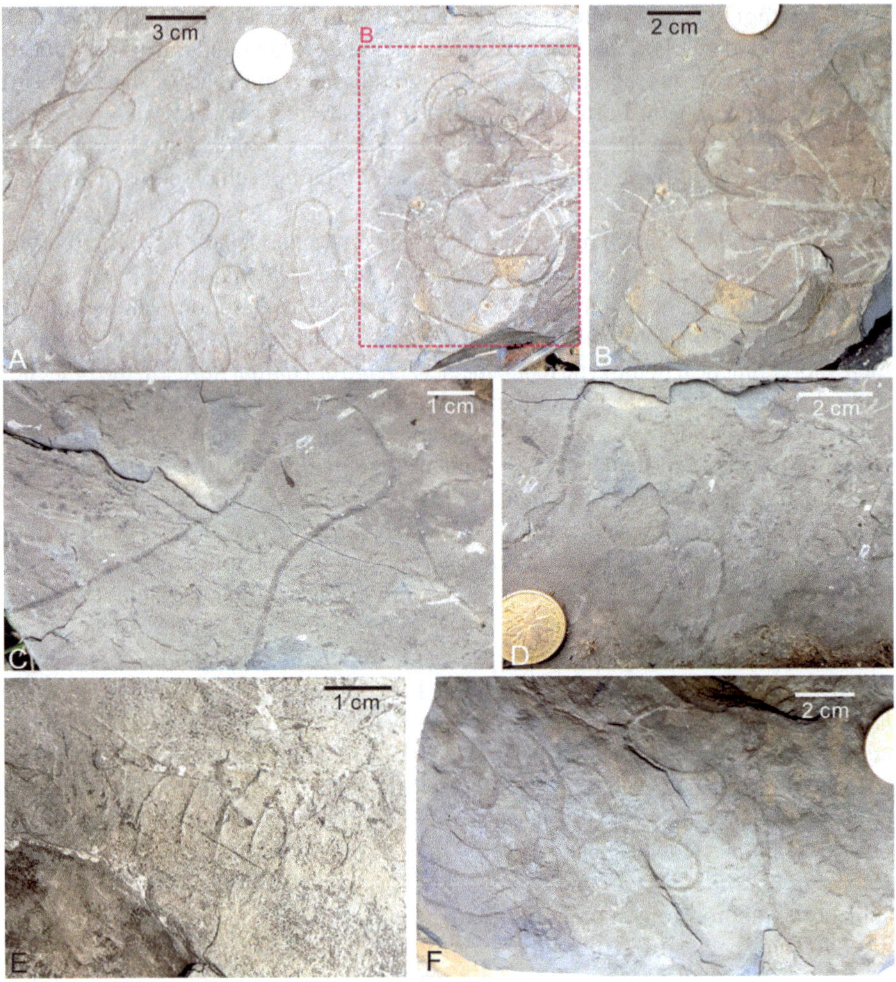

Fig. 3.18 *Dictyodora* in outcrop. (**a–f**), *D. zimmermanni* preserved in slate and siltstone beds, B is a close-up view of the dashed block in **a**. Changhsingian (late Permian) Maomaolong Formation, Malonggou section, Gansu Province, northwestern China

played menisci backfilling structures. Some specimens exhibit a complex three-dimensional structure that is roughly conical and vertical to the bedding plane. These meanders may represent the horizontal expressions of a vertical burrow wall. The limbs of the meanders are regular, subparallel, and slightly divergent (Pazos et al. 2015; Zhang et al. 2020). The trace appears as a three-dimensional meandering band corresponding to the intersection of the vertical spreite structure with the upper bedding surface. These traces are preserved in the siltstone beds of the Maomaolong Formation at the Malonggou section. Similar trace fossils were reported from the same locality by Zhang et al. (2020).

Remarks *Dictyodora* is a very complex trace fossil composed of a basal burrow and a vertical upward-tapering wall that originates from the dorsal line of the basal burrow. The most visible and commonly preserved features of *Dictyodora* in the field are the wall structures, which appear as epichnial or hypichnial meanders (Benton and Trewin 1980; Zhang et al. 2020). Zhang et al. (2020) reported and discussed several ichnospecies of *Dictyodora* from the same locality; they also discussed ichnostratigraphic and palaeoenvironmental evolution of this ichnogenus. The simplest ichnospecies of *Dictyodora* is *D. zimmermanni*, characterized by irregular, broad meanders with relatively high amplitudes and locally looping courses (Benton and Trewin 1980). *D. tenuis* is distinguished by secondary sinuosity superimposed on primary meandering burrows. *D. scotica* differs from *D. tenuis* in having regular first-order meanders with parallel limbs (Benton and Trewin 1980). *D. silurica* is represented by complex meanders with high amplitude and narrow width (Yang and Hu 1992). *D. liebeana* consists of complex spiral or helical meanders (Benton 1982; Orr et al. 1996; Baucon and Neto de Carvalho, 2008). *D. atuelica* is defined by nearly orthogonal angles between limbs and pointed turns of meanders (Pazos et al. 2015; Zhang et al. 2020).

Several organisms have been regarded as potential tracemakers of *Dictyodora*, and shell-less mollusks were one of the candidates (Neto de Carvalho 2001). However, the "snorkel-like" respiratory organ, which is currently used to explain the vertical wall, is dorsally located. This feature is inconsistent with primitive mollusks. Thus, soft-bodied worm-like organisms such as vermiform shell-less mollusks with snorkel-like organs may be plausible producers (Pazos et al. 2015). *Dictyodora* first appeared in shallow-marine environments during the early Cambrian, later migrated into deep-sea settings in the Ordovician, and adapted to marginal-marine environments by the Wenlock. By the Carboniferous and Permian, it became restricted to deep-marine environments. As a result, *Dictyodora* is not a reliable indicator of deep-marine environments in the Paleozoic (Zhang et al. 2020). Similar to *Psammichnites*, the tracemakers of *Dictyodora*, which possessed snorkel-like organs, were eliminated during the P-Tr mass extinction (Zhang et al. 2020).

Diplichnites Dawson 1873
Diplichnites isp.
Figure 3.19

Description Simple horizontal trackways, straight to slightly curved, composed of two approximately parallel rows of imprints. Individual imprints or tracks vary in shape from circular-shaped to ellipsoid-shaped. Ellipsoid tracks are oriented either perpendicular or oblique to the midline. Medial furrows or ridges are absent. The trackway width of *Diplichnites* is about 35 mm, with a minimum internal width of about 12 mm. Together with another form of trackway, *Kouphichnium*, *Diplichnites* was found in the muddy limestone of the Lower Triassic Daye Formation from the Dangwu section, Guizhou Province, South China. In addition, *Diplichnites* was found in the mudstone or shale beds of the Upper Changhsingian Maomaolong Formation from the Malonggou section, Xiqinling area, northwestern China.

Fig. 3.19 *Diplichnites* in outcrop. (**a**–**f**), *Diplichnites* isp., preserved as convex hyporeliefs on the sole bedding plane of siltstone beds. **a**–**c** are from the Lower Triassic Sunjiagou Formation, Houzhougongmiao section, Shaanxi Province, northwestern China. **d**–**f** are from the Changhsingian (late Permian) Maomaolong Formation, Malonggou section, Gansu Province, northwestern China

Diplichnites was also preserved together with another trackway of *Cruziana* in the Maomaolong Formation. All specimens were preserved as positive hyporeliefs.

Remarks Most specimens of *Diplichnites* are characterized by approximately parallel rows of circular or ellipsoid tracks. *D. gouldi* is widely reported in the literature and has been classified into several morphotypes. However, our specimens may not be attributed to this ichnospecies due to the lack of sufficient specimens and poor preservation. Previous works have proposed that the ichnotaxonomy of *Diplichnites* needs further revision (Minter et al. 2007). *Diplichnites* is a widespread trace fossil, with a record extending from the Early Cambrian to the Holocene. Notably, similar trackways have been reported from the terminal Ediacaran Shibantan Member (551–541 million years ago) of the Dengying Formation in the Yangtze Gorges area, South China, potentially representing the earliest known trackways. Their appearance in the terminal Ediacaran marked the beginning of bilaterian animals (Chen et al. 2018). *Diplichnites* has been reported in various environmental settings, from fully continental settings to marine settings, even in deep marine, just like the case from the Upper Changhsingian Maomaolong Formation. *Diplichnites* has been interpreted to be produced by various arthropods, most notably myriapods, as well as euthycarcinoids and trilobites (Johnson et al. 1994; Buatois et al. 1998b).

Diplocraterion Torell 1870
Diplocraterion parallelum Torell 1870
Figures 3.20, 3.21, and 3.22

Description Dumbbell-shaped structures on the upper surface of sandstone beds, featuring two circular holes resulting from the removal of sediment from the U-shaped limbs. In the vertical profile, it displays U-shaped burrows with spreite, a web-like structure located between the two arms of the U-shaped tube. The limbs are oriented perpendicular or subvertical relative to the bedding plane. The depth of the U-shaped limbs ranges from 30 to 200 mm below the bedding surfaces. The marginal tubes are 4–15 mm in diameter, and the limbs are 30–80 mm in width. Spreites are distinct and darker than the host rock, while the burrow fillings are the same as the host rock. The spreite can exhibit two forms: protrusive, where it lies between the paired tubes, or retrusive, positioned below the paired tubes. The former is more common in the studied horizons. No funnel-shaped openings are observed at the top of the burrow. The ichnogenus is very common in the sandstone beds of the Lower Triassic Xiahuancang Formation, Zhihema section, southern Qilian basin, northern China. They are extremely abundant on the beds where ripple marks develop.

Remarks Eight ichnospecies have been erection in *Diplocraterion*, among which the type ichnospecies, *D. parallelum* is the most common and important. *Diplocraterion* shares some morphological similarities with *Rhizocorallium*, but there are conspicuous differences. *Rhizocorallium* is oriented horizontally to the rock bedding, whereas *Diplocraterion* is vertical. This orientational difference reflects the different ecological niches occupied by these two ichnogenera. Incomplete *Diplocraterion* may also be confused with the spreite burrow *Teichichnus*, though the latter lacks a marginal tube (Fürsich, 1974b). *Diplocraterion* is regarded as a dwelling structure (Bromley 1996) produced by suspension feeders or predators in high-energy wave environments (Fürsich, 1974b, Fürsich and Palmer 1975). It is also recognized as an "equilibrium structure" (Bromley 1996), responding to sedimentation and erosion. Crustaceans and polychaete worms are the two

Fig. 3.20 *Diplocraterion* in outcrop. (**a**–**c**), *D. parallelum* preserved in sandstone. **a** is a bedding plane view of a rippled sandstone bed, and **b** and **c** are in vertical profile view. Lower Triassic Xiahuancang Formation, Zhihema section, Qinghai Province, northwestern China

main groups of potential producers (Knaust 2017). It has also been regarded as one of the ichnotaxa that first appeared after the P-Tr mass extinction (Knaust 2010), and this viewpoint perfectly matches our observation. We indeed observed *Diplocraterion* is only one ichnotaxa with high density that occurred in the basal of the Xiahuancang Formation. Other ichnotaxa all occurred in higher horizons, although there is no change in the lithology. *Diplocraterion* has been documented in marginal marine environments, including tidal settings and estuaries, as well as shoreface and offshore environments. It is a long-ranging trace fossil known from the Cambrian to the Holocene.

Fig. 3.21 *Diplocraterion* in outcrop. (**a–f**), *D. parallelum* preserved in sandstone. All are vertical profile views, except for **b**, which is a bedding plane view of a rippled sandstone bed. Lower Triassic Xiahuancang Formation, Zhihema section, Qinghai Province, northwestern China

Gordia Emmons 1844
Gordia marina Emmons 1844
Figure 3.23

Description Slender, worm-like trails that are mostly bent but not meandering, preserved in both positive and negative reliefs. The traces are freely curved to straight and often cross each other or self-crossing. The burrows maintain a uniform width throughout their length and display a rounded base in cross-section. Some specimens exhibit a tendency to meander. Burrow diameters range from 3 to 10 mm. The burrow fill is identical to the host rock. The burrow fill is the same as the host rock. This ichnogenus has been found in the limestone beds of the Daye Formation.

Remarks The type ichnospecies *G. marina* is most common and easily discriminated from other ichnospecies. *G. nodosa* is annulated rather than cylindrical, while *G. indianaensis* is a sharply angled trail (Pickerill and Peel 1991; Buatois et al. 1998a). *G. maendria* shows guided meanders. *Gordia* can be differentiated from

Fig. 3.22 *Diplocraterion* in outcrop. (**a–g**), *D. parallelum* preserved in sandstone. **a–e** are vertical profile views, and **f–g** are bedding plane views. Lower Triassic Xiahuancang Formation, Zhihema section, Qinghai Province, northwestern China

several similar ichnogenera, including *Helminthopsis* and *Helminthoidichnites*, by the presence of self-crossing. The occurrence of *Gordia* is abundant in different sedimentary environments. The tracemaker is likely a worm, as the burrow resembles that of the hair-worm *Gordius* (Häntzschel 1975). Książkiewicz (1977) posited

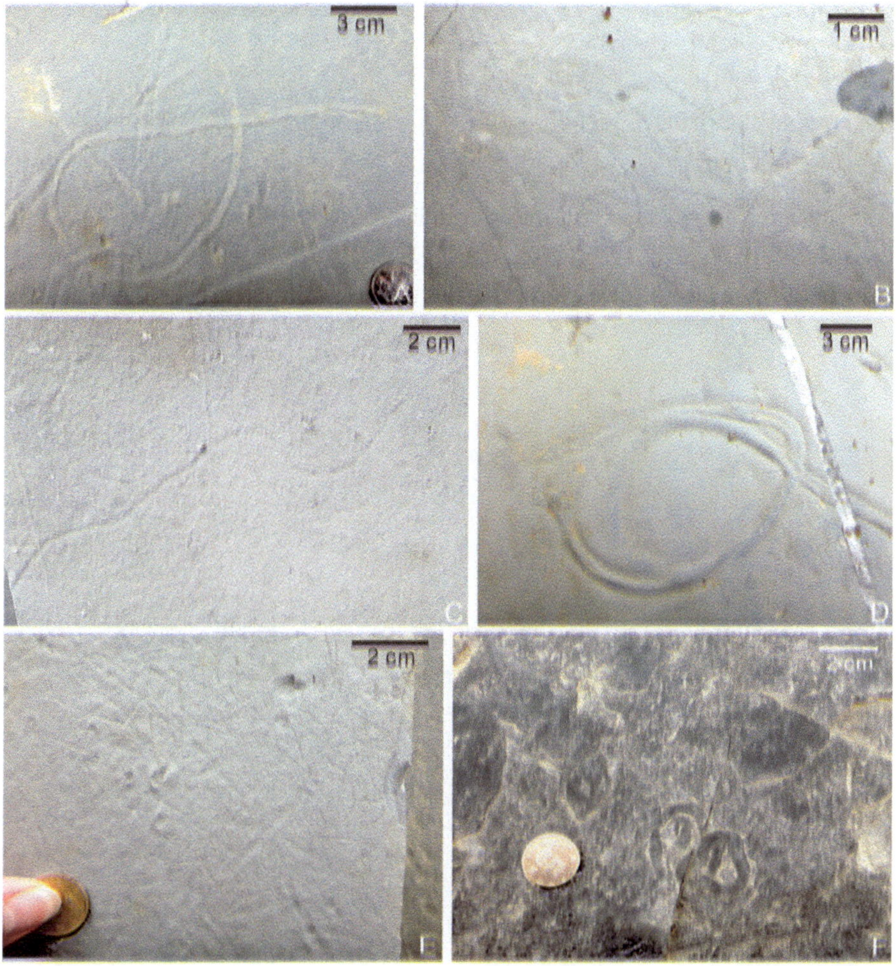

Fig. 3.23 *Gordia* in outcrop. (**a–e**), *G. marina* preserved in argillaceous limestone. Lower Triassic Daye Formation, Dangwu section, Guizhou Province, southwestern China. (**f**) *G. marina* preserved in siltstone beds. Lower Triassic Shangaling Group, Zhayougou section, Gansu Province, northwestern China

that it might be a feeding burrow or a trail resulting from polychaete activity, while some others viewed it as a locomotory trace made by worms or gastropods (Yang 1984; Gong 1987; Li et al. 1997). In addition, some authors also proposed this ichnogenus as a pasichnia or fodinichnia produced by worms or worm-like organisms (Aceñolaza and Buatois 1993; McCann 1993; Geyer and Uchman 1995). In particular, *Gordia* was preserved in association with large pelagic body fossils that had been reported from the Cambrian Burgess Shale-type Kaili Biota (Wang et al. 2009). *Gordia* is found in association with the discoidal animal *Pararotadiscus* and some bivalved arthropods. *Gordia* first occurred in the Ediacaran and has been a common ichnotaxon throughout the Phanerozoic.

Gyrochorte Heer 1865
Gyrochorte comosa, Heer 1865
Figures 3.24 and 3.25

Description Horizontal traces preserved on bedding planes. They are preserved in the form of plaited ridges, characterized by biserially arranged, obliquely aligned sediment pads. Burrow is divided into two parallel lobes by a median groove, visible in positive epirelief, and displayed biserial grooves separated by a median ridge in negative hyporelief. Traces are typically 6–12 mm wide, 30–180 mm long.

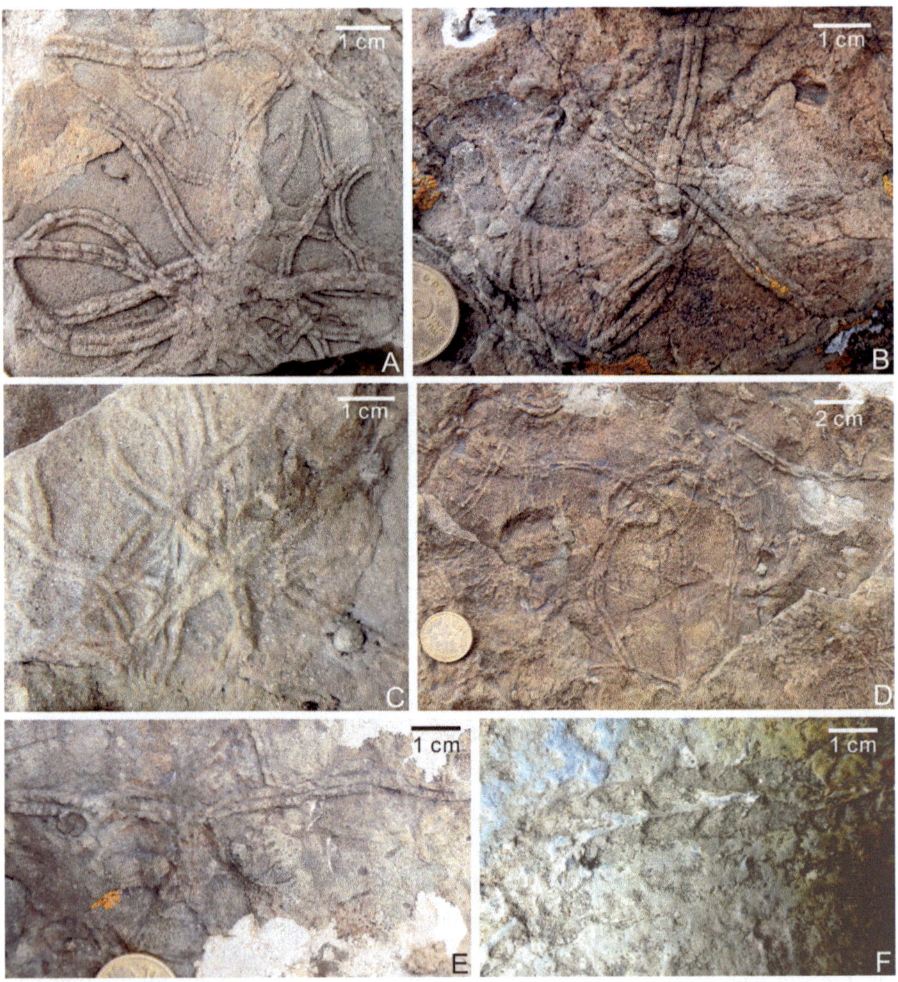

Fig. 3.24 *Gyrochorte* in outcrop. (**a–e**), *G. comosa* preserved as convex epireliefs in sandstone. Lower Triassic Xiahuancang Formation, Zhihema section, Qinghai Province, northwestern China. (**f**), *G. comosa* preserved as concave hyporeliefs. Lower Triassic Daye Formation, Lichuan section, Hubei Province, South China

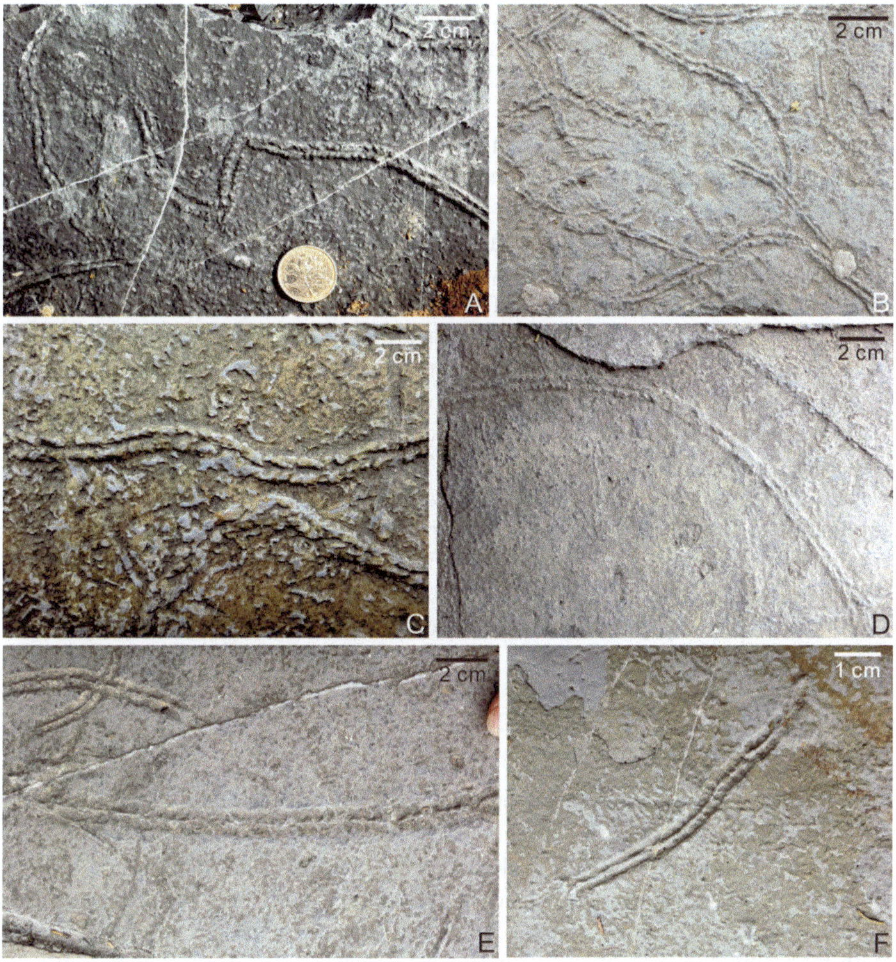

Fig. 3.25 *Gyrochorte* in outcrop. (**a–f**), *G. comosa* preserved in argillaceous limestone, **a–d** are preserved as convex epireliefs, and **e–f** are concave hyporeliefs. Lower Triassic Daye Formation, Lichuan section, Hubei Province, South China

Burrows can often crosscut each other, but self-crossing is not observed. They are the most distinct ichnotaxa preserved on limestone beds of the Daye Formation from the Lichuan section; both positive and negative forms can be observed. This ichnogenus has also been reported from the upper surface of sandstone beds of the Xiahuancang Formation, Zhihema section.

Remarks *Gyrochorte* exhibits strong winding and sharp directional changes, with its length showing significant variation across exposures. It often crosscut other traces on bedding planes. The specimens are identified as *G. comosa* according to the plaited ridges that are obliquely aligned. In some cases *Gyrochorte* can be easily confused with other bilobate trails such as *Cruziana*, *Archaeonassa*, *Psammichnites*, and *Didymaulichnus*, due to the preservation condition (Uchman 1998). *Cruziana* and *Didymaulichnus* are preserved as positive hypopireliefs, and the former is always characterized by herringbone-shaped or transverse striations. *Psammichnites* is preserved as positive epirelief with the bilobate feature but exhibit a clear median dorsal ridge or a groove and transverse meniscate backfill. *Archaeonassa* is marked by simple horizontal trails that are deformed by irregular levees on both sides, which never display obliquely aligned pads of sediment. The plaited ridges and biserial grooves of *Gyrochorte* provide insights into the movement and feeding strategies of the trace makers (Schlirf 2000). The consistent presence of these traces in specific sedimentary contexts helps reconstruct the dynamic processes and conditions of ancient shallow marine environments. *Gyrochorte* has been frequently found in shallow marine and nearshore environments with moderately high energy (Gibert and Benner 2002). The traces are commonly linked to fine-sand tempestites, indicating post-event burrowing activities. The producers of *Gyrochorte* are hypothesized to be worm-like animals, likely polychaetes, which moved obliquely through the sediment, displacing sediment upward and backward to form the plait-like ridges above sandy beds (Seilacher 2007; Fürsich et al. 2018). These behaviors indicate that *Gyrochorte* producers were surface deposit feeders that exploited the sediment for food during quieter periods between storm events.

Helminthoidichnites Fitch 1850
Helminthoidichnites tenuis Fitch 1850
Figure 3.26

Description Simple, unbranched, straight to gently curved trails, oriented parallel to the bedding plane. Cross-cutting among different individuals is common. Trail fillings are identical to the host rock. Trails are densely distributed on the bedding plane, and the diameters among different individuals from the same slab are always very similar. Trail sizes are about 0.5–1.5 mm. These trails were widely distributed with high abundance in many of the lowest Triassic strata from the studied area, especially from the thin-bedded limestone beds of the Daye Formation from South China. They always crosscut each other, forming a relatively high degree of bedding plane bioturbation. This ichnotaxon could be the first ichnogenera that reappeared after the P-Tr mass extinction in carbonate facies, and it has been confused with *Planolites* in some literature. In addition, the *Helminthoidichnites*-bearing limestone beds of the Daye Formation in many carbonate facies sections also developed vermicular fabrics, namely "vermicular limestone." In some cases it is difficult to distinguish the tiny trail of *Helminthoidichnites* from those fabrics of vermicular limestone, especially in vertical sections. The vermicular limestone seems to be very common in carbonate ramp and platform environmental settings in the aftermath of mass extinctions.

Remarks Although *Helminthoidichnites* was synonymized with *Gordia* by Häntzschel (1975), it is widely regarded as a valid ichnogenus in subsequent works. *Helminthoidichnites* differs from *Gordia* in its crosscutting among individuals

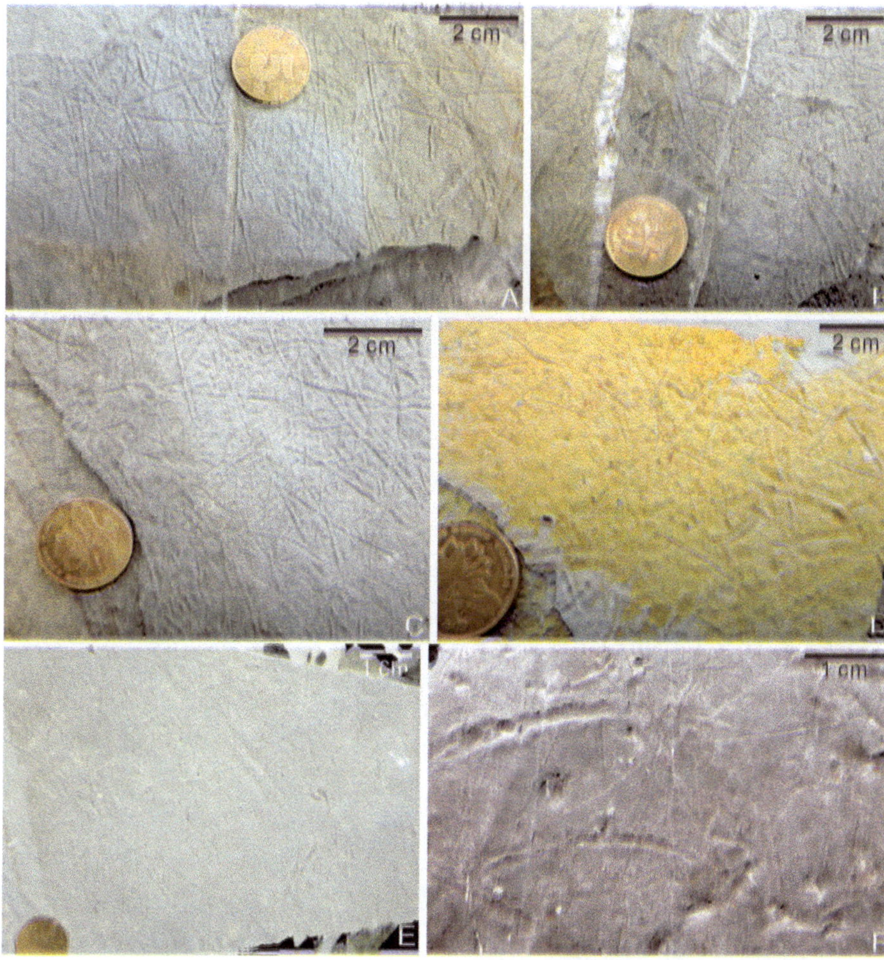

Fig. 3.26 *Helminthoidichnites* in outcrop. (**a–f**), *H. tenuis* preserved in argillaceous limestone. **a–d** are from the Lower Triassic Daye Formation, Daxiakou section, Hubei Province, South China. **e** is from the Lower Triassic Daye Formation, Dangwu section, Guizhou Province, southwestern China. **f** is from the Lower Triassic Guomugou Formation, Longwuxia section, Qinghai Province, northwestern China

rather than self-crossing, a specific feature of *Gordia* (Hofmann and Patel 1989). *Helminthoidichnites* is not meandering as *Helminthopsis* (Hofmann and Patel 1989; Buatois and Mángano 1993a, b). Most specimens in our material show straight, unbranched, without meanders or self-crossing, thus belonging to this ichnogenus. However, these traces are relatively similar to *Planolites*, as the actively backfilled sediments in *Planolites* were not clearly displayed in such limestone beds. *Helminthoidichnites* is a monospecific ichnotaxon with a single ichnospecies *H. tenuis*. *Helminthoidichnites* is interpreted as a grazing trail (pascichnia) produced by arthropods or vermiform animals (Buatois et al. 1998a). It is one of the most common ichnofauna during the Ediacaran and early Cambrian but is also widely distributed in the Phanerozoic.

Helminthopsis Heer 1877
Helminthopsis abeli Książkiewicz 1977
Figure 3.27

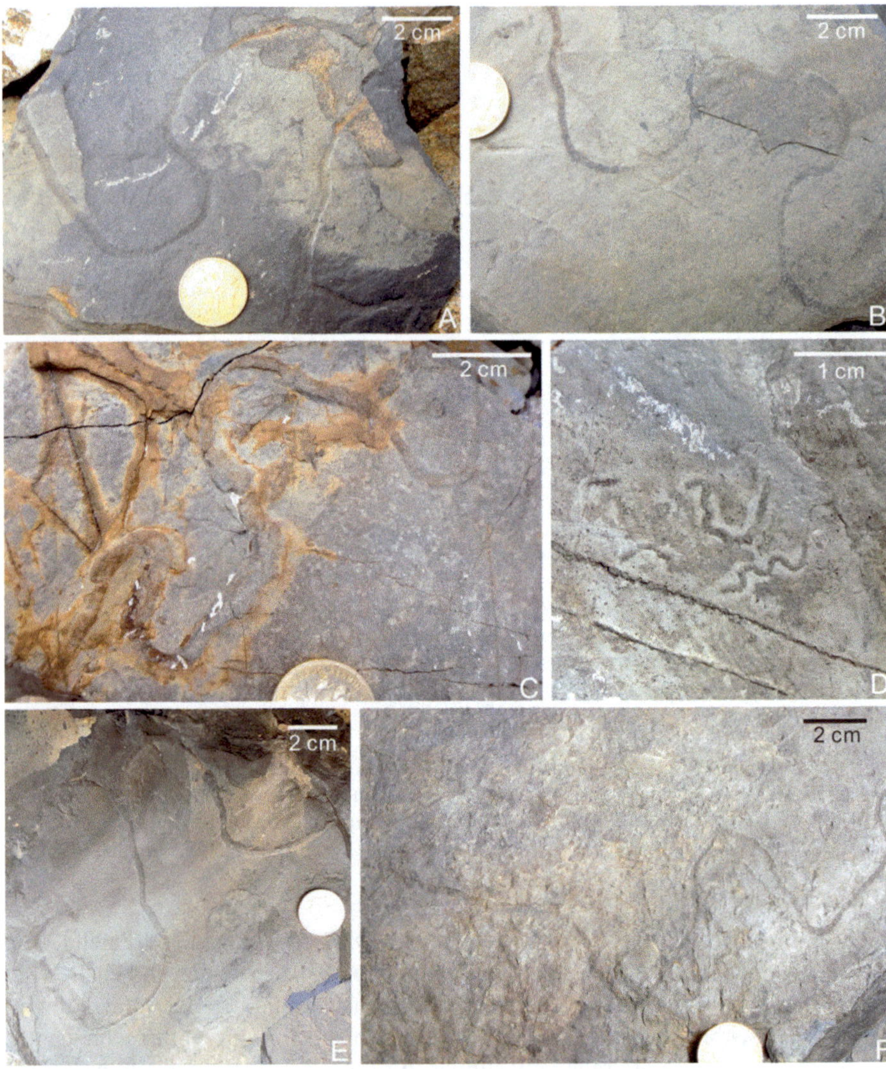

Fig. 3.27 *Helminthopsis* in outcrop. (**a**–**f**), *H. abeli* preserved in slate and siltstone beds. **a**–**b** and **e**–**f** are from Changhsingian (late Permian) Maomaolong Formation, Malonggou section, Gansu Province, northwestern China. **c** is from the Lower Triassic Shangaling Group, Zhayougou section, Gansu Province, northwestern China. **d** is from the Lower Triassic Guomugou Formation, Longwuxia section, Qinghai Province, northwestern China

Description Simple, unbranched, rope-like meanders that exhibit horseshoe-shaped turns. Tubes are irregular windings with low amplitude, without a tendency to loop or self-crossing. Meanders are preserved as full reliefs or positive hyporeliefs. The diameter varies between approximately 1 and 6 mm, and the length varies between 50 and 400 mm. The height and width of the horseshoe-shaped turns are about 10–20 mm and 20–40 mm. Burrow fillings are slightly darker and coarser than the host rock, but some specimens show the same with the host rock. This ichnotaxon was mostly documented from the siltstone beds of the Maomaolong Formation from the deep slope Malonggou section.

Remarks *Helminthopsis* differs from *Helminthoidichnites* by its meandering course (Hofmann and Patel 1989), from *Palaeophycus* by the absence of a distinct lining, and from *Gordia* by its tendency to meander and lack of self-overcrossing (Pickerill 1981; Crimes and Anderson 1985; Pickerill et al. 1987; Wetzel and Bromley 1996). There were once more ten ichnospecies of *Helminthopsis* since its first establishment, which made the ichnotaxonomy of this ichnogenus confusing. Wetzel and Bromley (1996) made a reasonable sorting and revision on this ichnogenus based on their re-examination of the type material of the *Helminthopsis* Heer. They confirmed the validity of the name of this ichnogenus and proposed *Helminthopsis hieroglyphica*, as the type ichnospecies of *Helminthopsis*. Besides *H. hieroglyphica*, there are now other valid ichnospecies, *H. abeli*, and *H. tenuis*, which were later erected by Książkiewicz (1968, 1977). According to Wetzel and Bromley (1996), the differences among the three ichnospecies were their geometrical patterns: *H. abeli* has open meanders and horseshoe-shaped turns with relatively smaller heights and widths, *H. tenuis* has high-amplitude windings and U-shaped turns, but without horseshoe-shaped turns, while *H. hieroglyphica* has low-amplitude meanders and straight segments, just like a box-shaped fold appearance. We therefore attribute our specimens to *H. abeli*, due to horseshoe-shaped turns, and relatively low-amplitude meanders. This ichnogenus has been widely distributed in marine environments ranging from deep sea and offshore to shallow marine, as well as in nearshore tidal flats, lagoons, and deltas. *Helminthopsis* is generally considered to be the grazing trace produced by deposit-feeding organisms such as polychaete worms (Crimes and McCall 1995; Fillion and Pickerill 1990; Książkiewicz 1977; Buatois et al 1998a). *Helminthopsis* occurs since the Ediacaran and throughout the Phanerozoic.

Kouphichnium Nopsca 1923
Kouphichnium isp.
Figures 3.28, 3.29, and 3.30

Description The trackways show successive sets of (or isolated) imprints characterized by a pair of well-preserved tetradactyl external imprints and a variable number (mostly three pairs) of straight thin intermediate imprints. A median imprint is often observed. Isolated external imprints are also very abundant. Most of them are preserved as convex hyporeliefs; few are concave epireliefs. These imprints and trackways are identified as *Kouphichnium*, which is characterized by strongly het-

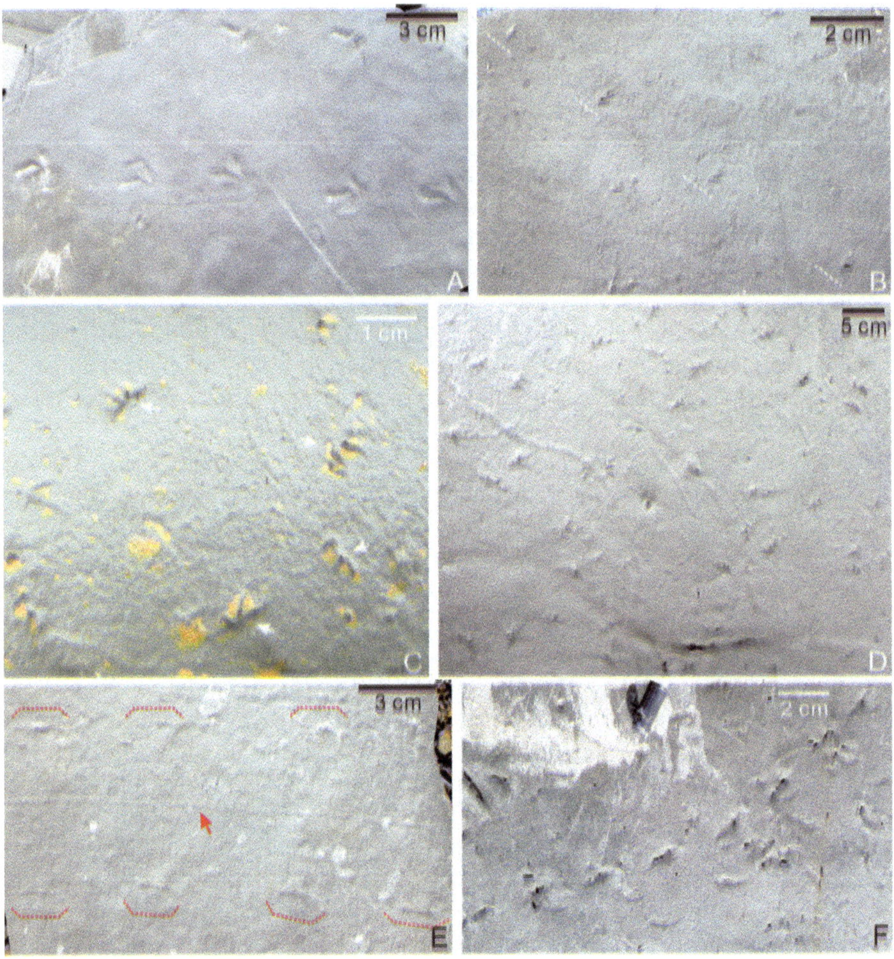

Fig. 3.28 *Kouphichnium* in outcrop. (**a–f**), *Kouphichnium* isp., preserved in argillaceous limestone. **a–b** and **e–f** are composed of one or two rows of imprints, and a central imprint produced by telson is observed in **e**. **c–d** are abundant isolated external imprints. **a** is preserved as concave epireliefs, and **b–f** are convex hyporeliefs. Lower Triassic Daye Formation, Dangwu section, Guizhou Province, southwestern China

eropodous tracks. Several parameters have been measured. The average stride/repetition distance is 52.4 mm. The average spacing of the middle imprint is 40 mm, and the angle between the middle imprint and the midline is ~23°. *Kouphichnium* is a common trackway preserved in the muddy limestone beds of the Daye Formation from the Dangwu section and in the sandstone beds of the Sunjiagou Formation of the Houzhougongmiao section, both of which are Early Triassic in age.

Remarks *Kouphichnium* is a well-known ichnogenus that has been widely accepted as being produced by limulids (horseshoe crabs) (Moreau et al. 2014; Shu et al.

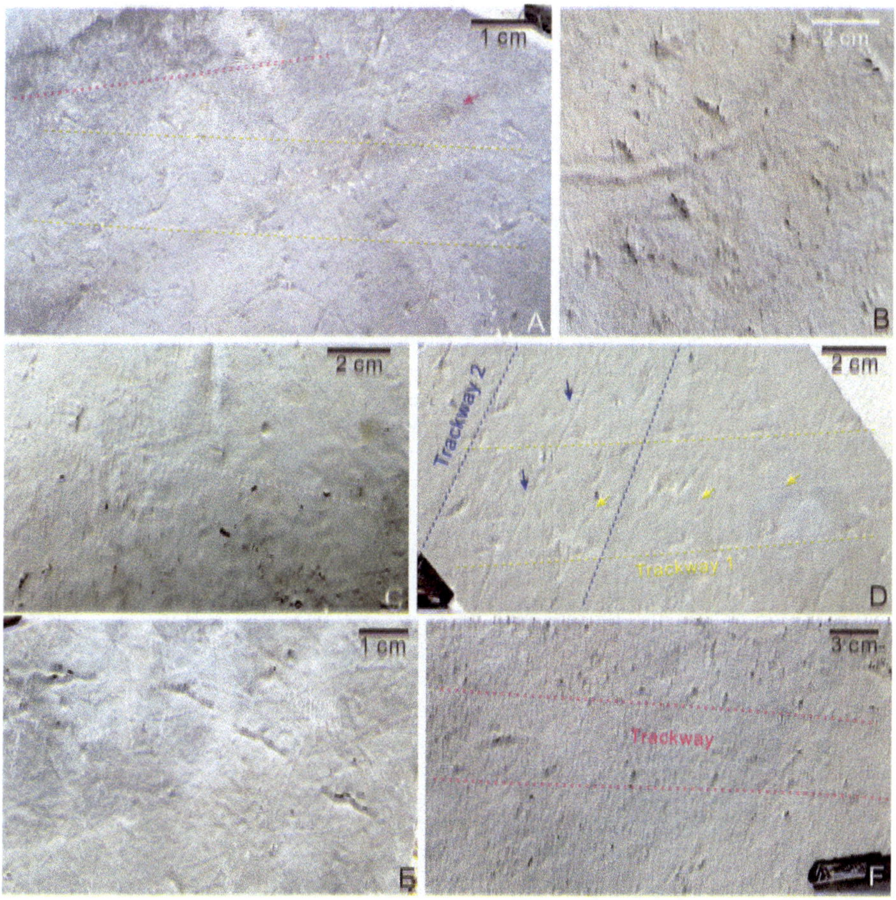

Fig. 3.29 *Kouphichnium* in outcrop. (**a**–**f**), *Kouphichnium* isp., preserved in argillaceous limestone. **a** and **c**–**f** are composed of one or two rows of imprints, and two sets of *Kouphichnium* trackways with central imprints produced by the telson are observed in **d**. **b** is composed of abundant isolated external imprints. **e** is preserved as concave epireliefs, and all others are convex hyporeliefs. Lower Triassic Daye Formation, Dangwu section, Guizhou Province, southwestern China

2018; King et al. 2019; Shibata and Varricchio 2020). *Kouphichnium* is featured with distinctly heteropodous trackways (Gaillard 2011; Moreau et al. 2014; Bicknell et al. 2018). The smallest pair of chelicers is used only for ingestion, while the other five pairs of legs are used for motion. The last pair is more complex in morphology and is an active pusher and digger when locomotion (Shu et al. 2018). Furthermore, in the Pennsylvanian-period Steven C. Minkin Fossil Site located in Alabama's Union Chapel Mine, USA, a wealth of *Kouphichnium* specimens has been collected. Research indicates that the formation of certain *Kouphichnium* ichnospecies might not be exclusively attributed to limulids, and alternative chelicerate ought to be incorporated as possible trace makers (King et al. 2019). However, the specimens

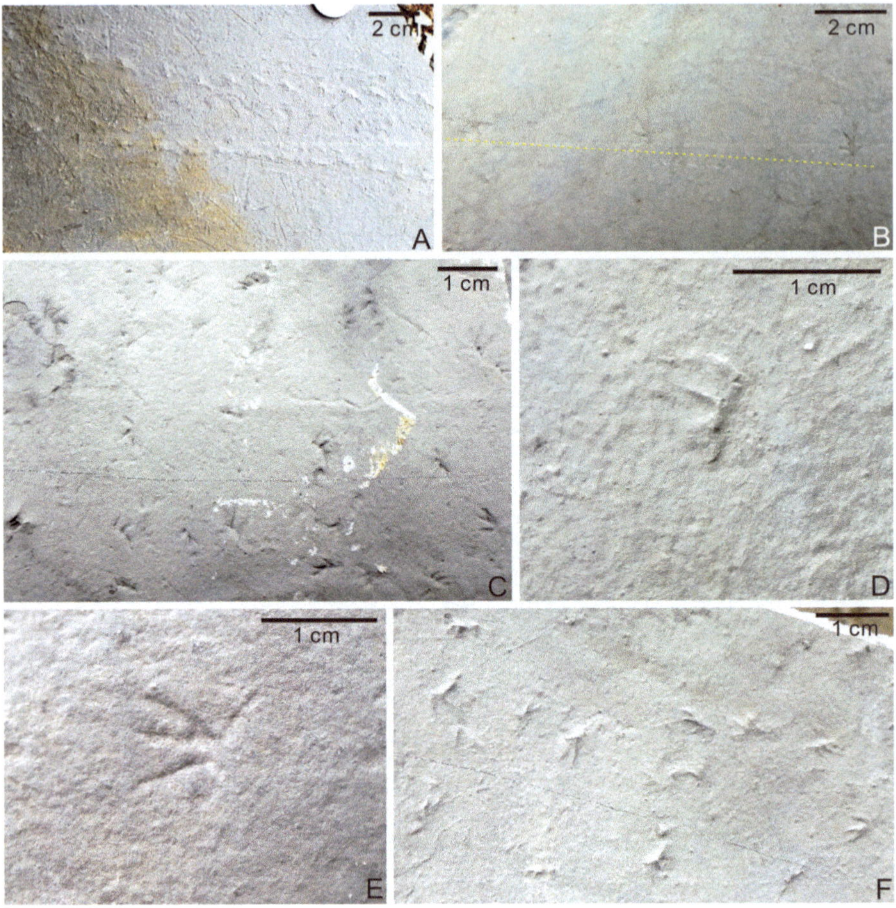

Fig. 3.30 *Kouphichnium* in outcrop. (**a–f**), *Kouphichnium* isp., preserved in argillaceous limestone. **a–b** are composed of one or two rows of imprints. **c–f** are composed of abundant isolated external imprints. **b** is preserved as concave epireliefs, and all others are convex hyporeliefs. Lower Triassic Daye Formation, Dangwu section, Guizhou Province, southwestern China

from Dangwu quarries show no such features as those from the Union Chapel Mine; thus, the trace-making animals are attributed to limulids. This is also supported by the fact that a prosoma imprint of limulid is preserved together with *Kouphichnium*. The distinctly defined imprints of the hind limbs highlight their primary function as propulsive structures during locomotion. The elongated straight morphology suggests small pressure exerted on the sea floor. The highly consistent repetition of the traces likely arose from standard locomotory behavior. The considerable distance between two sets of propulsive limb imprints implies that the movement was both efficient and potentially swift. The exceptionally straight line of the trackways is also evidence of rapid locomotion. These characteristics collectively suggest a typical yet rapid mode of locomotion along the seafloor.

Limulid legs can puncture relatively deep into the sediment and produce undertracks (Seilacher 2007). The preservation of undertracks indicates the stepwise disappearance of shallowly imprinted components of the trackway such as the telson imprint, because the telson may have the shallowest contact with the seafloor, and its marks may not be preserved in deeper laminae. Therefore undertracks may have higher potential for preservation than telson tracks or other surface tracks, as traces left on the seafloor surface were fragile and easily destroyed by water currents or waves. Thus the Dangwu limulid tracks were probable undertracks rather than true tracks, especially isolated specimens or sets of tetradactyl external imprints of leg VI (pushers). However, we examine the *Kouphichnium* specimens from Dangwu and find that many of them possess one axial median imprint produced by the telson, indicating they are not undertracks. *Kouphichnium* occurs from the Middle Ordovician to the Holocene and has been reported from restricted marine, nonmarine and freshwater palaeoenvironments.

Laevicyclus Quenstedt 1879
Laevicyclus parvus (Desio 1940)
Figure 3.31

Description Regular concentric circles with a central core, visible in bedding plane view. The whole shape indicates it is a cylindrical vertical burrow. The string diameter measures up to 6 mm, and the ring diameter reaches up to 50 mm. It is commonly present on the muddy limestone bedding surface of the Daye Formation from the Lichuan section. The penetration depth of the central core is not obtained as the trace fossils were on large bedding planes. Sampling and polishing vertical slabs seem to be impossible. This trace may penetrate up to approximately 40 mm below the top of the bed. The burrow fill of the central core is massive, suggesting a passive filling. The concentric area is smooth, without any grooves or radial imprints from the central core.

Remarks There have been debates on the status of this ichnogenus with several other ichnogenera, among which *Dolopichnus*, *Monocraterion*, and *Mammillichnis* are the most confused. Knaust (2015) suggested that *Dolopichnus* and *Monocraterion* were junior synonyms of *Laevicyclus*, and *Mammillichnis* can be maintained as a separate ichnogenus, but further work is still needed. Besides the Lichuan specimens, this ichnogenus has also been reported from several other Early Triassic localities such as the Spathian Virgin Limestone Member in the western United States (Pruss and Bottjer 2004) and the Smithian Kockatea Shale Formation of the Perth Basin in Western Australia (Chen et al. 2012), though none of these specimens were well illustrated. Seilacher (1953, 1955) first proposed a formation model of *Laevicyclus* in which the central teat was formed by a slender, vertical shaft encircled by a circular groove at the bedding plane. This interpretation was tested through observations of the activities of the polychaete worm *Scolecolepis squamata* (Knaust 2015). *Laevicyclus* has a broad stratigraphical and geographical distribution, ranging from the Lower Cambrian (Alpert and Moore 1975) to the Holocene (Uchman 1995; Mude 2012; Buatois and Mángano 2018).

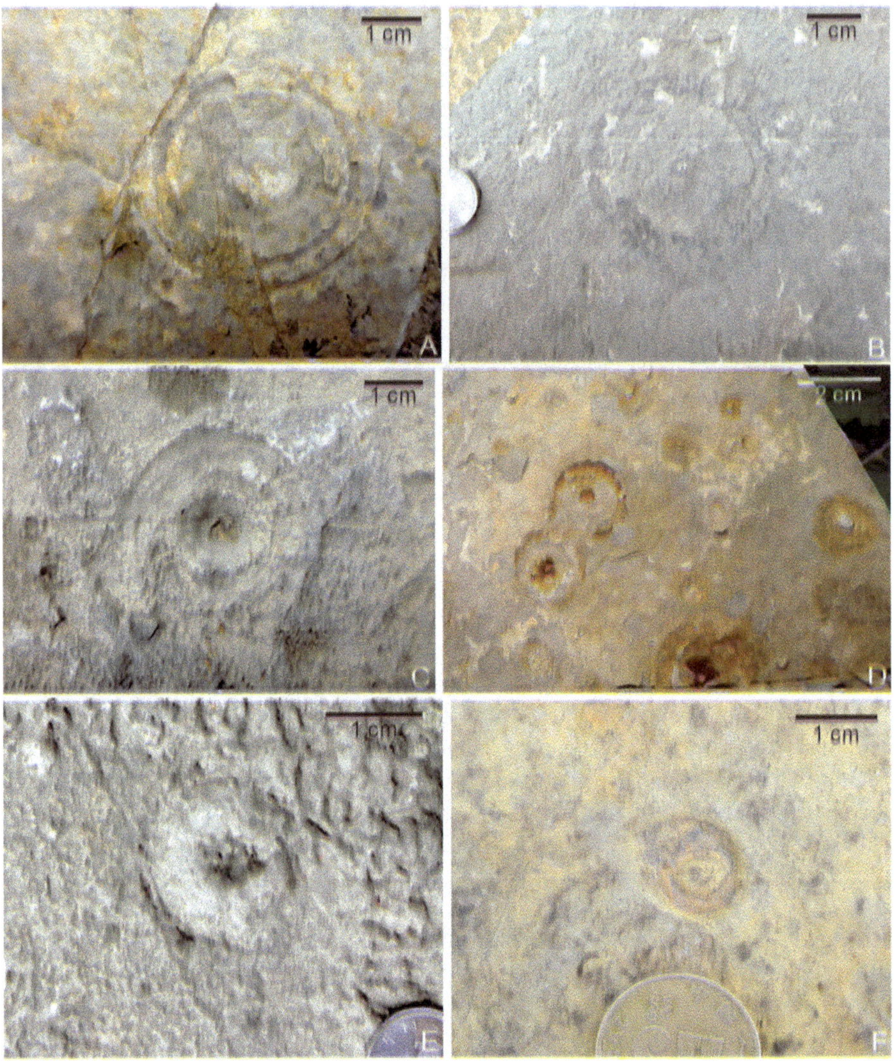

Fig. 3.31 *Laevicyclus* in outcrop. (**a–f**), *L. parvus* preserved in argillaceous limestone. **a–e** are from the Lower Triassic Daye Formation, Lichuan section, Hubei Province, South China. **f** is from the Lower Triassic Daye Formation, Shatian section, Hubei Province, South China

Lockeia James 1879
Lockeia siliquaria (James 1879)
Figure 3.32

Description Almond-shaped traces preserved as convex hyporelief. Traces are roughly symmetrical along the longitudinal median line, tapering at both ends with well-developed median ridges. Most specimens are 7–15 mm in length and 3–5 mm in width. Trace surfaces are typically smooth but may exhibit slight irregularities.

Fig. 3.32 *Lockeia* in outcrop. (**a, e**) *L. siliquaria* preserved as convex hyporeliefs in argillaceous limestone. Lower Triassic Jialingjiang Formation, Xiejiacao section, Sichuan Province, southwestern China. (**b**) *L. siliquaria* preserved as concave epireliefs in sandstone. Lower Triassic Xiahuancang Formation, Zhihema section, Qinghai Province, northwestern China. (**c–d, f**) *L. siliquaria* preserved as convex hyporeliefs in sandstone. Changhsingian Talung Formation, lower Penglaitan section, Guangxi Province, southwestern China

Overlaps between specimens are not observed. The burrow fill is identical to the host rock, suggesting a passive filling of the structure. These features agree with *L. siliquaria.* This ichnogenus has been observed in the sole of sandstone beds in the Lower Triassic Xiahuancang Formation, southern Qilian area, northwestern China.

These traces were preserved together with bivalve fauna *Unionites*, *Eumorphotis*, *Promyalina*, and *Neoschizodus*, all of which were very common during the Early Triassic worldwide. *Lockeia* also co-occurred with other burrows such as *Diplocraterion* and *Skolithos*.

Remarks *Lockeia* shares certain morphological characteristics with other similar ichnogenera such as *Pelecypodichnus*, but also possesses distinct diagnostic features that set it apart. The latter is now considered a junior synonym of *Lockeia* (Maples and West 1989). There are several key differences between *Lockeia* and other similar ichnogenera. *Lockeia* is characterized by its almond-shaped morphology that tapers at both ends, with a smooth or slightly irregular surface. In contrast, other ichnogenera may exhibit similar shapes but lack the characteristic tapering ends or smooth surfaces (Hammersburg et al. 2018). In addition, *Lockeia* is typically preserved as separated or linear patterns of hypichnial mounds, with individual segments frequently exhibiting a prominent median ridge (Schlirf et al. 2001). This unique combination of features helps to distinguish *Lockeia* from other similar ichnogenera. There are more than ten ichnospecies of *Lockeia*, among which the almond-shaped *L. siliquaria* is the most common one, while the validity of other ichnospecies needs more work and further examination. *Lockeia* is one of the most common cubichnia ichnotaxa, which indicates the resting behavior of the tracemakers. *Lockeia* has often been reported to co-occur with *Protovirgularia*, a burrow form characterized by chevronate or spicate marks separated by a median axis (Ekdale and Bromley 2001). *Lockeia* may also intergrade with *Protovirgularia* in some cases. *Protovirgularia* and *Lockeia* are generally regarded as locomotion and resting traces of cleft-foot bivalves (Seilacher and Seilacher 1994). Moreover, exquisitely preserved trace fossils demonstrating transitional features between clusters of *Protovirgularia-Lockeia* structures have been illuminated as a compound biological structure, embodying the movement, break, and feeding activities of burrowing protobranch bivalve (Cabrera et al. 2019). *Lockeia* has been found in diverse aquatic environments, ranging from marine to brackish water to freshwater fluvial and lacustrine habitats, indicating a broad ecological tolerance of the trace-making organisms. The stratigraphical occurrence of *Lockeia* spans from the Cambrian to the Holocene (Buatois and Mángano 2018).

Lorenzinia Gabelli 1900
Lorenzinia isp.
Figure 3.33

Description Radiating, daisy-shaped ridges extending from a central area; ridges are short and smooth. The central area has a diameter of 0.3 cm, and the ridges vary in length from less than 1.0 to 1.5 cm. The trace possesses about 7–10 ridges arranged in a circle around a blank central area, with the overall diameter ranging from 10 to 40 mm. The ridges are cylindrical in cross-section and vary in length and spacing, not forming a perfectly uniform circle. The separate ridges have no surface

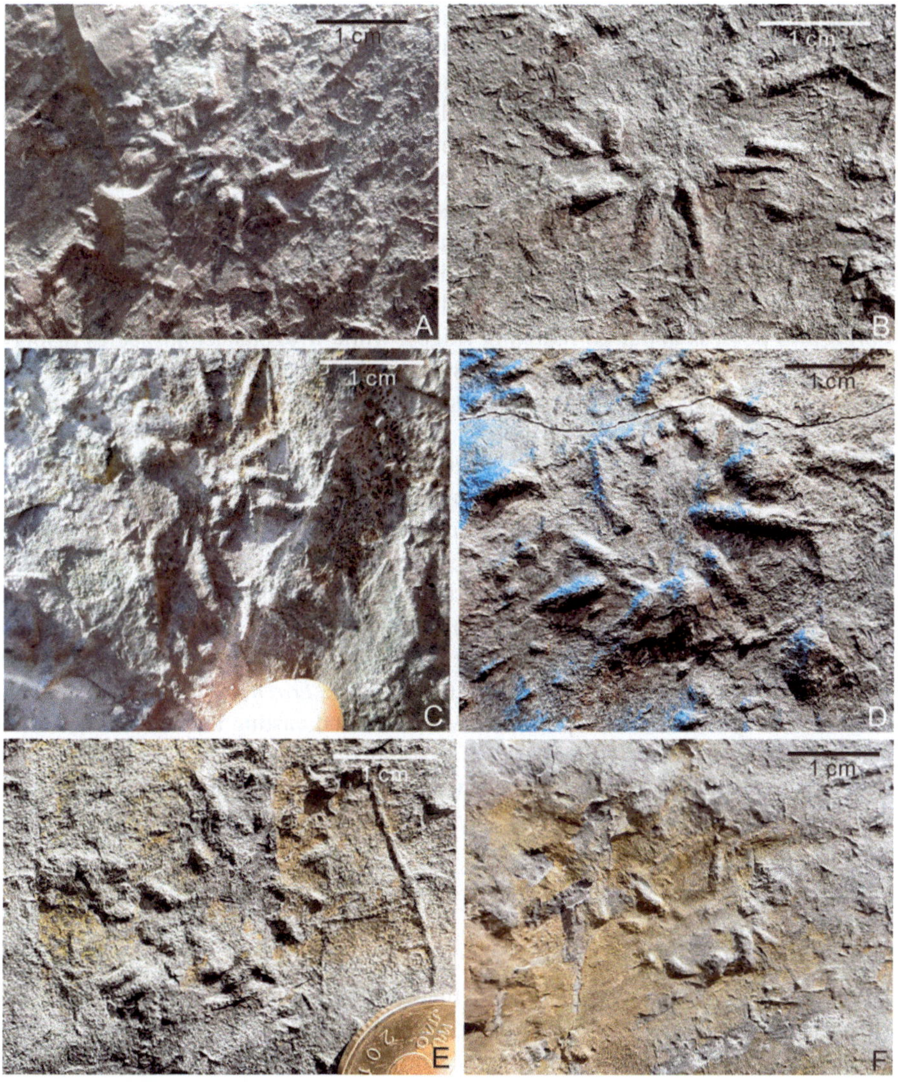

Fig. 3.33 *Lorenzinia* in outcrop. (**a–f**), *Lorenzinia* isp., preserved as convex hyporeliefs in sandstone beds. Lower Triassic Guomugou Formation, Longwuxia section, Qinghai Province, northwestern China

ornamentation and are preserved in convex hyporelief. Burrow fill is indistinct from the host rock, indicating passive filling. These traces were only found on the sole beds of the siltstone layers of the Lower Triassic Guomugou Formation from the Longwuxia section. This ichnotaxon was preserved on the same beds with other graphoglyptids, such as *Dendrotichnium*, *Acanthoraphe*, and *Ubinia*, but was less common.

Remarks The ichnogenus *Lorenzinia* was discussed by Uchman et al. (1998). The similar ichnogenus *Sublorenzinia* was included in *Lorenzinia* according to the similar pattern of the radiating ridges. There are several ichnospecies of *Lorenzinia*. The type ichnospecies, *L. apenninia* Gabelli 1900, is characterized by radiating traces forming a circular or elliptical pattern around a central blank area. These radiating traces consist of 14–26 spindle-shaped protruding ridges of nearly equal length, typically arranged in a regular or equidistant manner. Uchman et al. (1998) proposed a reconstruction model of *Lorenzinia* based on abundant materials and suggested that *Lorenzinia* most probably existed as a three-dimensionally branching structure characterized by systematic radiating branches. Ichnospecies of regular *Lorenzinia* possess about 20 radiating ridges, while irregular ichnospecies display only about 10 ridges. Most specimens from the Guomugou Formation of the Longwuxia section display only 10 or less than ten radiating ridges, indicating they are irregular ichnospecies. *Lorenzinia* was regarded as one of the graphoglyptids by Seilacher (1977). Demircan and Toker (2003) identified *L. pustulosa* (Książkiewicz 1977) in the middle and outer fan sediments of the Miocene Cingöz Formation. Uchman et al. (2004) documented *Lorenzinia* in siliciclastic flysch of the Eocene Kusuri Formation from the Sinop-Boyabat Basin (Nielsen et al. 2012). It has been interpreted as produced by holothurians, crabs, annelids, or sipunculoids. Książkiewicz (1977) regarded *Lorenzinia* as a feeding trace of hydromedusae, and *Sublorenzinia* (=*Lorenzinia*) as produced by polychaetes. Uchman et al. (1998) argued against the kinship to hydromedusae because *Lorenzinia* represents a complex burrow system that was preserved via scour and casting and is different from surface markings created by tentacles of hydromedusae. *Lorenzinia* has been found in flysch deposits from the Lower Cambrian (Crimes and Anderson 1985) to the Holocene. *Lorenzinia* belongs to the *Nereites* ichnofacies (Pemberton et al. 1992).

Monomorphichnus Crimes 1970
Monomorphichnus isp.
Figure 3.34

Description A set of horizontal scratch marks, preserved as convex hyporelief. The ridges are parallel or subparallel to each other. Each set consists of 5–7 ridges, with one ridge typically wider than the others. The ridges are smooth, straight to slightly curved, equidistant, and measure approximately 15–30 mm in length. The spacing between ridges is about 2 mm. The ridges vary from flat to deep and are highly convex, with widths of up to 1 mm (Hofmann et al. 2012). These traces were found on the sole bedding plane of siltstone beds of the Lower Triassic Sunjiagou Formation, Houzhougongmiao section.

Remarks *Monomorphichnus* is a controversial ichnogenus. The key debate may be its distinction with another similar trackway *Dimorphichnus*. In his original description of *Monomorphichnus*, Crimes (1970) attributed the scratch marks to trilobites raking the sediment surface with their endopodites. Crimes excluded scratch marks lacking diagnostic pusher imprints from the ichnogenus *Dimorphichnus*, indicating

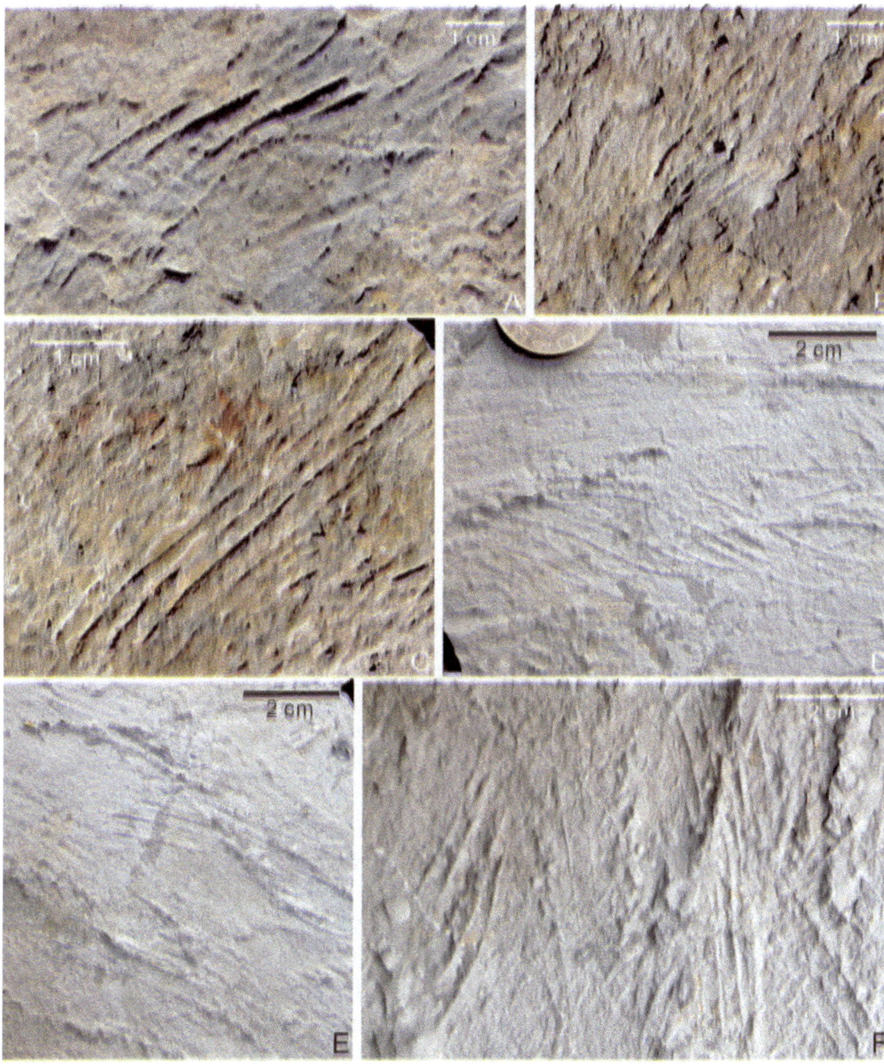

Fig. 3.34 *Monomorphichnus* in outcrop. (**a–f**), *Monomorphichnus* isp., preserved as convex hyporeliefs on the sole bedding plane of siltstone. Lower Triassic Sunjiagou Formation, Houzhougongmiao section, Shaanxi Province, northwestern China

that *Monomorphichnus* represents a distinct behavioral pattern. Seilacher maintained that the type specimen of *Monomorphichnus* featured pusher marks. He suggested that *Monomorphichnus* should be treated as a junior synonym of *Dimorphichnus*, as it was viewed to represent an alternative mode of preservation for *Dimorphichnus* (Seilacher 1985, 1990). However, Fillion and Pickerill (1990) later dismissed this view, contending that pushing marks are absent in the type specimens of *M. bilinearis*. Because the differences do exist between *Monomorphichnus* and *Dimorphichus*, it is not easy to include all *Monomorphichnus*-type material into

Dimorphichus or in other available ichnogenera, Mángano and Buatois (2004) suggested that *Monomorphichnus* may be provisionally retained. Mángano and Buatois (2004) also noted that some earlier described specimens of *M. bilinearis* were likely a fragments of a shallow rusophycid structure, and some other specimens figured as *Monomorphichnus* isp. were actually partially preserved shallow *Cruziana*. Jensen (1997) noted that similar structures can form as physical tool marks and *Monomorphichnus* specimens were more often found on current-marked beds. This statement seems to be consistent with our specimens as they were preserved together with tool marks on the sole bedding plane of siltstone. Within shallow-marine deposits of the lower Paleozoic, *Monomorphichnus* is widely considered to represent the trace of trilobites (Hofmann et al. 2012). Other arthropods could also be the possible trace-makers. *Monomorphichnus* occurs in marine deposits from the Lower Cambrian to the Holocene (Buatois and Mángano 2018).

Nereites MacLeay 1839
Nereites missouriensis Weller 1899
Figure 3.35

Description The burrows are curved to serpentine-shaped, consisting of a central backfill zone and disturbed zones on either side. The central zone is 5–8 mm wide, and the envelope zone is 4–6 mm wide. The diameter of the entire trace measures approximately 15–25 mm and is consistent throughout the burrow. The variably preserved envelope zones were arranged asymmetrically on both sides along the median groove. Some lobate zones form a discrete uniserial chain of pustules, resembling the initial descriptions of *Scalarituba missouriensis* by Conkin and Conkin (1968), while others form evenly imbricate sediment pads. Several slabs containing this ichnofauna were documented from the sandstone beds of the Lower Triassic Shangaling Group, Zhayougou section, Hezuo City, western Qinling area.

Remarks It is now widely accepted that *Neonereites*, *Scalarituba*, and *Helminthoida* are synonyms of *Nereites* (Uchman 1995; Mángano et al. 2000). The differentiation of *Nereites* trace fossil species is primarily based on the relative proportion of the central backfill zone to the disturbed zone and the morphology of the disturbed zone (Uchman 1995; Mángano et al. 2000, 2002). *N. missouriensis* has a well-developed central backfill zone, which is typically equal to or slightly wider than the marginal disturbed zone. The distinct median tunnel, surrounded by a lobate zone of reworked sediments, agrees well with Weller's ichnospecies. Although 30 ichnospecies have been described and ascribed to *Nereites*, many of them are now regarded as preservational variants of a few ichnospecies (Uchman 1995; Mángano et al. 2000, 2002; Knaust 2017). In addition, Conkin and Conkin (1968) mentioned that some preservational variants of *Scalarituba missouriensis* (*Nereites missouriensis*) possess a scalariform backfill menisci of the central tunnel and a uniserial row of small pustules of the envelope zone. The sole expression of this form corresponds to its subjective synonym *Neonereites biserialis* (Seilacher 1960). The exterior of the central tunnel of *Scalarituba* (*Nereites*) sometimes shows a uniserial row of pustules. Moreover, the uniserial chain passes into or is covered by a biserial chain of

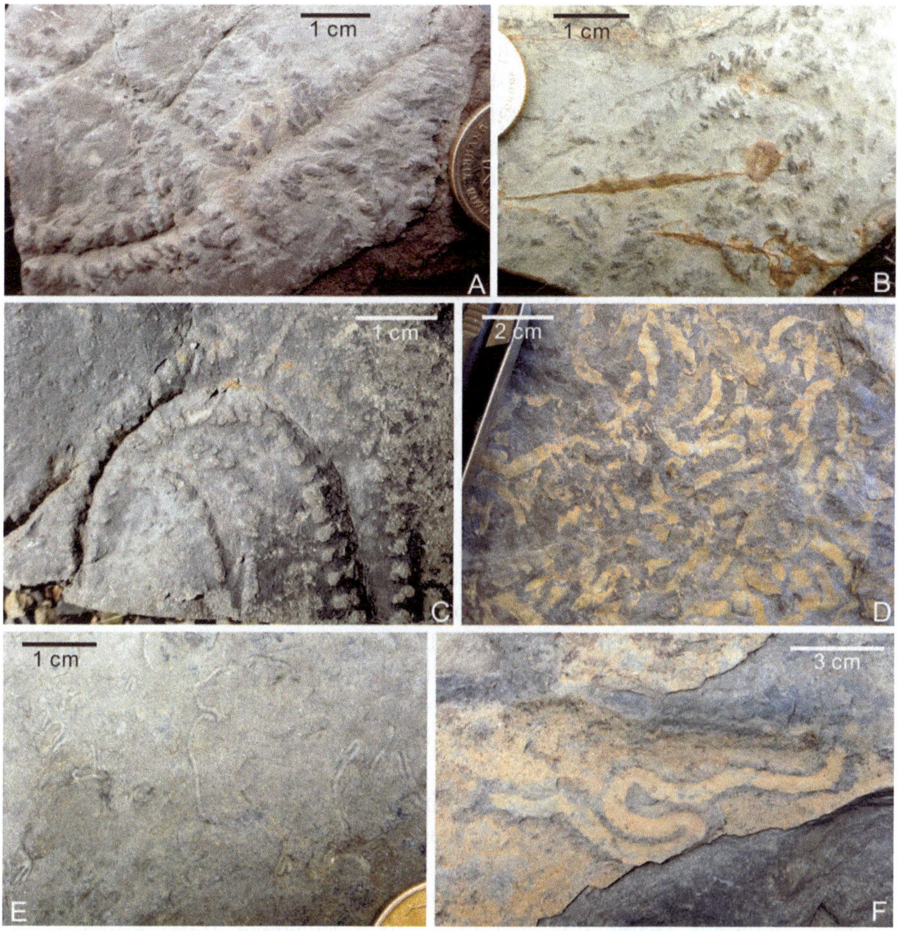

Fig. 3.35 *Nereites* in outcrop. (**a–c**), *N. missouriensis* preserved as full reliefs in sandstone. Lower Triassic Shangaling Group, Zhayougou section, Gansu Province, northwestern China. (**d**, **f**), *N. missouriensis* preserved as full reliefs in silty slate. Lower Triassic Guomugou Formation, Longwuxia section, Qinghai Province, northwestern China. (**e**), *N. missouriensis* preserved as full reliefs in silty slate. Changhsingian (late Permian) Maomaolong Formation, Malonggou section, Gansu Province, northwestern China

sediment pustules on the illustration of the holotype of *Neonereites uniserialis*. Thus there was no important difference between the type material of *N. uniserialis* and *N. biserialis*. Uchman (1995) further proposed that the ichnospecies previously classified under *Neonereites* (*N. uniserialis*, *N. biserialis*, and *N. multiserialis*) should instead be informally regarded as ichnosubspecies or preservational variants of *Nereites missouriensis*. However, Mángano et al. (2000) suggested that caution should be exercised in cases where the internal structure is not visible and the assignment to an ichnospecies is uncertain. *Nereites* is a cross-stratigraphic trace fossil but is commonly found in deep-sea turbidite facies (Wetzel 2002). In the

Palaeozoic, *Nereites* is a common component of both shallow-marine and deep-sea deposits (Fillion and Pickerill 1990; Rindsberg 1994; Orr 1995). *Nereites* may be categorized as the trace of a deposit-feeder (fodinichnia) or as a grazing trace (pascichnia) resulting from both locomotion and feeding activities (Orr 1995; Mángano et al. 2000; Knaust 2017), but it is most likely produced by a wormlike sediment-feeder (Mángano et al. 2000). *Nereites* has a distribution ranging from the early Cambrian to the Holocene (Crimes 1987; Wetzel 2002).

Palaeophycus Hall 1847
Palaeophycus heberti Saporta 1872-1873
Figures 3.36 and 3.37

Description Positive hyporeliefs or full reliefs; distinctly lined, smooth-walled, unornamented cylindrical burrows. Burrows are straight to slightly curved, branching or unbranching. Vertical or oblique shafts are occasionally observed on bedding surfaces. The diameter of individual burrows tends to remain more or less constant. Most burrows, 12–20 mm in diameter, are rather robust. Few specimens are very long, up to 50 cm in length; others are 10–20 cm long. Wall linings constitute a substantial portion of the overall diameter, with actual thickness varying following burrow size. Collapsed structures are not commonly present in the studied materials. The burrow fill is nearly identical to the host rock, while the wall linings are typically composed of agglutinated sediments that are coarser and better sorted than the surrounding rock. This ichnofauna is very common in muddy limestone beds from many localities of carbonate ramp environmental settings.

Remarks *Palaeophycus* is characterized by passive, gravity-driven sedimentation within open, lined burrows and differs from *Planolites* due to its wall linings and burrow fill, which are the same as the host rock. Taxonomic differences between various ichnospecies of *Palaeophycus* were based primarily upon wall linings and burrow sculptings (Pemberton and Frey 1982; Buckman 1995; Keighley and Pickerill 1995; Mángano et al. 1996; Jensen 1997; Kim et al. 2000). Of these, *P. heberti* is characterized by its thick wall lining and the absence of striations (Pemberton and Frey 1982), which match our specimens well. The categorization of *Palaeophycus* holds significant consequences for explaining sedimentology and diagenesis. Characteristics of *Palaeophycus* indicate passive sediment accumulation in the dwelling burrow of a predatory or suspension-feeding animal. In contrast, *Planolites* displays features of active sediment refilling in a temporary burrow made by a mobile, deposit-feeding animal (Pemberton and Frey 1982). *Palaeophycus* is considered to represent dwelling burrows (dominichnia) created by suspension feeders, such as polychaetes or annelids. It is a facies-crossing ichnotaxon, occurring in nearly all types of depositional environments (Pemberton and Frey 1982). The age range of *Palaeophycus* spans from the Ediacaran to the Holocene.

Phycodes Richter 1850
Phycodes circinatus Richter 1853
Figures 3.38 and 3.39

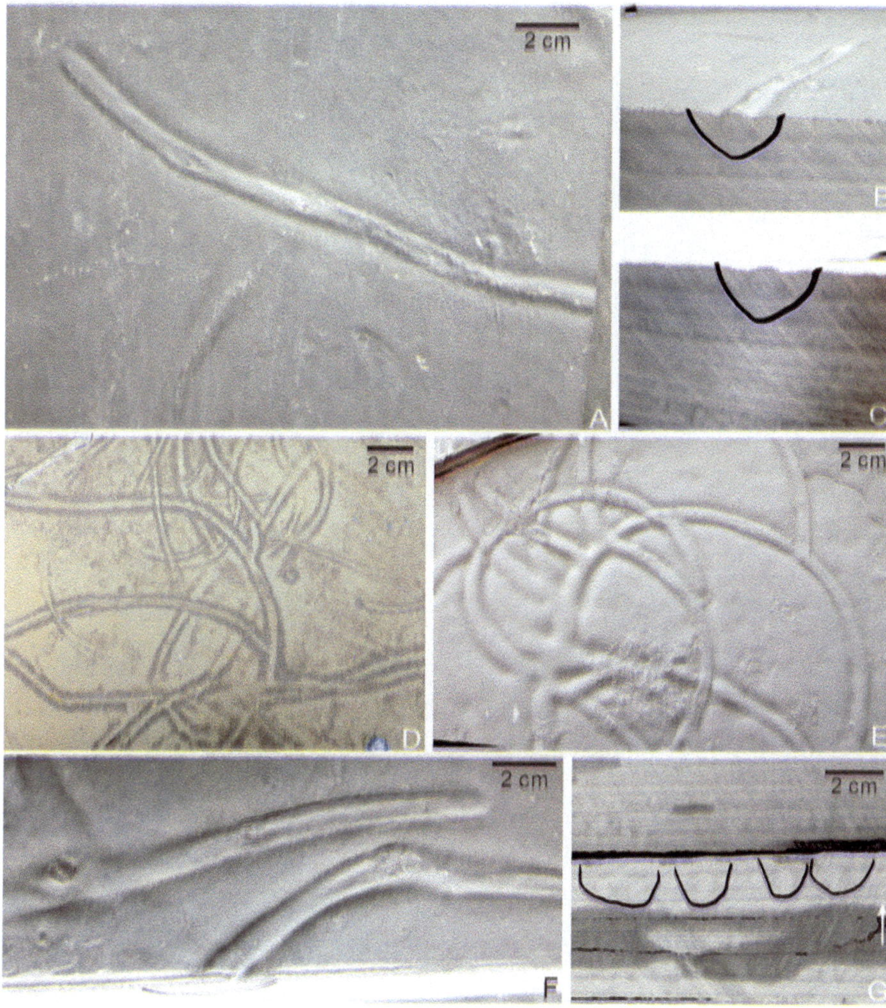

Fig. 3.36 *Palaeophycus* in outcrop. (**a–f**), *P. heberti* preserved as full reliefs in argillaceous limestone. **b–c** and **g** are vertical profile views; others are upper bedding plane views. Lower Triassic Daye Formation, Dangwu section, Guizhou Province, southwestern China

Description Horizontally grouped burrows preserved as convex hyporeliefs. The burrow systems display a range of shapes, such as reniform, fasciculate, flabellate, broom-like, ungulate, linear, falcate, or circular outlines. Most forms feature one or several main branches with a spreite-like structure, from which numerous free branches extend distally. Other forms lack spreiten and show more irregular branching patterns. In some cases, lateral tubes are aligned along both sides of the main tube. The main tube is straight, while the lateral tube curves. Individual branches are terete and finely annulate or smooth and embrace distinct fine scratch marks on burrow surfaces in some specimens. Individual lateral tubes are about 5–10 mm

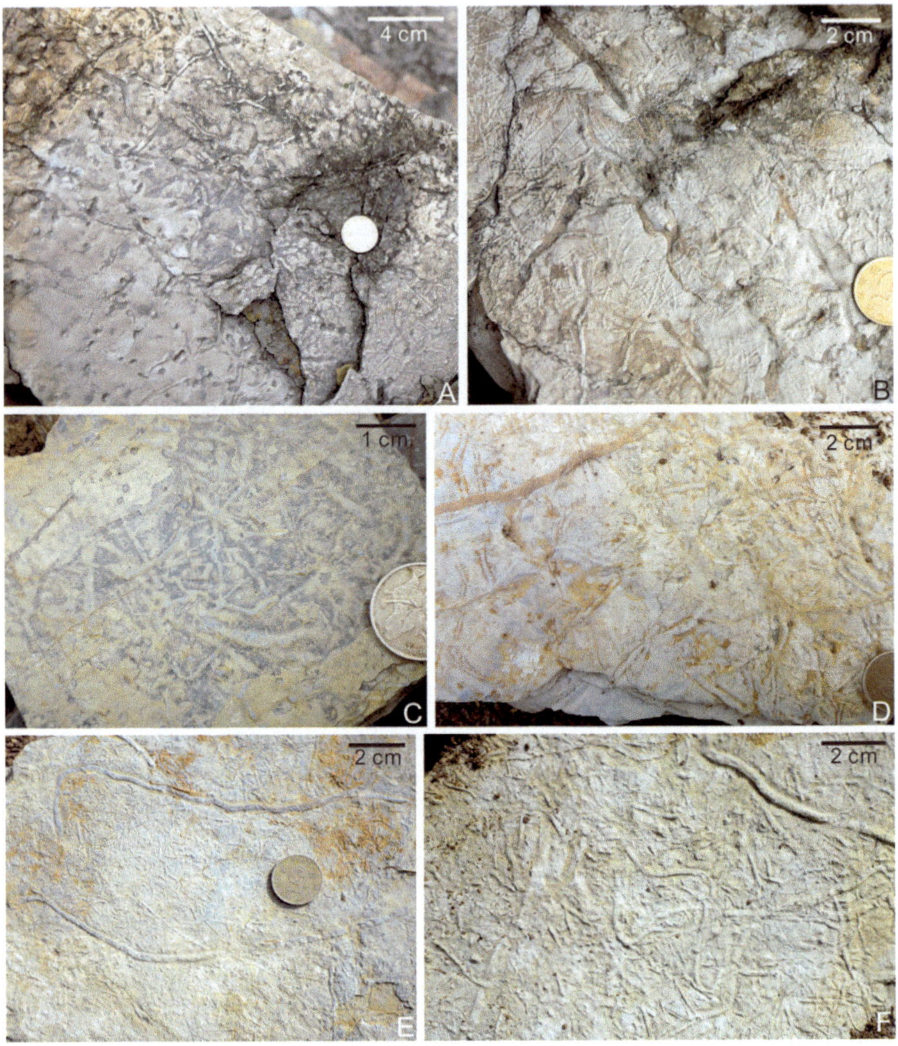

Fig. 3.37 *Palaeophycus* in outcrop. (**a**–**f**), *P. heberti* preserved in argillaceous limestone. All are upper bedding plane views. **a**–**b** are from the Lower Triassic Daye Formation, Daxiakou section, Hubei Province, South China. **c**–**f** are from Lower Triassic Daye Formation, Shatian section, Hubei Province, South China

wide, whereas the main tube has a diameter of 20 mm, and the maximum preserved length is 30 cm. The burrow fill is lithologically similar to the host rock.

Remarks *Phycodes* is one of the oldest ichnogenera established in the 1850s (i.e., Richter 1850, 1853). Mägdefrau (1934) modified the name of the type ichnospecies *P. circinatum*. Later, the concept and nomenclatures of *Phycodes* have been extensively discussed (Seilacher 1955; Osgood 1970; Bradley 1980, 1981; Fillion and

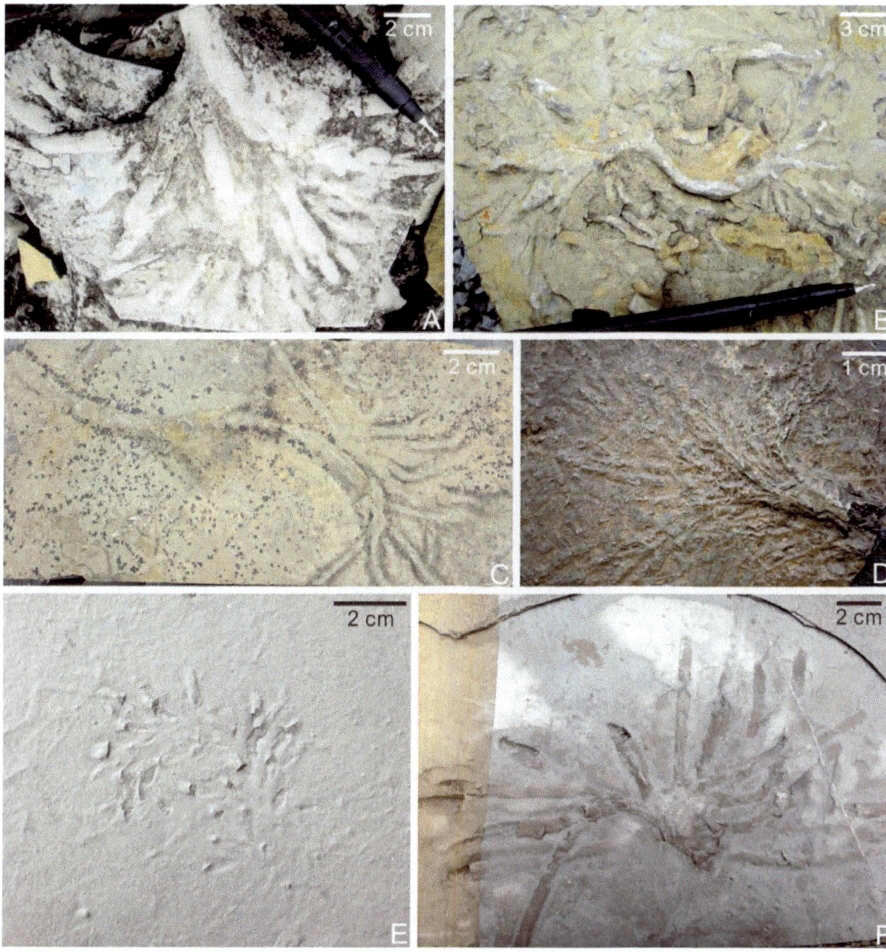

Fig. 3.38 *Phycodes* in outcrop. (**a–f**), *P. circinatus* preserved in argillaceous limestone. **a–e** are preserved as convex hyporeliefs, and **f** is full relief. Lower Triassic Daye Formation, Dangwu section, Guizhou Province, southwestern China

Pickerill 1990). Richter (1850) established this ichnogenus based on the specimens of the Early Ordovician in Germany and later described *P. circulatum* as the type ichnospecies of the ichnogenus by the same author (Richter 1853). This ichnospecies name was later corrected as *P. circulatus* (Richter 1853). Seilacher (1955) pointed out that the starting point of the burrow bundle is composed of a few channels, and the burrow has a receding web-like structure inside. The orthographic history of *P. circinatus* was discussed in detail by Fillion and Pickerill (1990). The taxonomy of *Phycodes* has been well studied by Dam (1990), Han and Pickerill (1994), and Pickerill et al. (1995). The specimens examined agree well with *P. circinatus* in all observed aspects.

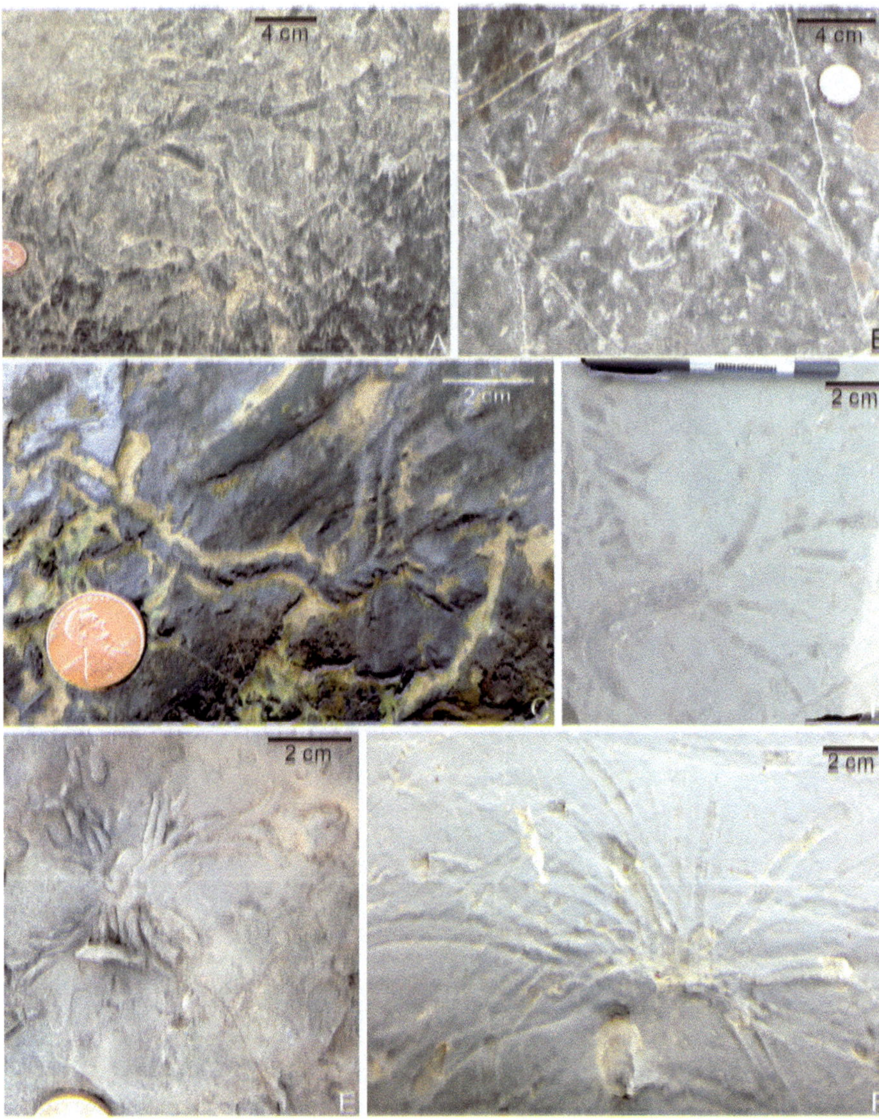

Fig. 3.39 *Phycodes* in outcrop. (**a–c**), *P. circinatus* preserved as convex hyporeliefs in sandstone. Changhsingian Talung Formation, lower Penglaitan section, Guangxi Province, southwestern China. (**d, f**), *P. circinatus* preserved as full reliefs in argillaceous limestone. Lower Triassic Daye Formation, Dangwu section, Guizhou Province, southwestern China. (**e**), *P. circinatus* preserved as convex hyporeliefs in sandstone. Lower Triassic Xiahuancang Formation, Zhihema section, Qinghai Province, northwestern China

Up to now, a total of 20 ichnospecies assigned to *Phycodes* are valid, namely *P. circinnatus* Richter 1853 (type ichnospecies of *Phycodes*), *P. palmatus*, *P. triadicum*, *P. flabellum*, *P.? antecedens*, *P. gregarius*, *P. bilix*, *P. reniforme*, *P. curvipalmatum*, *P. bromleyi*, *P. auduni*, *P. ungulatus*, *P. wabanensis*, *P. noha*, *P. templus*, *P. parallelum*, *P. fusiforme*, *P. dentatus*, *P. rarus*, and *Phycodes canelensis*.

The producer of *Phycodes* has been interpreted as a sediment-feeding vermiform annelid (Seilacher 1955; Häntzschel 1975), a Pennatulacean (Bradley 1980), an anthoptiloid sea pen (Bradley 1981), or possibly even arthropods (Rindsberg and Martin 2003). *Phycodes* was traditionally considered a reliable indicator of shallow marine environments and a characteristic trace fossil of the *Cruziana* ichnofacies of Seilacher (1964), typically found in low-energy settings with relatively stable substrates. However, this ichnogenus has also been reported from non-marine, deltaic, and deep-sea environments (Bradley 1981; Narbonne 1984; Pollard 1985; Greb and Chesnut 1992), implying a much broader environmental range than previously thought. The earliest occurrence of *Phycodes* is known from the base of the Cambrian (Clausen and Vilhjálmsson 1986; Jensen 1997; Jensen et al. 1998; Seilacher 2007; Desai et al. 2010). Their stratigraphical occurrences range from the basal part of the Cambrian to Holocene.

Phycosiphon Fischer-Ooster 1858
Phycosiphon incertum Fischer-Ooster 1858
Figures 3.40 and 3.41

Description Small, narrow, and sinuous to U-shaped tubes that curve into a series of lobes, creating protrusive spreiten structures. The most notable characteristic of Phycosiphon, whether preserved as epirelief or on split bedding surfaces, is the shape and arrangement of its marginal tube. The burrow fill of the marginal tube is darker than the surrounding host rock. The diameter of the marginal tube ranges from 0.6 to 1 mm and remains uniform throughout the entire trace. The Spreiten structure in the small U-shaped lobe is poorly preserved and is always not observed. Abundant U-shaped lobes are always densely arranged and looped together, forming antler-shaped, complex burrow systems. Individual lobes are hairpin-shaped, resembling miniature *Rhizocorallium*, while larger arrays may appear similar to meandering burrows. The antler-shaped burrow systems are mainly parallel to the bedding plane. Some specimens are inclined and cut the upper and neighboring lower layers, indicating they are complex three-dimensional burrow systems. It is hardly observed from vertical sections. A large number of these U-shaped lobes form an antler-shaped burrow system on the bedding plane, which may result in a high bioturbation degree on the bedding plane. Crosscutting relationships among different U-shaped lobes or antler-shaped burrow systems, as well as with other horizontal burrows, are relatively common. This ichnotaxon has only been found in the muddy limestone beds of the Lower Triassic Daye Formation, Dangwu section, Guizhou Province, South China.

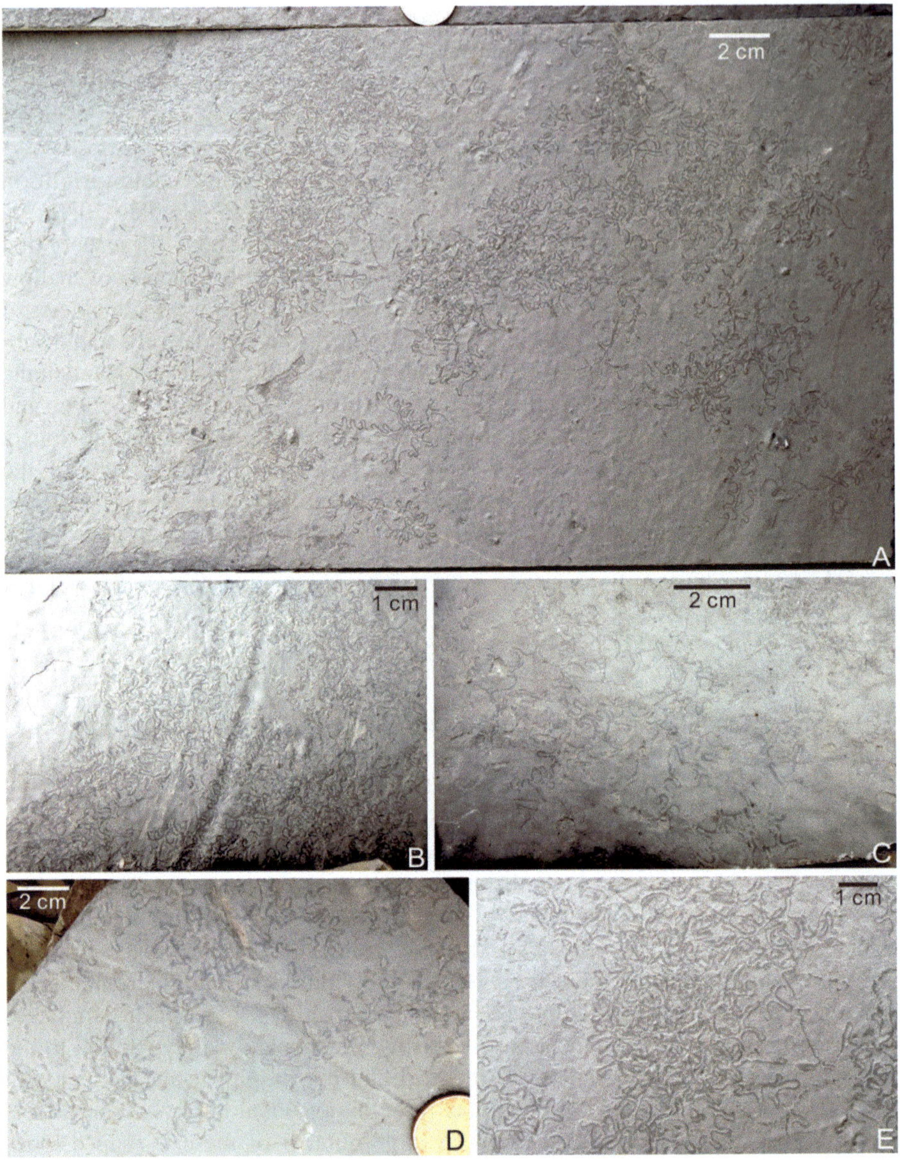

Fig. 3.40 *Phycosiphon* in outcrop. (**a–e**), *P. incertum* preserved in argillaceous limestone. Lower Triassic Daye Formation, Dangwu section, Guizhou Province, southwestern China

Remarks There is only one ichnospecies of *Phycosiphon*, *P. incertum*. The three-dimensional *Anconichnus horizontalis* is now considered a junior synonym of *P. incertum* (Wetzel and Bromley 1994). Three-dimensional reconstructions of *P. incertum* burrow systems were conducted, revealing morphological differences from the type material as redescribed by Wetzel and Bromley (1994) (Bednarz and

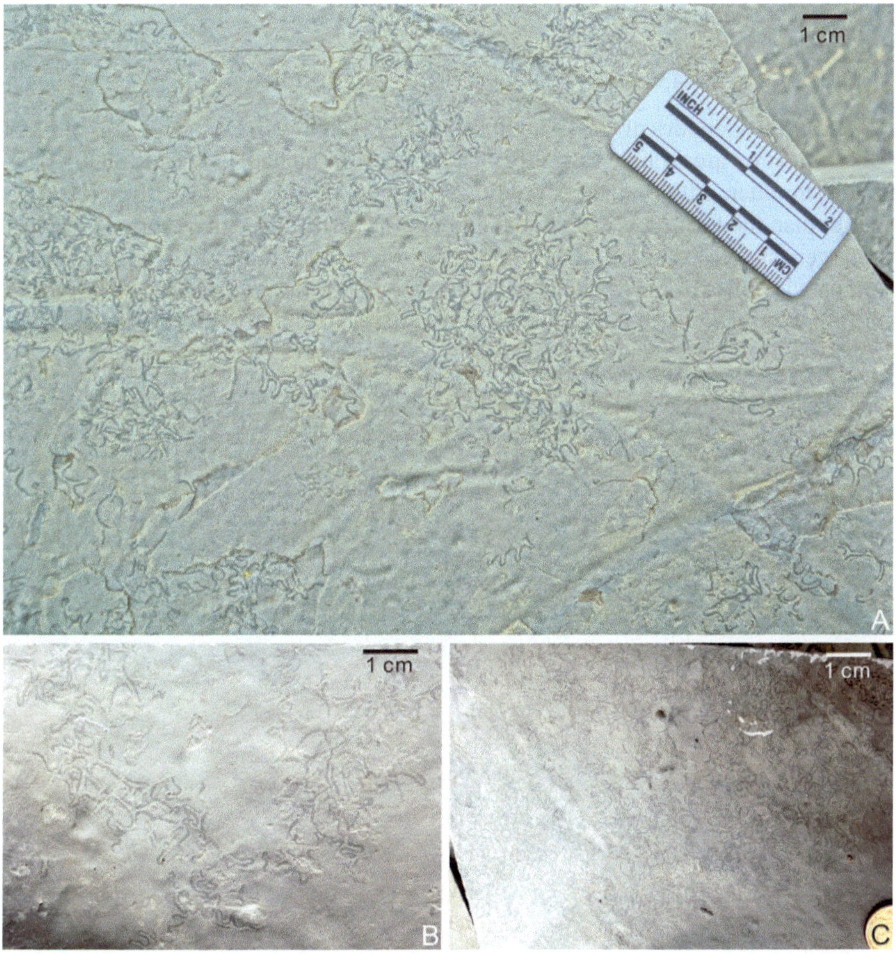

Fig. 3.41 *Phycosiphon* in outcrop. (**a–c**), *P. incertum* preserved in argillaceous limestone. Lower Triassic Daye Formation, Dangwu section, Guizhou Province, southwestern China

McIlroy 2009). *Phycosiphon* is best explained as the result of deposit-feeding activity by small vermiform organisms that exploit organic-rich matter (Wetzel 2010; Izumi 2014). These small, worm-like bulldozers moved through the sediment without a clear connection to the sediment surface. *Phycosiphon* is a common component of offshore (shelf) to lower shoreface settings (Goldring et al. 1991; Pemberton et al. 2012). It also occurs in slope (e.g., Savrda et al. 2001; Naruse and Nifuku 2008). Additionally, *Phycosiphon* is common in deep-marine settings (Callow et al. 2013). Callow and McIlroy (2011) noted a dominance of phycosiphoni-form ichnotaxa in Mesozoic and Cenozoic rocks, while *Nereites* prevails in Paleozoic rocks. They then speculated that a shift from a *Nereites*-dominated Paleozoic to a *Phycosiphon*-dominated Mesozoic and Cenozoic ecological niches. The age range of *Phycosiphon* spans from the Early Devonian to the Holocene (Buatois and Mángano 2018).

Planolites Nicholson 1873
Planolites beverleyensis Billings 1862
Figures 3.42 and 3.43

Description Cylindrical or subcylindrical horizontal burrows with smooth surfaces. Burrows are straight to gently curved, rarely branched, but occasionally intersect each other, with diameters ranging from 6 to 14 mm. Burrow fill is relatively darker than the host rock, which may indicate active filling. Small burrows, as well as larger ones in crowded conditions, tend to be more highly curved or tortuous compared to uncrowded large burrows. In some cases, burrows are densely arranged, and crosscutting among individuals is common. In siliciclastic rocks, the burrow fill

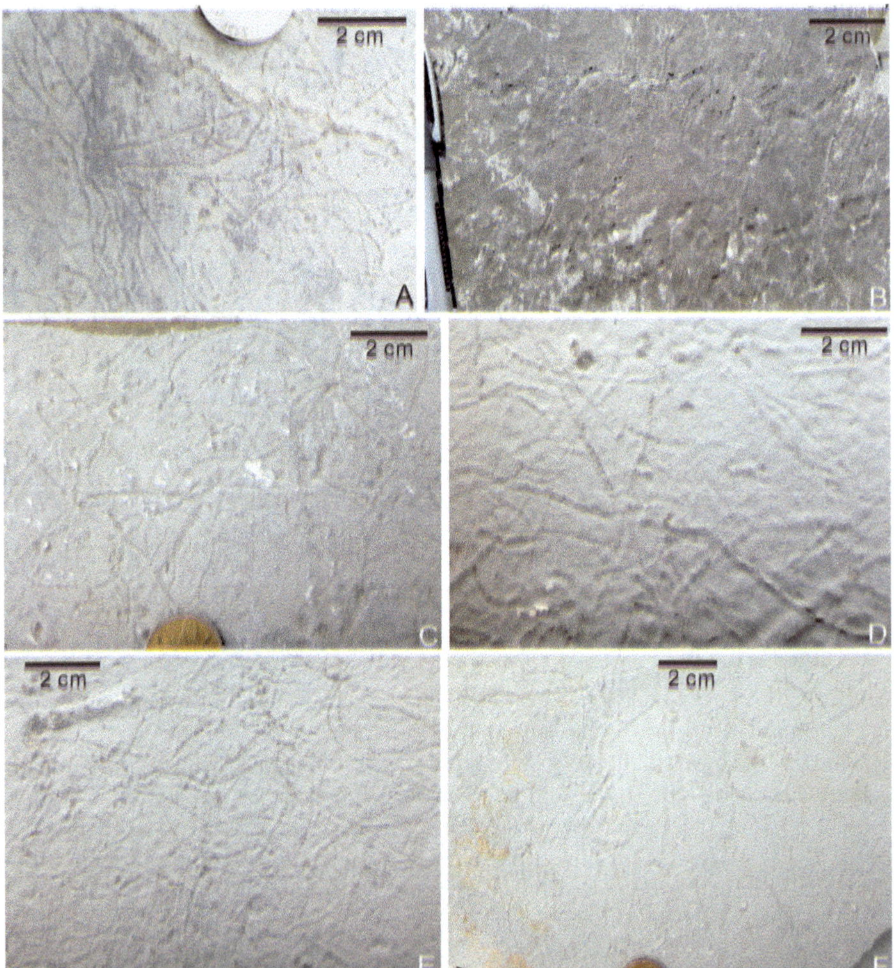

Fig. 3.42 *Planolites* in outcrop. (**a–f**), *P. beverleyensis* preserved in argillaceous limestone. Lower Triassic Daye Formation, Dangwu section, Guizhou Province, southwestern China

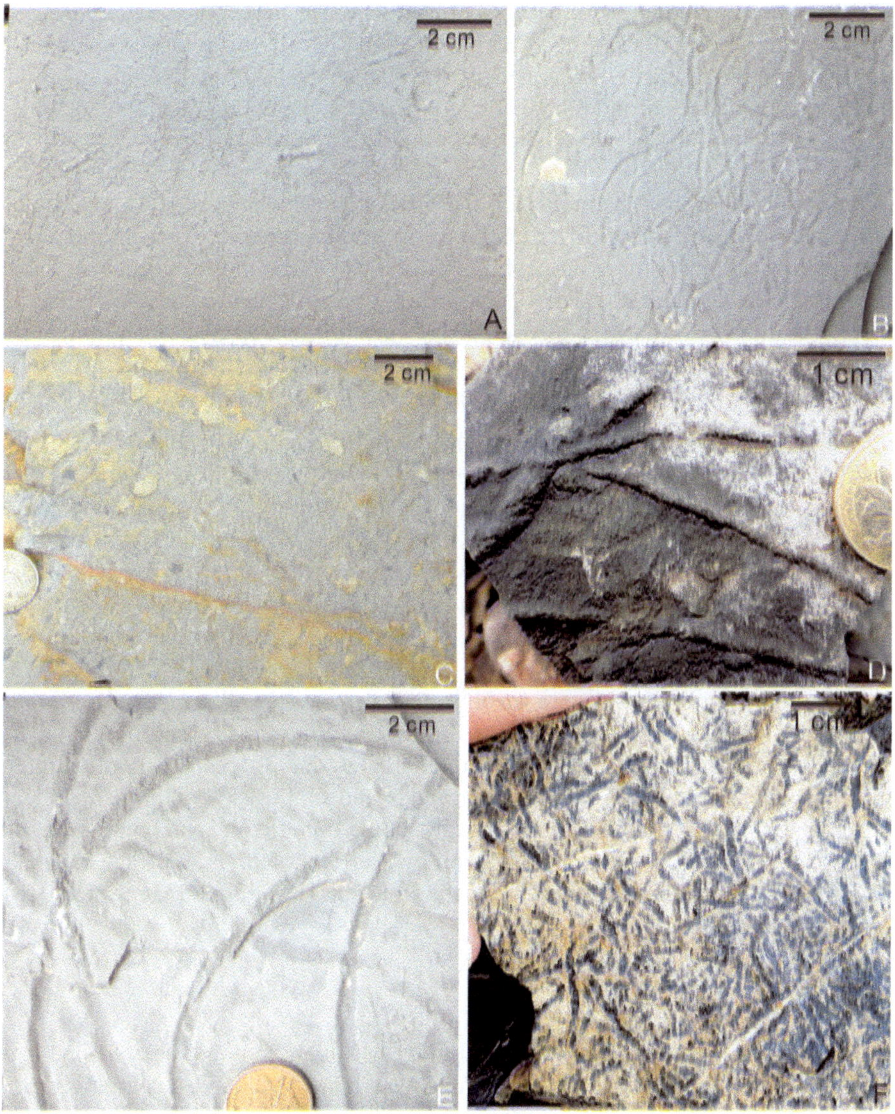

Fig. 3.43 *Planolites* in outcrop. (**a–b**, **e**), *P. beverleyensis* preserved in argillaceous limestone. Lower Triassic Daye Formation, Dangwu section, Guizhou Province, southwestern China. (**c**), *P. beverleyensis* preserved in argillaceous limestone. Lower Triassic Daye Formation, Shatian section, Hubei Province, South China. (**d**), *P. beverleyensis* preserved in sandstone. Lower Triassic Shangaling Group, Zhayougou section, Gansu Province, northwestern China. (**f**), *P. beverleyensis* preserved in silty slate. Changhsingian (late Permian) Maomaolong Formation, Malonggou section, Gansu Province, northwestern China

of *Planolites* is distinct with the host rock, while in carbonates, the sediments between the burrow fill and host matrix are lithologically difficult to distinguish. *Planolites* is one of the first ichnogenera that reappeared after the P-Tr mass extinction in carbonate facies, and it has been reported from many Early Triassic localities from South China and elsewhere in the world.

Remarks *Planolites* is distinguished from *Palaeophycus* primarily by its unlined walls and burrow fills that differ in texture from the host rock. The fills may also vary in fabric, composition, and color. The infills of *Planolites* represent sediment processed by the tracemaker, particularly through the deposit-feeding activities of mobile animals. Similarities also exist between *Planolites* and *Macaronichnus*. However, although *Macaronichnus* also has an active filling, its surrounding mantle differs from the smooth margin characteristic of *Planolites*.

The ichnospecies of *Planolites* has been well discussed (Pemberton and Frey 1982; Stanley and Pickerill 1994; Keighley and Pickerill 1995). According to Pemberton and Frey (1982), three ichnospecies, namely *Planolites annularius*, *P. beverleyensis*, and *P. montanus*, were valid. *P. annularis* is an obviously annulate ichnospecies, *P. beverleyensis* is a relatively large unornamented form, and *P. montanus* is rather small (less than 5 mm in diameter) and unornamented. In addition, new ichnospecies of *Planolites*, namely, *P. constriannulatus*, exhibiting both striation and annulation, has been reported (Stanley and Pickerill 1994). *Planolites* is only one of several ichnogenera involving fecal materials. *Planolites* is a simple trace fossil that can be produced by a wide range of organisms from different phyla. Wormlike animals, such as annelids, hemichordates, and priapulids, are often interpreted as its producers, as well as arthropods and mollusks. *Planolites* has been documented in all aquatic environments, including both marine and nonmarine settings. It is a common component of shallow-tier ichnofabrics. The age range of Planolites extends from the Ediacaran to the Holocene (Buatois and Mángano 2018).

Protovirgularia M'Coy 1850
Protovirgularia dichotoma M'Coy 1850
Figures 3.44 and 3.45

Description Horizontal, unbranched, straight or slightly curved, keel-shaped trail or burrow featuring a median ridge or furrow and lateral appendages arranged in a chevron pattern. The lateral appendages are symmetrically arranged along the main branch. They are perpendicular or inclined to the central main branch. The appendages are short, with only 2–4 mm in length and 0.5–1.5 mm in diameter. The central main branch is just a narrow furrow with only 0.5–1 mm in diameter when preserved as negative epirelief and is a tiny intermittent ridge when preserved as positive hyporelief. The sediments in the main branch and the lateral appendages are observed to be disturbed obviously. The whole trace is rather tiny, with only 5–8 mm in diameter, but quite long, with about 20–30 cm in length. This trace has been found in several localities in both Late Permian and Lower Triassic strata. The most abundant specimens of this ichnogenus were documented from the mudstone or

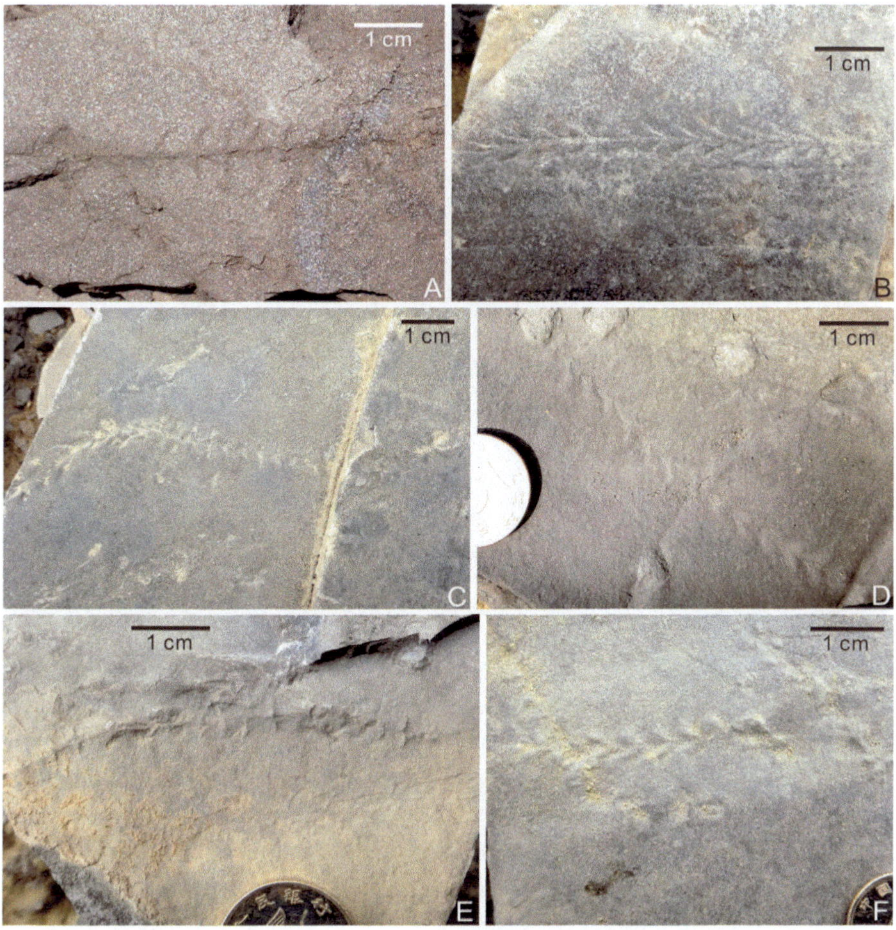

Fig. 3.44 *Protovirgularia* in outcrop. (**a**), *P. dichotoma* preserved in siltstone. Lower Triassic Sunjiagou Formation, Houzhougongmiao section, Shaanxi Province, northwestern China. (**b–f**), *P. dichotoma* preserved in siltstone. Upper Changhsingian Talung Formation, upper Penglaitan section, Guangxi Province, southwestern China. All preserved as convex hyporeliefs

muddy limestone beds of the Late Permian Talung Formation from the Penglaitan section, Guangxi Province, South China. In addition, several specimens were also collected from the siltstone beds of the Lower Triassic Sunjiagou Formation, Houzhougongmiao section, North China.

Remarks *Protovirgularia* is a quite complex trace fossil. It has long been debated among *Protovirgularia* and many other similar ichnogenera, but it has not been well understood, especially due to the confusion of its ichnospecies, until the systematic review by Knaust (2023). A total of fourteen ichnogenera have now been regarded as its junior synonyms. According to the carding by Knaust (2023), five ichnospe-

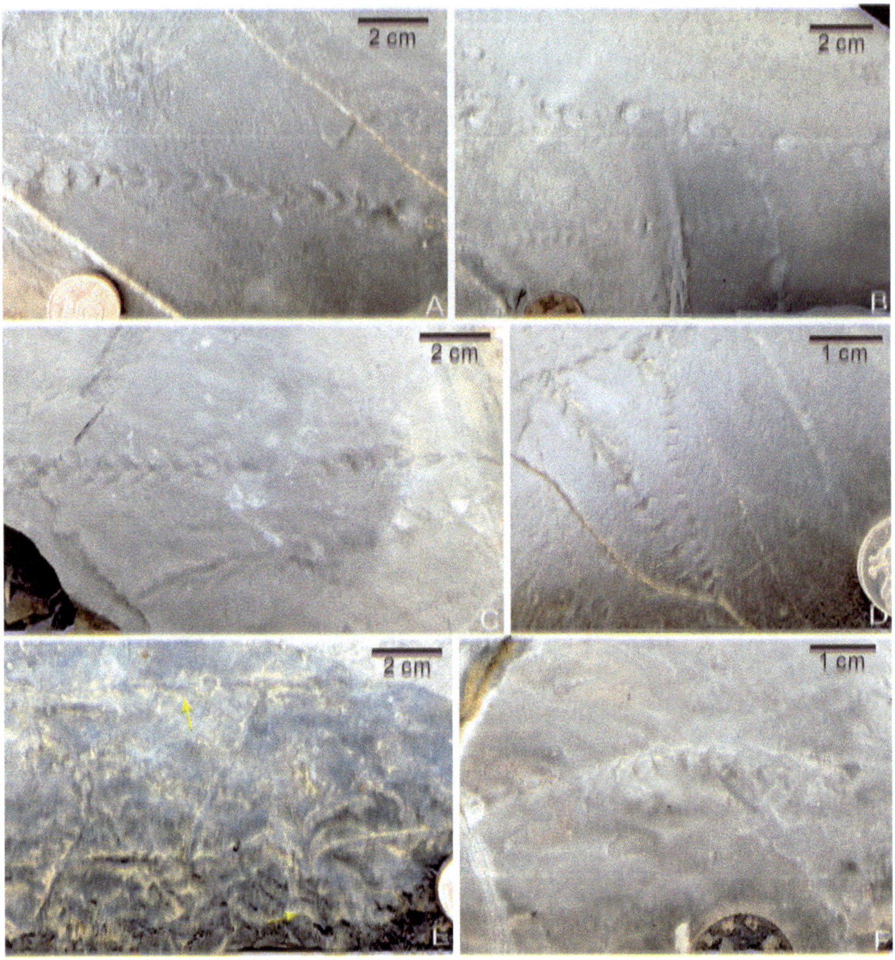

Fig. 3.45 *Protovirgularia* in outcrop. (**a–f**), *P. dichotoma* preserved in siltstone. Upper Changhsingian Talung Formation, upper Penglaitan section, Guangxi Province, southwestern China. (All preserved as convex hyporeliefs)

cies, namely, *Protovirgularia dichotoma*, *P. pennata*, *P. rugosa*, *P. longespicata* and *P. bifurcate*, are considered valid due to their distinct morphological features. Among these, *P. dichotoma* is the type ichnospecies. The Penglaitan specimens from South China match well with the morphological features of *P. dichotoma*. In contrast to other ichnospecies, which represent shallow burrows, *P. dichotoma* is better interpreted as a trackway (Knaust 2023). Seilacher and Seilacher (1994) demonstrated by experiments that protobranch bivalves with a cleft foot were the primary tracemakers of *Protovirgularia*. The intergradation of *Protovirgularia-Lockeia* in a fan-shaped cluster of ridges has been interpreted as a compound trace, representing the locomotion and resting activities of burrowing protobranch bivalves

(Cabrera et al. 2019). This interpretation has been widely accepted by other workers (Rindsberg 1994; Uchman 1998; Mángano et al. 1998; Gibert and Domènech 2008; Carmona et al. 2010; Fernández et al. 2010; Nara and Ikari 2011; Luo and Shi 2017; Hammersburg et al. 2018; López Cabrera et al. 2019). However, many earlier interpretations in fact did not support this conclusion that other clades such as annelids and arthropods were also potential trace-makers of *Protovirgularia* (Richter 1941; Volk 1961; Claus 1965; Greiner 1972; Han and Pickerill 1994). Additionally, in continental environments, isopods, amphipods, and odonatans have also been identified as producers of *Protovirgularia* (Lima et al. 2017; Metz 2020; Knaust 2022). Most Palaeozoic records of *P. dichotoma* are from deep-sea environments, while younger occurrences are predominantly found in shallow-marine to continental environments (Knaust 2023). *Protovirgularia* occurs throughout the Phanerozoic.

Protovirgularia longespicata (De Stefani 1885)
Figure 3.46

Description Horizontal, straight, imbricated burrow with a chevron pattern, and the central furrow or ridge is not obvious. The burrow is quite robust, with about 20–30 mm in diameter. The burrow may be interrupted by smooth sections and include a vertical spreite. Burrow is preserved in hyporelief. The burrow is smooth, unbranching, without a papillate structure. The lateral imbricated sediment pads are

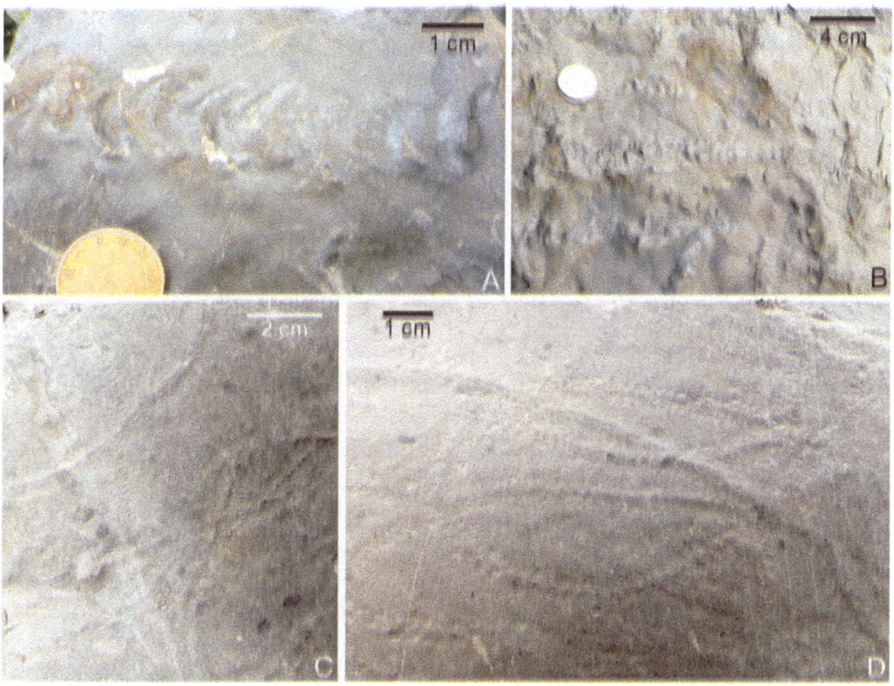

Fig. 3.46 *Protovirgularia* in outcrop. (**a–d**), *P. longespicata* preserved in sandstone. Changhsingian (late Permian) Maomaolong Formation, Malonggou section, Gansu Province, northwestern China

symmetrically arranged along the central area. The imbricated sediment pads were filled with the same sediments as the host rock, indicating passive filling. The most distinct feature of this trace is the chevron pattern and its relatively strong burrow size. It differs from *P. dichotoma* in its larger size, imbricated sediment pads, and lack of central furrow or ridge, but it has a vertical spreite. This ichnofauna was only collected from the sole bedding plane of turbidity sandstone beds of the Late Permian Maomaolong Formation, Malonggou section, western Qinling area, northwestern China.

Remarks When compared with *P. dichotoma*, *P. longespicata* has been only occasionally reported (Seilacher and Seilacher 1994; Uchman 1998; Luo and Shi 2017; Kuwazuru and Nakadawa 2018; Ding et al. 2021). Knaust (2023) systematically reviewed and amended this ichnospecies; many were previously ambiguous ichnospecies or those that had been classified into other ichnospecies, or other ichnogenera were included in this ichnospecies. For example, *Protovirgularia bidirectionalis* was once a new ichnospecies erected from the Carboniferous of Kansas, which was a shallow, arcuate burrow with basal V-shaped markings (Mángano et al. 2002). This ichnospecies was also reported from the Middle Triassic strata in Germany (Knaust 2022). Knaust (2023) argued that all the features of *P. bidirectionalis* can be observed in the type material of *Pennatulites longespicata* (now *Protovirgularia longespicata*). This justifies the inclusion of *P. bidirectionalis* in *P. longespicata*. *P. longespicata* is the most complex ichnospecies of *Protovirgularia*, distinguished from other ichnospecies by its branched forms, smooth burrow segments, spreiten, and densely imbricated sediment pads (Knaust 2023). In addition, previous works have described the combination or association of *Protovirgularia* with *Rusophycus*-like resting traces (Mángano et al. 2002; Knaust 2022), suggesting an arthropod trace-maker. *P. longespicata* primarily occurs in shallow marine environments and, to a lesser extent, in marginal marine settings from the Cambrian to the Jurassic. From the Cretaceous to the Neogene, it shifts into deep-sea environments.

Rhizocorallium Zenker 1836
Rhizocorallium jenense Zenker 1836
Figure 3.47

Description Horizontal to oblique, U-shaped spreite burrows. The burrows measure 10–30 mm in width and are unbranched. They exhibit abundant scratches on both marginal limbs and protrusive spreiten. The marginal tube has a diameter of 6–10 mm. The burrows have a relatively steep inclination relative to the bedding plane, ranging from 30° to 60°. The spreiten is well defined in many specimens, though occasionally unclear in some individuals due to the steep inclination. Some burrows show variable U-shaped structures, including straight and short, slightly acute or sinuous, and fan-shaped forms. Scratch marks are very common, appearing as continuous and parallel lines on both the marginal tubes and the spreiten. The distal of the U-shaped burrows broaden and thicken occasionally when the burrows are orientated with respect to the bedding plane at a low angle. Fecal pellets are

Fig. 3.47 *Rhizocorallium* in outcrop. (**a–c**), *R. jenense* preserved on the upper surface of sandstone. Lower Triassic Xiahuancang Formation, Zhihema section, Qinghai Province, northwestern China. (**d**), *R. jenense* preserved in argillaceous limestone. Lower Triassic Jialingjiang Formation, Shawan section, Sichuan Province, southwestern China

more common in the distal part of the marginal limbs than in the spreiten with concentric disposition of ridges and grooves. This ichnospecies of *Rhizocorallium* has been obtained from the Lower Triassic of the Daxiakou and Yashan sections, South China.

Remarks *Rhizocorallium* is among the earliest described ichnotaxa. Following the introduction of the type ichnospecies, *Rhizocorallium jenense*, by Zenker (1836), many additional ichnospecies have been established. However, in a review by Fürsich (1974a), only three ichnospecies, *R. jenense*, *R. irregulare* Mayer 1954, and *R. uliarense* Firtion 1958, were considered valid ichnotaxa. Later, Knaust (2013) argued that only *R. jenense* and *R. commune* are valid, distinguished by the substrate type and the presence of scratches. *R. jenense* is a relatively small, inclined burrow with heavy scratch marks. It is found in firm-ground substrates and features passive fills. Most of the new materials fit the concept very well. Scratches are not easily observed in both marginal limbs and poorly preserved spreiten. However, the small size, steep inclination from the bed-tops, and fecal pellets in marginal limbs of the U-shaped burrows are characteristic of *R. jenense*.

Rhizocorallium commune Schmid 1876

Comment: This ichnospecies is well-established and widely used. Many researchers have realized that size variations of *R. commune* exist (Worsley and Mørk 2001; Bann and Fielding 2004; Jaglarz and Uchman 2010; Hofmann et al. 2011), a fact with significant paleoenvironmental implications. Since small and large forms exhibit gradual transitions, a subdivision at the ichnosubspecies level cannot be justified. However, distinguishing varieties appear meaningful for establishing different ichnosubspecies. Therefore, *R. commune* var. *auriforme* was proposed to include the relatively smaller forms, which are discriminated from the relatively larger forms that are assigned to *R. commune* var. *irregulare*.

R. commune var. *irregulare* Mayer 1954
Figures 3.48, 3.49, and 3.50

Description Winding, long, and horizontal U-shaped burrows were collected from all of these ten South Chinese localities but are extremely abundant in the Luoping and Gaimao sections. The burrows are primarily preserved as negative epirelief, though occasionally a combination of negative and positive epirelief occurs. The burrows are mostly horizontal but can occasionally incline relative to the sediment. Most burrows measure 18–80 cm in length and 3–12 cm in width, with maximum dimensions reaching up to 120 cm in length and over 15 cm in width. The limb tubes have diameters ranging from 0.5 to 3.6 cm. The spreiten is consistently well-preserved, and most specimens exhibit complete protrusive spreiten. Vertically retrusive forms are also frequently present. Of these, these Luoping burrows (28–80 cm long, 5–12 cm wide, with maximum length and width up to 120 cm and >15 cm, respectively) having tube diameters of 1.6–3.6 cm (averagely ~2.0 cm) are larger than the Gaimao ones (mostly 18–46 cm long, 3–10 cm wide, with maximum length and width up to 70 cm and 12 cm, respectively) having tube diameters of 0.5–2.4 cm (averagely ~1.2 cm). All features documented here agree well with *R. commune* var. *irregulare* (Knaust 2013; Feng et al. 2018). This ichnosubspecies often occurs in the lower Middle Triassic but can extend down to the uppermost Spathian (Lower Triassic) in Gaimao.

Remarks Knaust (2013) proposed that only *R. jenense* and *R. commune* are valid ichnospecies of *Rhizocorallium*, distinguished by substrate type and the presence of scratches. *R. commune* comprises extensive, predominantly horizontal burrows with occasionally scratched marginal tubes and an actively filled spreite between the limb tubes. *Rhizocorallium* is linked to the dwelling and feeding activities of deposit-feeding crustaceans, annelids, or insects (Häntzschel 1975; Pemberton 1992; Knaust 2007, 2013; Feng et al. 2018). *Rhizocorallium* is found in a wide range of environments, from continental to marine depositional systems (Rodríguez-Tovar and Pérez-Valera 2008; Hofmann et al. 2011; Rodríguez-Tovar et al. 2012; Knaust 2013; Zhang et al. 2016; Feng et al. 2018). It is one of the long-ranging trace fossils, with a record extending from the early Cambrian to the Holocene (Knaust 2013).

Fig. 3.48 *Rhizocorallium* in outcrop. (**a–d**), *R. commune* var. *irregulare* preserved on the upper surface of sandstone, they preserved together with wrinkle structures (a kind of microbially induced sedimentary structures) and bivalves in **a** and **b**. Lower Triassic Xiahuancang Formation, Zhihema section, Qinghai Province, northwestern China. (**e–f**), *R. commune* var. *irregulare* preserved on the upper surface of argillaceous limestone. Lower Triassic Anshun Formation, Gaimao section, Guizhou Province, southwestern China

Rhizocorallium, particularly *R. commune* var. *irregulare* is very abundant in Lower and Middle Triassic strata from South China, which has been used for the analysis of biotic recovery processes following the P–Tr biotic crisis (Feng et al. 2017, 2018).

Rhizocorallium was sporadically present during the Griesbachian-Smithian interval, experienced significant proliferation and spatial expansion in the Spathian,

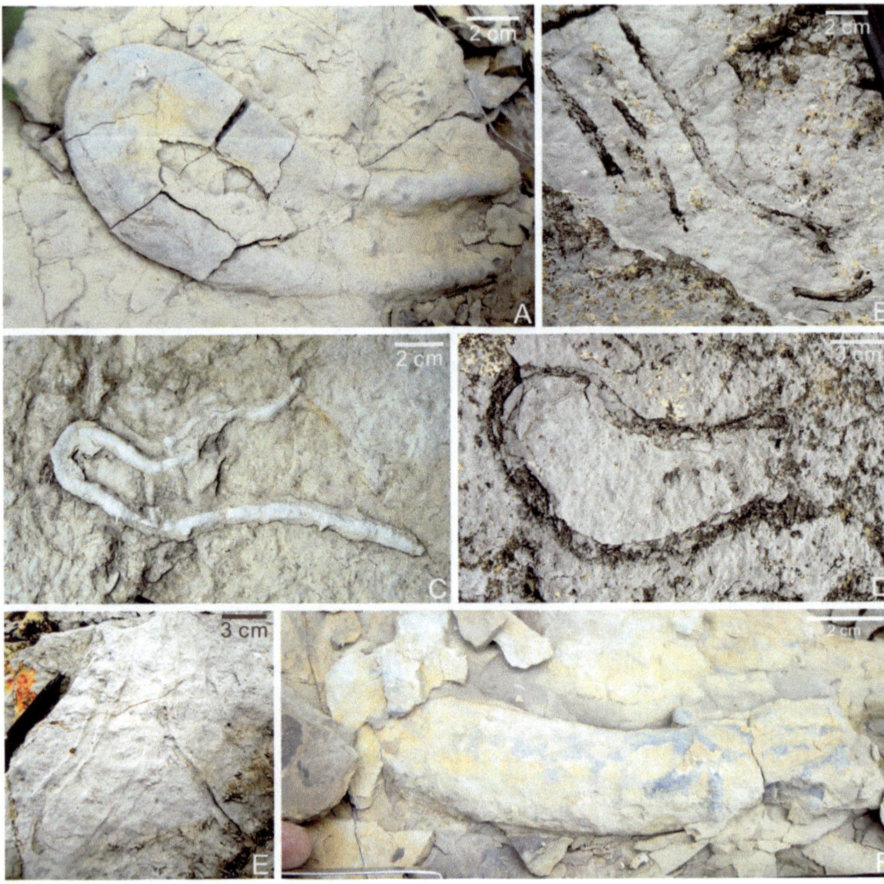

Fig. 3.49 *Rhizocorallium* in outcrop. (**a–f**), *R. commune* var. *irregulare* preserved on the upper surface of argillaceous limestone. Lower Triassic Anshun Formation, Gaimao section, Guizhou Province, southwestern China

and became highly abundant during the Anisian (Middle Triassic). The Spathian and Pelsonian (middle-late Anisian) periods marked peaks in ichnospecies richness and morphological complexity. Griesbachian-Smithian *Rhizocorallium* was characterized by relatively smaller sizes, shallower depth penetration, fewer ichnospecies, and limited geographical distribution. While Spathian *Rhizocorallium* displayed increased ichnospecies diversity, morphological complexity, and expanded environmental and geographical ranges, most ichnospecies remained relatively small, suggesting stages 2–3 of ecological recovery. By comparison, middle-late Anisian *Rhizocorallium* not only reached the highest ichnospecies richness and deepest penetration depth but also demonstrated the greatest morphological complexity and largest burrow sizes, with some forms featuring tube diameters over 20 mm. This marks the final stage (stage 4) of ecological recovery after the P–Tr crisis. According

Fig. 3.50 *Rhizocorallium* in outcrop. (**a–f**), *R. commune* var. *irregulare* preserved on the upper surface of dolomitic limestone. Lower Triassic Anshun Formation, Gaimao section, Guizhou Province, southwestern China

to ichno-ecological proxies, Early-Middle Triassic *Rhizocorallium* exhibits a stepwise recovery pattern: an initial rebound in the Griesbachian-Smithian, improved ecological recovery in the Spathian, and full recovery by the middle-late Anisian (Feng et al. 2018).

Rosselia Dahmer 1937
Rosselia socialis Dahmer 1937
Figure 3.51

Fig. 3.51 *Rosselia* in outcrop. (**a–d**), *R. socialis* preserved in sandstone (blue dashed circles), they preserved together with pictograph-like sand cracks/crackfills (a kind of microbially induced sedimentary structures, MISS) (red arrows) in **a** and **c**. Lower Triassic Xiahuancang Formation, Zhihema section, Qinghai Province, northwestern China. (**e–f**), *R. socialis* preserved in siltstone. Upper Changhsingian Talung Formation, upper Penglaitan section, Guangxi Province, southwestern China

Description The funnel-shaped burrow has a central sandy tunnel and surrounding well-defined concentric laminae. Spindle- or bulbous-shaped burrows are not observed from the studied material. Concentric laminae are typically 1–3 cm in diameter, and the central core is 3–8 mm in diameter. Height is about 40–50 mm of the whole funnel-shaped burrow, with a greatest depth of ~15 cm. Alternating layers of light and dark sediments composed the inner concentric. Most specimens are vertical, while only few are highly inclined. This ichnofauna was collected from the sandstone beds of the Xiahuancang Formation, southern Qilian, North China.

Remarks Only few ichnospecies of *Rosselia* have been erected or well inspected. Knaust (2017) noted that *R. socialis* is the most common ichnospecies and probably the only valid one. The funnel-shaped structure of *R. socialis* may represent the truncated lower portion of the spindle-shaped *Rosselia* (Nara 1995). Thus well-preserved *R. socialis* is spindle- or bulbous-shaped. The funnel-shaped morphology always appears below the erosive bedding plane or the outcrop surface parallel to the bedding, while the spindle-shaped morphology occurs in sediments without sedimentary discontinuities (Nara 1995). The specimens observed here are all preserved as funnel-shaped structures, suggesting they may occur just below an erosional surface. Another ichnogenus with concentric laminations is *Cylindrichnus*. Unlike *Rosselia*, it exhibits bow-shaped morphology and characteristic concentric, cone-in-cone arrangement of the laminae (Goldring 1996; Goldring et al. 2002; Ekdale and Harding 2015). *Rosselia* is explained as the dwelling structure (domichnion) of detritivorous polychaetes (Nara 1995, 2002). *Rosselia* has been recorded mainly in shallow marine such as lower and middle shoreface, tidal flats, and lagoons (Howard and Frey 1984; Frey and Howard 1985; Uchman and Krenmayr 1995; Carmona et al. 2008), and ranges in age from early Cambrian to Holocene (Jensen 1997; Nara and Haga 2007).

Sinusichnus Gibert 1996
Sinusichnus sinuosus Gibert 1996
Figure 3.52

Description The burrow system composed of horizontal, sinusoidal tunnels with frequent branching and few short shafts. Tunnels are 4–8 mm in diameter. Vertical shafts are short and relatively common in the burrow system. The branching points most frequently involve three tunnels meeting at 120° angles, creating a three-armed, helix-like structure. Burrows are preserved as positive epireliefs or full relief. The diameter of tunnels in a burrow system remains consistent, with 15–22 mm in diameter. The burrow systems are densely distributed on a large bedding plane, with a square of 5 m × 10 m, forming complex burrow networks; the vertical shafts are always distributed at the branching point of the sinusoidal tunnel and the beginning of an individual tunnel. The branching point of the burrow system always forms a Y- or T-shaped junction. The burrow networks are found from several large bedding surfaces, exposed at the Dangwu quarry near Guiyang City, Guizhou Province, and the bearing strata is thinly bedded muddy limestone of Member II of the Daye Formation, and the age is Smithian of the Early Triassic.

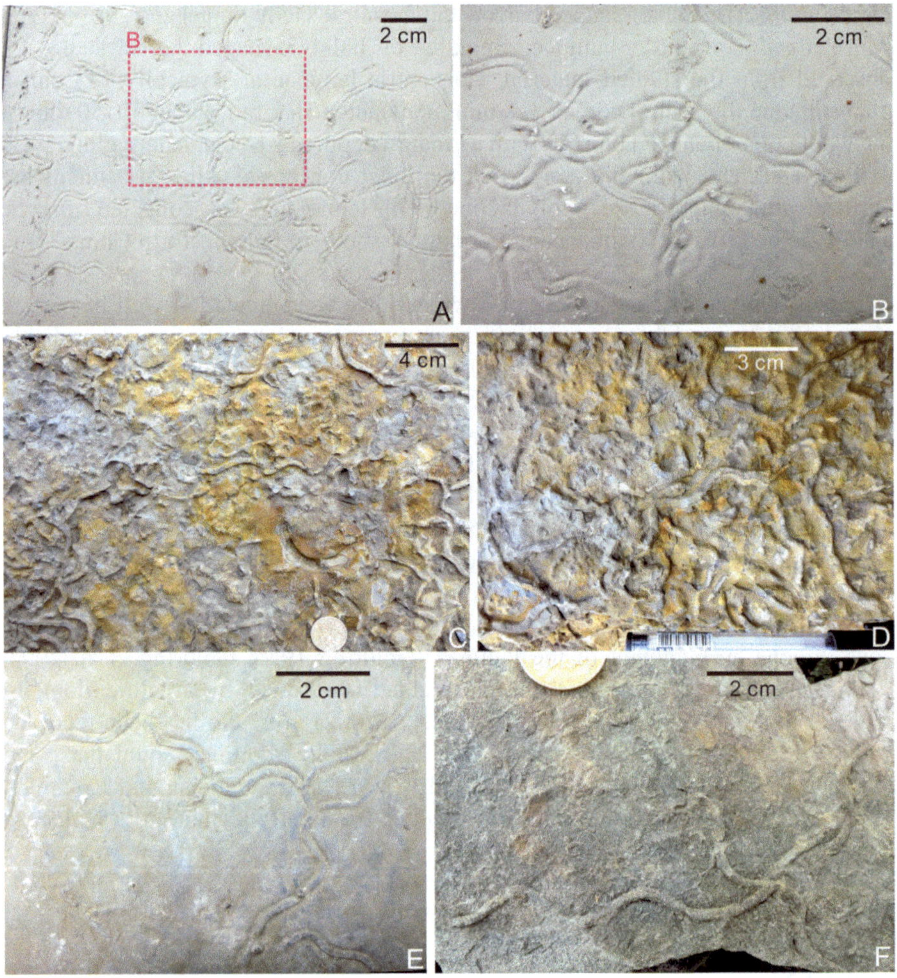

Fig. 3.52 *Sinusichnus* in outcrop. (**a**–**b**, **e**), *S. sinuosus* preserved in argillaceous limestone, B is a close-up view of the dashed block in **a**. Lower Triassic Daye Formation, Dangwu section, Guizhou Province, southwestern China. (**c**–**d**), *S. sinuosus* preserved in argillaceous limestone. Lower Triassic Daye Formation, Gaimao section, Guizhou Province, southwestern China. (**f**), *S. sinuosus* preserved in sandstone. Lower Triassic Xiahuancang Formation, Zhihema section, Qinghai Province, northwestern China

Remarks The main difference between the first two ichnospecies is that *S. sinuosus* has smooth walls, whereas *S. priesti* has a crisscrossed pattern of scratches on the walls. Buatois et al. (2009) discussed some doubts about specimens described by Kappel (2003), proposing that they could be partially eroded *Protopaleodictyon*. According to Belaústegui et al. (2014), some of the Cretaceous specimens described by Kappel (2003) and Buatois et al. (2009) are not clear. The *S. sinuosus* figured by Kappel (2003) is less regular and lacks any branching. Afterward, Buatois et al. (2009) described this trace as displaying some branches, irregular sinuosity and low amplitude/wavelength ratio. *S. seilacheri* exhibits a slight sinuosity, a bilobate floor with perpendicular to oblique striations and pronounced elliptical cross sections (Knaust et al. 2016).

The ichnogenus *Sinusichnus* was erected by Gibert (1996) according to specimens from the Pliocene of Spain and France, with the establishment of the ichnospecies *S. sinuosus*. Kappel (2003) proposed a new ichnospecies, *S. priesti*, from the Upper Cretaceous of the Münster Basin, Germany. Currently, *Sinusichnus* comprises three ichnospecies: *S. sinuosus*, *S. priesti*, and *S. seilacheri*, the last of which was described from the Middle Triassic Udelfangen Formation from Germany (Knaust et al. 2016). The key distinction between *S. sinuosus* and *S. priesti* is that the former has smooth walls, while the latter features a crisscrossed pattern of scratches on the walls. Buatois et al. (2009) expressed skepticism about the specimens described by Kappel (2003), proposing they might represent partially eroded *Protopaleodictyon*. Belaústegui et al. (2014) noted that some of the Cretaceous specimens described by Kappel (2003) and Buatois et al. (2009) are unclear. *Sinusichnus* is thought to have been produced by an animal combining deposit-feeding and farming strategies (Gibert et al. 1999) and is commonly interpreted as the work of decapod crustaceans (Gibert 1996) or isopod crustaceans (Knaust and Desrochers 2020). *Sinusichnus* has been reported in numerous environmental settings worldwide, including deltaic, coastal, shallow-marine and deep-marine environments (Luo et al. 2019). Its occurrence in the Early Triassic is rare (Zhang et al. 2019).

Skolithos Haldeman 1840
Skolithos linearis Haldeman 1840
Figure 3.53

Description Burrows are straight, vertical to steeply inclined, and filled with homogeneous sediments. On bedding planes, they are preserved as shafts or holes. In lateral view, burrows are usually 2–4 mm in diameter and 5–10 cm deep. Some specimens penetrated from the top to the bottom surface of the sandstone bed. Lining structures in many specimens are not observed from the lateral direction as a result of the strong weathering. A funnel-shaped aperture at the top may be present (Schlirf 2000). However, we have not observed such a funnel-shaped structure in our specimens. This ichnogenus has been well documented from the Lower Triassic successions of the studied area. It is commonly found in the carbonate successions and siliciclastic sandstone beds. *Skolithos* burrows are one the most common traces of the Zhihema section, southern Qilian, and occur together with *Diplocraterion* in sandstone beds of the lower member of the Xiahuancang Formation.

Remark *Skolithos* is a simple, nearly vertical cylindrical tubular structure that may or may not have a lining, and its fill is passive (Alpert 1974; Knaust 2017; Knaust et al. 2018). It is one of the earliest defined valid ichnotaxa and has been widely reported elsewhere (Knaust et al. 2018). Despite being widely reported, the type ichnospecies *S. linearis* of the ichnogenus remains ambiguous in terms of holotype and definition because no original description, holotype and illustrations were pub-

Fig. 3.53 *Skolithos* in outcrop. (**a–e**), *S. linearis* preserved in sandstone. **a–d** are vertical profile view. Lower Triassic Xiahuancang Formation, Zhihema section, Qinghai Province, northwestern China. (**f–g**), *S. linearis* preserved in dolomitic limestone (arrows). Vertical profile view. Lower Triassic Anshun Formation, Gaimao section, Guizhou Province, southwestern China. (**h**), *S. linearis* preserved in argillaceous limestone (arrows). Vertical profile view. Lower Triassic Jialingjiang Formation, Shawan section, Sichuan Province, southwestern China

lished when *S. linearis* was first established. Recently, Knaust et al. (2018) designated, described and illustrated a neotype of *Skolithos* based on new material collected from the type locality. Knaust et al. (2018) also gave a brief review of *S. linearis*. Taxonomic difference between *Skolithos* and its allies has been discussed by many authors (i.e., Schlirf and Uchman 2005; Knaust et al. 2018). However, a thorough ichnospecific revision within *Skolithos* is still needed, and thus, the assignment to *S. linearis* of our materials is tentative. *Skolithos* has been widely accepted as a common indicator of shallow, relatively high-energy, near-shore to marginal-marine environments, but it also occurs in deep-marine and continental environments (Melchor et al. 2012). *Skolithos* can be produced by a variety of organisms from different phyla, including polychaetes, phoronids, priapulids, anthozoans, crustaceans, holothurians, insects, and so on (Knaust 2017; Knaust et al. 2018). *Skolithos* ranges in age from early Cambrian to Holocene (Nara and Haga 2007; Knaust 2017). The high abundance of *Skolithos* in the strata may form piperocks during the Palaeozoic. Mángano and Buatois (2014) proposed that the occurrence of this *Skolithos*-dominated piperock might have played an important role in the functioning of the Early Cambrian ecosystem, and an evolutionary breakthrough in ecological structure was embodied by deeply-burrowing suspension feeders.

Taenidium Heer 1877
Taenidium barretti (Bradshaw 1981)
Figure 3.54

Fig. 3.54 *Taenidium* in outcrop. (**a–b**), *T. barretti* preserved in sandstone. Changhsingian (late Permian) Maomaolong Formation, Malonggou section, Gansu Province, northwestern China

Description Straight or slightly curved cylindrical burrows are equipped with well-spaced, arcuate meniscate packets. Distance between menisci is equal to or less than the width of the burrow. Burrow fill is homogeneous and has almost the same component in lithology as the host rock. Single burrows are ~8 mm in diameter. Meniscate packets are 4–6 mm long. Burrows are preserved as full reliefs or positive epireliefs on the upper bedding surface. Branching and intersecting are rare. This kind of burrow has been found in the reddish sandstone beds of the basal Xiahuancang Formation, as well as in the limestone beds of the Daye Formation.

Remarks Several authors have discussed the taxonomic issues of *Taenidium* and other similar meniscate trace fossils such as *Scoyenia*, *Beaconites* and *Ancorichnus* (Frey et al. 1984; D'Alessandro and Bromley 1987; Keighley and Pickerill 1994; Retallack 2001; Smith et al. 2008). *Taenidium* differs from *Rhizocorallium* in lacking a U-shaped marginal tube and differs from *Nereites* in lacking an envelope zone of reworked sediment along meniscate segments (Knaust 2013; Feng et al. 2018). Less common ichnogenera, *Parataenidium* superficially resembles *Taenidium* but has more complex, actively back-filled meniscate and possesses two levels of backfill structures (Buckman 2001). The well-developed two-level construction of *Parataenidium*, the distinct characteristics of each level, and the sharp partitioning between the levels suggest that each level represents a separate ethological activity. Generally, the studied specimens are cylindrical, straight to curved, unbranched meniscate traces. They have no linings, outer membranes or longitudinal stripes; these characteristics classify them into the ichnogenus *Taenidium*. In addition to the type ichnospecies *T. serpentinum*, three other ichnospecies namely *T. satanassi*, *T. cameronensis*, and *T. barretti*, are currently valid. All features observed on our burrows agree well with *T. barretti*. *Taenidium* often occurs in shallow- and deep-marine deposits, as well as in alluvial, fluvial, and other continental settings (Hembree and Hasiotis 2008; Bedatou et al. 2009). This ichnogenus is interpreted to be produced by a combination behavior such as deposit-feeding, locomotion, and dwelling by arthropods or worm-like organisms in marine environmental settings (Knaust 2017), while beetle larvae, insects, and earthworms are the most likely tracemakers in continental settings (Smith et al. 2008; Hembree and Hasiotis 2008). *Taenidium* ranges in age from Cambrian to Holocene.

Teichichnus Seilacher 1955
Teichichnus rectus Seilacher 1955
Figure 3.55

Description Vertical burrow composed of a horizontal, passively filled terminal causative burrow and a set of convex down laminae. The whole burrow looks like an arcuate spreite burrow, with spreite structures beneath the terminal burrow. The length of the terminal burrow and laminae is almost the same. The trace is unbranched, about 20–50 mm in length. The terminal burrow is about 10–15 mm in diameter. The height of the whole trace is 30–45 mm. *Teichichnus* crosscut other horizontal burrows such as *Palaeophycus*. This ichnotaxon can be found in the argillaceous limestone beds of the Daye Formation from quarries of the Shawan and

Fig. 3.55 *Teichichnus* in outcrop. (**a–b**), *T. rectus* preserved in argillaceous limestone (arrows). Vertical profile view. Lower Triassic Jialingjiang Formation, Shawan section, Sichuan Province, southwestern China. (**c–d**), *T. rectus* preserved in argillaceous limestone. **d** is a bedding plane view of **c**. Lower Triassic Daye Formation, Dangwu section, Guizhou Province, southwestern China

Dangwu sections, South China. Specimens were mainly observed from cut slabs, which were roughly polished by workers.

Remarks Seilacher (1955) erected the ichnogenus *Teichichnus* from the Lower Cambrian of the Salt Range, Pakistan. Since then, *Teichichnus* has been debated and amended to sustain its ichnotaxonomic stability (Frey and Howard 1985; Corner and Fjalstad 1993; Jensen 1997; Mángano et al. 2002; Mángano and Buatois 2011; Buatois et al. 2017; Knaust 2018). Knaust et al. (2018) revised the ichnotaxonomy of *Teichichnus* and proposed that only four out of eighteen erected ichnospecies are valid, namely *T. rectus*, *T. zigzag*, *T. patens* and *T. duplex*. Since the Early Cambrian and throughout the Phanerozoic, *T. rectus* has been the most common ichnospecies (Orłowski 1989; Pacześna 1996; McIlroy and Brasier 2017). *T. zigzag* is also a well-studied ichnospecies. The terminal burrow of *T. rectus* maintains the same size range or width as its spreite, while the width of the spreite of *T. zigzag* exceeds that of its terminal burrow (Frey and Bromley 1985). *T. patens* is differentiated from other ichnospecies by its branched morphological structure (Schlirf 2000). *T. duplex* was established based on materials from the Middle Cambrian of Sweden (Martinsson 1965). It has a bilobed structure and transverse to oblique scratches and therefore has an affinity with the hypothesized arthropod burrows (such as *Cruziana* and *Trichophycus*) (Jensen 1997). Not all parts of the *Teichichnus* burrow can be equally well-developed or well-preserved, due to the high degree of morphological variability.

Teichichnus is commonly interpreted as a feeding structure, representing a combined feeding and dwelling behavior (Chisholm 1970; Frey and Howard 1985). Other interpretations include suspension feeding (Ekdale and Bromley 1984), gardening (Mikuláš 2001) and equilibrium movement in response to sedimentation (Corner and Fjalstad 1993). Superficial deposit-feeding or suspension-feeding has been considered (Knaust 2018). The producers of *Teichichnus* have been considered to be polychaetes (bristle worms) (Seilacher 1957), echiurans (Reineck et al. 1967, 1968; Hertweck 1970), and holothurians. Moreover, some groups of bivalves can also generate similar spreite-like structures with their feet (Bromley 1996; Werner 2002; Winn 2006). *Teichichnus* is a marine trace fossil, which is distributed in environments ranging from marginal seas to deep seas. Studies indicate that in marginal sea areas under environmental stress, such as lagoons, estuaries or deltas, *T. rectus* occurs in monoichnospecific associations (Knaust 2018).

Thalassinoides Ehrenberg 1944
Thalassinoides suevicus Rieth 1932
Figures 3.56, 3.57, and 3.58

Description Predominantly horizontal, cylindrical burrows with more or less regular branching; Y-shaped bifurcations are more common than T-shaped branches. The burrow surface is smooth; ornamentation is not present. Obvious swellings are displayed at the points of bifurcation. The Y-shaped segments typically are 8–20 mm in diameter. Vertical shafts are not observed. Burrow fill has no internal structure and is consistent with the lithology of the surrounding rock. The Y-shaped burrows are mainly distributed on the bedding surface and are not observed in a vertical profile. Burrows are often connected by the Y-shaped bifurcations that form complex networks. This ichnogenus is relatively common in most carbonate successions of the Lower and Middle Triassic of South China. The Lower Triassic specimens are always simple, small and without complex networks, while the Middle Triassic specimens such as those from the Luoping Biota and neighboring areas are quite robust and form complex networks on large bedding surfaces.

Remarks Several allies, *Spongeliomorpha*, *Ophiomorpha*, and *Thalassinoides* (SOT group), are often confused with one another (Bromley and Frey 1974; Schlirf 2000), but each of them represents an independent ichnogenus (Fürsich 1973; Uchman 1995). *Thalassinoides* shows a high degree of similarity to the burrows of certain crustaceans. Studies have shown that burrow wall is the key discriminative feature for differentiating *Thalassinoides* from similar burrows created by crustaceans, such as *Ophiomorpha*, *Balanoglossites*, and *Spongeliomorpha* (Kern and Warme in 1974; Bromley in 1990 and 1996). *Ophiomorpha* is a trace fossil that

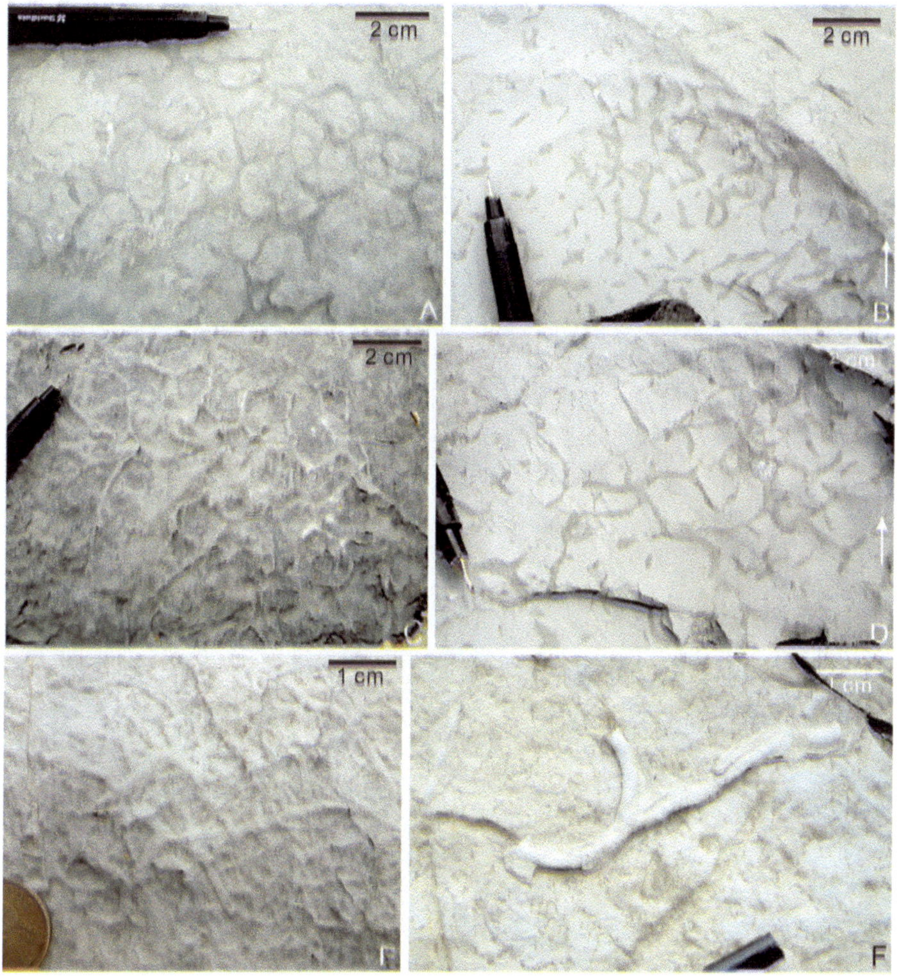

Fig. 3.56 *Thalassinoides* in outcrop. (**a–f**), *T. suevicus* preserved in dolomitic limestone. **b** and **d** are vertical profile view, others are bedding plane view. Lower Triassic Anshun Formation, Gaimao section, Guizhou Province, southwestern China

occurs almost exclusively in sandy substrates. It is characterized by a nodular surface structure formed by a thick lining of fecal pellets, and this structure is helpful for the reinforcement of the burrow (Shinn 1968; Frey et al. 1978). The characteristics that distinguish *Balanoglossites* from *Thalassinoides* lie in its highly irregular shape and the phenomenon of the blunt termination of its lateral branches (Knaust 2008; Knaust and Costamagna 2012). The ichnogenus *Spongeliomorpha* can be distinguished from *Thalassinoides* by its prominent surface ridges or groove structures in different directions and the massive filling characteristics (Melchor et al. 2010).

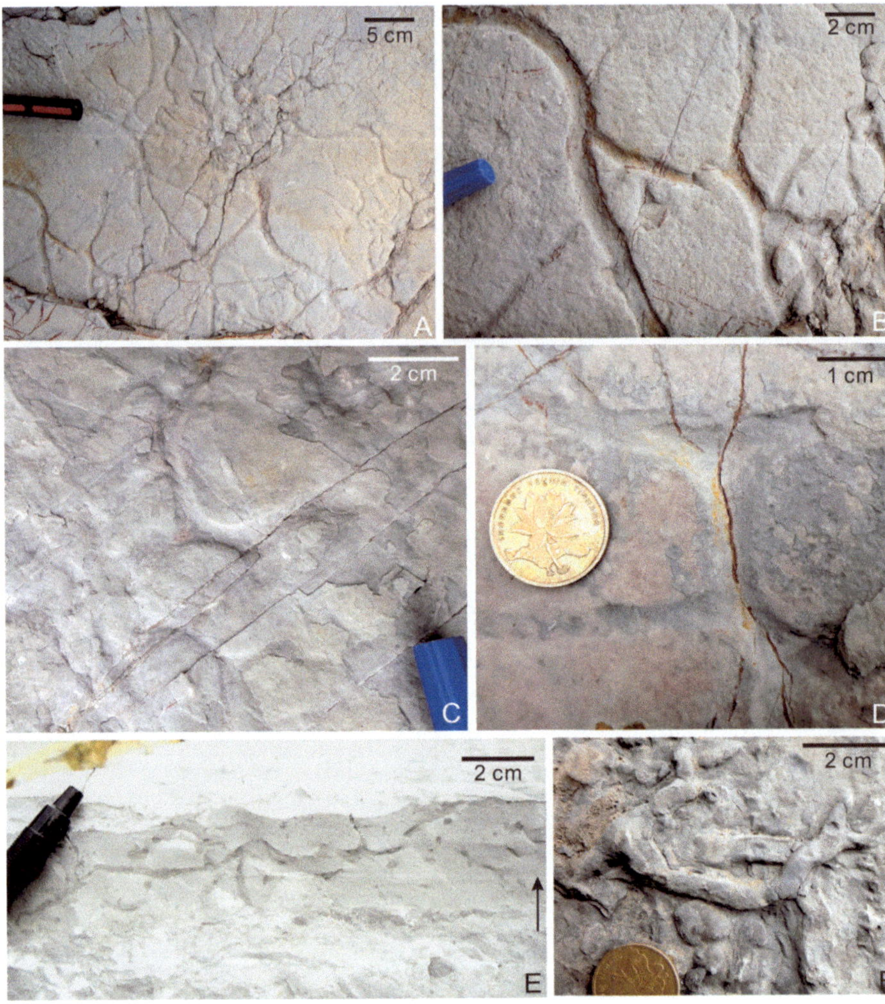

Fig. 3.57 *Thalassinoides* in outcrop. (**a–f**), *T. suevicus* preserved in dolomitic limestone. **e** is vertical profile view, others are bedding plane view. Lower Triassic Anshun Formation, Gaimao section, Guizhou Province, southwestern China

Fifteen ichnospecies of *Thalassinoides* have been erected in total, and the following are most common: *T. callianassa*, *T. horizontalis*, *T. bacae*, *T. suevicus*, *T. paradoxicus*, *T. saxonicus*. *T. horizontalis* is characterized as a horizontal, branching network of smooth-walled, unlined burrows lacking vertically oriented shafts (Neto de Carvalho and Bayet-Goll 2023). *T. bacae* is defined as horizontal burrows with short and vertical shafts densely distributed along the burrows (Ekdale and Bromley 2003). *T. suevicus* is dominated by a horizontal morphology, and enlarged structures may develop at the Y-shaped bifurcations. *T. paradoxicus* is characterized by a reticular burrow system exhibiting a high degree of irregularity in terms of size

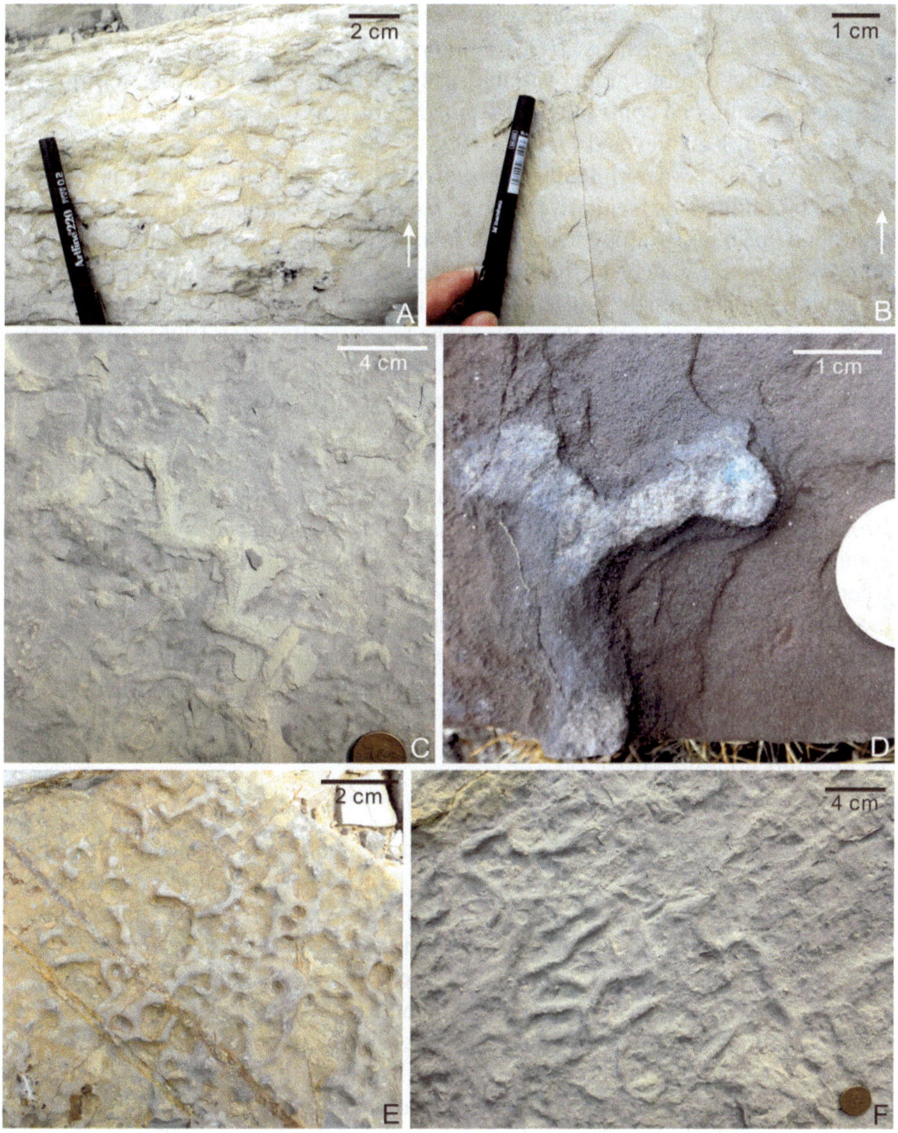

Fig. 3.58 *Thalassinoides* in outcrop. (**a**–**b**), *T. suevicus* preserved in dolomitic limestone. Vertical profile view. (**c**, **f**), *T. suevicus* preserved in argillaceous limestone. Lower Triassic Jialingjiang Formation, Shawan section, Sichuan Province, southwestern China. (**d**), *T. suevicus* preserved in sandstone. Lower Triassic Xiahuancang Formation, Zhihema section, Qinghai Province, northwestern China. (**e**), *T. suevicus* preserved in limestone. Lower Triassic Daye Formation, Shatian section, Hubei Province, South China. **c**–**f** are bedding plane view

and geometric shape. *T. saxonicus* is a very large form with tunnels measuring 5–20 cm in diameter (Knaust 2008). *T. suevicus* is distinguished from *T. paradoxicus*, as the latter exhibits highly variable, irregularly branched systems. Additionally, *T. suevicus* cannot be confused with *T. horizontalis*, as the latter lacks burrow walls (Myrow 1995; Keighley and Pickerill 1997).

Thalassinoides was originally interpreted as a feeding trace, which is typically formed by obligate or facultative deposit-feeding endobenthic crustaceans or other arthropod species (Bromley and Frey 1974; Ekdale 1992). Studies have indicated that *Thalassinoides* may be burrows for cultivating microorganisms. The trace-making organisms obtain nutrients by cultivating the microbial communities that grow on the surface or inside the walls of the burrows (Bromley 1996). Comparative studies with modern organisms suggest that the trace-making organisms of *Thalassinoides* are thalassinidean shrimps, especially callianassidae. This conclusion is basically applicable to *Thalassinoides* from the Permian to the modern era. However, the trace-making organisms of *Thalassinoides* in the Paleozoic include other arthropods (such as trilobites), sea anemones, and worm groups (Ekdale and Bromley 2003; Knaust 2008, 2017). *Thalassinoides* produced by shrimp is most commonly found in shallow-marine environments (Keighley and Pickerill 1997; Chen et al. 2011, 2012, 2015; Luo et al. 2016; Feng et al. 2017). It also occurs in brackish environments, such as estuaries and fan deltas, and spans a wide range of settings, from marginal marine to deep marine (Monaco et al. 2009). *Thalassinoides* has been widely documented from the early Cambrian to the Holocene (Knaust 2008).

Treptichnus Miller 1889
Treptichnus bifurcus Miller 1889
Figures 3.59 and 3.60

Description Horizontal, straight to curved, zigzag burrow segments. Vertical shafts are observed at the ends of burrow segments, indicating a three-dimensional burrow structure. Burrow segments feature projections extending from the angles of juncture, interpreted as the bedding plane expression of oblique burrow segments. These projections are irregularly positioned on alternate sides of the segments (Buatois and Mángano 1993a, b; Schlirf 2000). The entire burrow measures approximately 15–40 cm in length, with individual segments about 3–6 mm wide. The burrow fill is structureless and identical to the host rock. The burrows are preserved as negative epireliefs or full reliefs. This ichnogenus is relatively common in many carbonate successions of the Lower Triassic from South China, but is rare in siliciclastic rocks.

Remarks Miller (1889) first established the ichnogenus *Treptichnus* to describe ichnofossils with a bifurcated structure, a zigzag extension, and protrusions at the end of the burrow. While studying the ichnogenus *Treptichnus*, Miller also described and discussed the ichnogenera *Haplotichnus* and *Plangtichnus*, believing that both were highly similar to *Treptichnus* in terms of burrow size, the type of trace-making organisms, and morphological characteristics. Initially, Miller (1889) described *Plangtichnus* as a zigzag trail with pits. Subsequent works proposed that the key feature distinguishing *Plangtichnus* from *Treptichnus* lies in the lack of the exten-

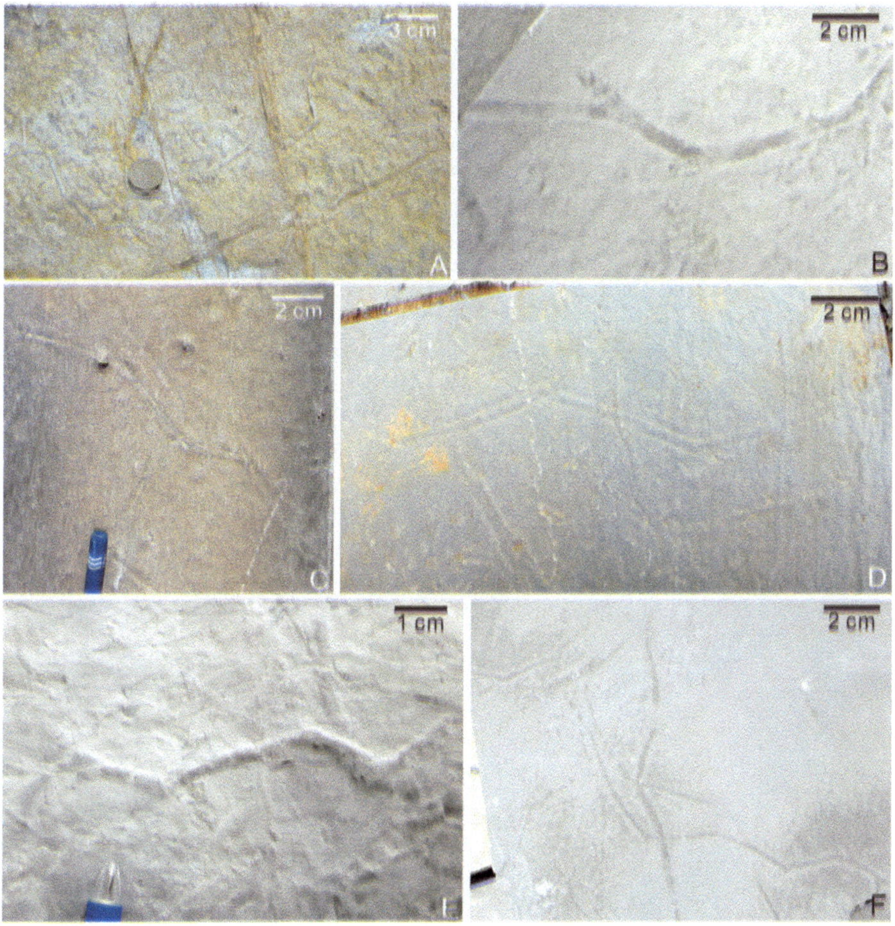

Fig. 3.59 *Treptichnus* in outcrop. (**a**), *T. bifurcus* preserved in limestone. Lower Triassic Daye Formation, Shatian section, Hubei Province, South China. (**b–f**), *T. bifurcus* preserved in argillaceous limestone. Lower Triassic Daye Formation, Dangwu section, Guizhou Province, southwestern China

sion structure at the end of the burrow. This absence gives *Plangtichnus* a highly angular zigzag shape (Archer and Maples 1984; Maples and Archer 1987).

However, Buatois and Mángano (1993a, b) argued that the projections of *Treptichnus* and the pits of *Plangtichnus* represented morphologically similar vertical shafts along the burrow system and that the lack of the burrow–end projections was likely caused by erosion. Buatois and Mángano (1993b) claimed that since both ichnogenera had similar morphology and represented similar behaviors, *Plangtichnus* and *Treptichnus* should be considered synonymous. Buatois and Mángano (1993b) proposed that the terminal extension structure of *Treptichnus* and the pits of *Plangtichnus* are actually vertically shaft structures with similar

Fig. 3.60 *Treptichnus* in outcrop. (**a–b, d**), *T. bifurcus* preserved in limestone, **b** is a close-up view of the boxed area in **a**. Lower Triassic Daye Formation, Lichuan section, Hubei Province, South China. (**c**), *T. bifurcus* preserved in argillaceous limestone. Lower Triassic Jialingjiang Formation, Shawan section, Sichuan Province, southwestern China. (**e–f**) *T. bifurcus* preserved in argillaceous limestone. Lower Triassic Daye Formation, Gaimao section, Guizhou Province, southwestern China

morphology within the burrow system, and the difference between the two may be caused by the loss of the terminal structure of the burrow. They believed that *Plangtichnus* and *Treptichnus* should be regarded as synonyms. Considering the stability of nomenclature, they chose to retain *Treptichnus* and regarded *Plangtichnus* as an abandoned name (Buatois and Mángano 1993b). However, this suggestion was not followed by subsequent work (Schlirf 2000). Ten ichnospecies assigned to *Treptichnus* are valid. *T. bifurcus*, *T. coronatum*, *T. lublinensis*, *T. pedum*, *T. pollardi*, *T. rectangularis*, *T. tripleurum*, *T. triplex* and *T. vagans* have been reported in the early Cambrian (Mángano and Buatois 2016; Hammersburg et al. 2018), while

T. meandrinus was established from Eocene flysch in Carpathians, Poland (Uchman et al. 1998).

Treptichnus is typically interpreted as a deposit-feeding trace (Buatois and Mángano 1993a, b; Uchman et al. 1998), though it has also been linked to agricultural, grazing, reproductive, and predatory or scavenging activities (Rindsberg and Kopaska-Merkel 2005; Seilacher 2007). It is believed to be created by marine annelid worms (Buatois and Mángano 1993b; Uchman et al. 1998; Buatois et al. 2013) and, since the Pennsylvanian, by certain insect larvae in continental environments (Miller 1889; Rindsberg and Kopaska-Merkel 2005). *Treptichnus* has been recorded in both shallow- and deep-marine environments, as well as in continental proximal floodplain and proximal lacustrine deposits (Archer and Maples 1984; Buatois and Mángano 1993b; Jensen 1997; Uchman et al. 1998; Getty et al. 2016). Its occurrence spans from the Cambrian to the present day (Buatois and Mángano 1993a, b; Uchman et al. 1998). However, some *Treptichnus* traces have been identified in the Ediacaran, indicating a gradual increase in ichnofossil complexity culminating in the first appearance of *T. pedum* at the Precambrian-Cambrian boundary (Jensen et al. 2000; Droser et al. 2002; Buatois et al. 2013; Buatois 2018).

References

Abed AM, Schneider W (1980) A general aspect in the genesis of nodular limestones documented by the Upper Cretaceous limestones of Jordan. Sediment Geol 26:329–335

Aceñolaza FG, Buatois LA (1993) Nonmarine perigondwanic trace fossils from the late Paleozoic of Argentina. Ichnos 2:183–201

Alpert SP (1974) Systematic review of the genus *Skolithos*. J Paleontol 48:661–669

Alpert SP, Moore WS (1975) Lower Cambrian trace fossil evidence for predation on trilobites. Lethaia 8:223–230

Archer AW, Maples CG (1984) Trace fossil distribution across a marine-to-nonmarine gradient in the Pennsylvanian of southwestern Indiana. J Paleontol 58:448–466

Bann KL, Fielding CR (2004) An integrated ichnological and sedimentological comparison of non-deltaic shoreface and subaqueous delta deposits. In: Permian reservoir units of Australia, Special publications 228. Geological Society, London, pp 273–310

Baucon A, De Carvalho CN (2016) Stars of the aftermath: asteriacites beds from the Lower Triassic of the Carnic Alps (Werfen Formation, Sauris di Sopra), Italy. Palaios 31:161–176

Baucon A, Neto de Carvalho C (2008) From the river to the sea: Pramollo, a new ichnolagerstätte from the Carnic Alps (Carboniferous, Italy–Austria). Studi Trentini di Scienze Naturali Acta Geologica 83:87–114

Baucon A, Ronchi A, Felletti F, Neto de Carvalho C (2014) Evolution of crustaceans at the edge of the end-Permian crisis: ichnonetwork analysis of the fluvial succession of Nurra (Permian–Triassic, Sardinia, Italy). Palaeogeogr Palaeoclimatol Palaeoecol 410:74–103

Baucon A, Bednarz M, Dufour S, Felletti F, Malgesini G, Neto de Carvalho C, Niklas KJ, Wehrmann A, Batstone R, Bernardini F, Briguglio A, Cabella R, Cavalazzi B, Ferretti A, Zanzerl H, McIlroy D (2020) Ethology of the trace fossil *Chondrites*: form, function and environment. Earth Sci Rev 202:102989

Beach DK (1995) Controls and effects of subaerial exposure on cementation and development of secondary porosity in the subsurface of Great Bahama Bank. In: Budd DH, Saller AH, Harris PM (eds) Unconformities and porosity in carbonate strata, AAPG memoir 63. American Association of Petroleum Geologists, Tulsa, pp 1–33

Beales FW (1953) Dolomitic mottling in Palliser (Devonian) Limestone, Banff and Jasper National Parks, Alberta. AAPG Bull 37:2281–2293

Bedatou E, Melchor RN, Genise JF (2009) Complex palaeosol ichnofabrics from Late Jurassic-Early Cretaceous volcaniclastic successions of central Patagonia, Argentina. Sediment Geol 218:74–102

Bednarz M, McIlroy D (2009) Three-dimensional reconstruction of "phycosiphoniform" burrows: implications for identification of trace fossils in core. Palaeontol Electron 12(13A):15

Belaústegui Z, Gibert JM, López-Blanco M, Bajo I (2014) Recurrent constructional pattern of the crustacean burrow *Sinusichnus sinuosus* from the Paleogene and Neogene of Spain. Acta Palaeontol Pol 59:461–474

Belaústegui Z, Muñiz F, Nebelsick JH, Domènech R, Martinell J (2017) Echinoderm ichnology: bioturbation, bioerosion and related processes. J Paleontol 91:643–661

Bell CM (2004) Asteroid and ophiuroid trace fossils from the Lower Cretaceous of Chile. Palaeontology 47(1):51–66

Benton MJ (1982) *Dictyodora* and associated trace fossils from the Palaeozoic of Thuringia. Lethaia 15:115–132

Benton MJ, Trewin NH (1980) *Dictyodora* from the Silurian of Peeblesshire, Scotland. Palaeontology 23:501–513

Bicknell RDC, Klinkhamer AJ, Flavel RJ, Wroe S, Paterson JR (2018) A 3D anatomical Atlas of appendage musculature in the chelicerate arthropod *Limulus polyphemus*. PLoS ONE 13(2):e0191400

Billings E (1862) New species of fossils from different parts of the Lower, Middle and Upper Silurian rocks of Canada. In Palaeozoic Fossils, Volume 1 1861-1865. Geological Survery of Canada Advance Sheets, 96–168

Bradley J (1980) *Scolicia* and *Phycodes*, trace fossils of Renilla (Pennatulacea). Pac Geol 14:73–86

Bradley J (1981) *Radionereites*, *Chondrites* and *Phycodes*; trace fossils of anthoptiloid sea pens. Pac Geol 15:1–16

Bradshaw MA (1981) Palaeoenvironmental interpretations and systematics of Devonian trace fossils from the Taylor Group (lower Beacon Supergroup), Antarctica. New Zealand Journal of Geology and Geophysics, 24:615–652

Bradshaw MA (2010) Devonian trace fossils of the Horlick Formation, Ohio Range, Antarctica: systematic description and palaeoenvironmental interpretation. Ichnos 17:58–114

Bromley RG (1990) Trace fossils. Biology and taphonomy. Unwin Hyman, London

Bromley RG (1996) Trace fossils: biology, taphonomy and application, 2nd edn. Chapman and Hall, London

Bromley RG, Ekdale AA (1984) *Chondrites*: a trace fossil indicator of anoxia in sediments. Science 224:872–874

Bromley RG, Frey RW (1974) Redescription of the trace fossil *Gyrolithes* and taxonomic evaluation of *Thalassinoides*, *Ophiomorpha* and *Spongeliomorpha*. Bull Geol Soc Den 23:311–335

Brongniart AT (1828) Histoire des ve´ge´taux fossiles ou recherches botaniques et ge´ologiques sur les ve´ge´taux renferme´s dans les diverses couches du globe, volume 1. G. Dufour & E. d'Ocagne, Paris, 136

Buatois LA (2018) *Treptichnus pedum* and the Ediacaran–Cambrian boundary: significance and caveats. Geol Mag 155:174–180

Buatois LA, Mángano MG (1993a) Trace fossils from a Carboniferous turbiditic lake: implications for the recognition of additional continental ichnofacies. Ichnos 2:237–258

Buatois LA, Mángano MG (1993b) The ichnotaxonomic status of *Plangtichnus* and *Treptichnus*. Ichnos 2:217–224

Buatois LA, Mángano MG (2002) Trace fossils from Carboniferous floodplain deposits in western Argentina: implications for ichnofacies models of continental environments. Palaeogeogr Palaeoclimatol Palaeoecol 183:71–86

Buatois LA, Mángano MG (2018) The other biodiversity record: innovations in animal-substrate interactions through geologic time. GSA Today, p 28

Buatois LA, Jalfin G, Aceñolaza FG (1997) Permian nonmarine invertebrate trace fossils from southern Patagonia, Argentina: ichnologic signatures of substrate consolidation and colonization sequences. J Paleontol 71:324–336

Buatois LA, Mángano MG, Maples CG, Lanier WP (1998a) Ichnology of an Upper Carboniferous fluvio-estuarine paleovalley: the Tonganoxie Sandstone, Buildex quarry, eastern Kansas, USA. J Paleontol 72:152–180

Buatois LA, Mángano MG, Maples CG, Lanier WP (1998b) Taxonomic reassessment of the ichnogenus *Beaconichnus* and additional examples from the Carboniferous of Kansas, USA. Ichnos 5:287–302

Buatois LA, Netto RG, Mángano MG, Balistieri PRMN (2006) Extreme freshwater release during the late Paleozoic Gondwana deglaciation and its impact on coastal ecosystems. Geology 34:1021–1024

Buatois LA, Macsotay O, Quiroz LI (2009) *Sinusichnus*, a trace fossil from Antarctica and Venezuela: expanding the dataset of crustacean burrows. Lethaia 42:511–518

Buatois LA, Almond J, Germs GJB (2013) Environmental tolerance and range offset of *Treptichnus pedum*: implications for the recognition of the Ediacaran–Cambrian boundary. Geology 41:519–522

Buatois LA, Wisshak M, Wilson MA, Mángano MG (2017) Categories of architectural designs in trace fossils: a measure of ichnodisparity. Earth-Science Reviews 164:102–181

Buckman JO (1994) *Archaeonassa* Fenton and Fenton 1937 reviewed. Ichnos 3:185–192

Buckman JO (1995) A comment on annulate forms of *Palaeophycus* Hall 1847: with particular reference to *P.* '*annulatus*' sensu Pemberton and Frey 1982, and the erection of *P. crenulatus* ichnosp. nov. Ichnos 4:131–140

Buckman JO (2001) *Parataenidium*, a new *Taenidium*-like ichnogenus from the carboniferous of Ireland. Ichnos 8(2):83–97

Cabrera LMI, Mángano GM, Buatois LA, Olivero EB, Maples CG (2019) Bivalves on the move: the interplay of extrinsic and intrinsic factors on the morphology of the trace fossil *Protovirgularia*. Palaios 34:349–363

Callow RHT, McIlroy D (2011) Ichnofabrics and ichnofabric-forming trace fossils in Phanerozoic turbidites. Bull Can Petrol Geol 59:103–111

Callow RHT, McIlroy D, Kneller B, Dykstra M (2013) Integrated ichnological and sedimentological analysis of a Late Cretaceous submarine channel-levee system: the Rosario Formation, Baja California, Mexico. Mar Pet Geol 41:277–294

Carmona NB, Buatois LA, Mángano MG, Bromley RG (2008) Ichnology of the Lower Miocene Chenque Formation, Patagonia, Argentina: animal-substrate interactions and the modern evolutionary fauna. Ameghiniana 45:93–122

Carmona NB, Mángano MG, Buatois LA, Ponce JJ (2010) Taphonomy and paleoecology of the bivalve trace fossil *Protovirgularia* in deltaic heterolithic facies of the Miocene Chenque Formation, Patagonia, Argentina. J Paleontol 84:730–738

Chamberlain CK (1971) Morphology and ethology of trace fossils from the Ouachita Mountains, southeast Oklahoma. J Paleontol 45:212–246

Chen ZQ, Tong JN, Fraiser ML (2011) Trace fossil evidence for restoration of marine ecosystems following the end-Permian mass extinction in the Lower Yangtze region, South China. Palaeogeography, Palaeoclimatology, Palaeoecology 299:449–474

Chen ZQ, Fraiser ML, Bolton C (2012) Early Triassic trace fossils from Gondwana interior sea: implication for ecosystem recovery following the end-Permian mass extinction in south high-latitude region. Gondwana Res 22:238–255

Chen X, Ling HF, Vance D, Shields-Zhou GA, Zhu M, Poulton SW, Och LM, Jiang SH, Li D, Cremonese L, Archer C (2015) Rise to modern levels of ocean oxygenation coincided with the Cambrian radiation of animals. Nat Commun 6:7142

Chen Z, Chen X, Zhou CM, Yuan XL, Xiao SH (2018) Late Ediacaran trackways produced by bilaterian animals with paired appendages. Sci Adv 4:eaao6691

Chisholm JI (1970) *Teichichnus* and related trace-fossils in the Lower Carboniferous at St. Monance, Scotland. Bull Geol Surv Great Br 32:21–51

Claus H (1965) Eine merkwürdige Lebensspur (*Protovirgularia*? sp.) aus dem oberen Muschelkalk NW-Thüringens. Senckenb Lethaea 46:187–191
Clausen CK, Vilhjálmsson M (1986) Substrate control of Lower Cambrian trace fossils from Bornholm, Denmark. Palaeogeogr Palaeoclimatol Palaeoecol 56:51–68
Conkin JE, Conkin BM (1968) *Scalarituba missouriensis* and its stratigraphic distribution. Univ Kansas Paleontol Contrib 31:1–7
Corner GD, Fjalstad A (1993) Spreite trace fossils (*Teichichnus*) in a raised Holocene fjord-delta, Breidvikeidet, Norway. Ichnos 2:155–164
Crimes TR (1970) Trilobite tracks and other trace fossils from the Upper Cambrian of North Wales. Geological Journal, 7:47–68
Crimes TR (1987) Trace fossils and correlation of late Precambrian and early Cambrian strata. Geological Magazine, 124:98–119
Crimes TP, Anderson MM (1985) Trace fossils from late Precambrian-early Cambrian strata of southeastern Newfoundland (Canada): temporal and environmental implications. J Paleontol 59:890–907
Crimes TP, McCall GJH (1995) A diverse ichnofauna from Eocene-Miocene rocks of the Makran Range (SE Iran). Ichnos 3(4):231–258
D'Alessandro A, Bromley RG (1987) Meniscate trace fossils and the *Muensteria–Taenidium* problem. Palaeontology 30:743–763
Dahmer G (1937) Lebensspuren aus dem Taunusquarzit und aus den Siegener Schichten (Unterdevon). Jahrbuch der Preußisch- Geologischen Landesanstalt 57:523–539
Dam G (1990) Taxonomy of trace fossils from the shallow marine Lower Jurassic Neill Klinter Formation, East Greenland. Bull Geol Soc Denmark 38:119–144
Dawson JW (1873) Impressions and footprints of aquatic animals and imitative markings on Carboniferous rocks. American Journal of Science and Arts, 105:16–24
De Stefani C (1885) Studî paleozoologici sulla creta superiore e media dell' Apennino settentrionale. In: Atti della R. Academia dei Lincei, della Classe Scienze Fisiche, Matematiche e Naturali, Memorie, Serie 4, 1:73–121 pls. 1–2
Demircan H, Toker V (2003) Trace fossils in the western fan of the Cingöz Formation in the northern Adana Basin (southern Turkey). Bull Miner Res Explor (MTA) Turkey 127:15–32
Desai BG, Shukla R, Saklani RD (2010) Ichnology of the Early Cambrian Tal Group, Nigalidhar Syncline, Lesser Himalaya, India. Ichnos 17:233–245
Desio A (1940) Vestigia problematiche paleozoiche della Libia. Pubblicazioni dell'Istituto di Geologia, Paleontologia e Geografia Fisica della Università di Milano, Serie P, 47–92 Pubblicazione 20
Ding Y, Duan Y, Wu Y, Cao C (2021) Trace fossils from the Permian Lopingian Talung Formation at the northern Penglaitan section of Laibin area, South China: ichnology, palaeoenvironment, and palaeoecology. Geol J 56:6117–6134
d'Orbigny A (1842) Voyage dans l'Amérique méridionale (le Brésil, la République oriental de l'Uruguay, la République Argentine, la Patagonie, la République du Chili, la République de Bolivia, la République du Péron) exécuté pendant les années 1826, 1827, 1829, 1830, 1831, 1832, et 1833. v.3, pt. 4 (Paléontologie), 188
Droser ML, Jensen S, Gehling JG, Myrow PM, Narbonne GM (2002) Lowermost Cambrian ichnofabrics from the Chapel Island Formation, Newfoundland: implications for Cambrian substrates. Palaios 17:3–15
Ehrenberg K (1944) Ergänzende Bemerkungen zu den seinerzeit aus dem Miozän von Burgschleinitz beschriebenen Gangkernen und Bauten dekapoder Krebse. Paläontologische Zeitschrift 23:354–359
Ekdale AA (1992) Muckraking and mudslinging: the joys of deposit feeding. Trace fossils. Paleontol Soc Short Course 5:145–171
Ekdale AA, Bromley RG (1984) Comparative ichnology of shelf-sea and deep-sea chalk. J Paleontol 58:322–332
Ekdale AA, Bromley RG (2001) A day and a night in the life of a cleft-foot clam: *Protovirgularia-Lockeia-Lophoctenium*. Lethaia 34:119–124

Ekdale AA, Bromley RG (2003) Paleoethologic interpretation of complex *Thalassinoides* in shallow-marine limestones, Lower Ordovician, southern Sweden. Palaeogeogr Palaeoclimatol Palaeoecol 192:221–227

Ekdale AA, Harding SC (2015) *Cylindrichnus concentricus* Toots in Howard, 1966 (trace fossil) in its type locality, Upper Cretaceous, Wyoming. Ann Soc Geol Pol 85:427–432

Emmons E (1844) The Taconic System: Based on Observations in New York, Massachusetts, Maine, Vermont and Rhode Island. Caroll and Cook, Albany, 68

Farrés MF (1967) Los "*Dendrotichnium*" de España. Notas y Communications, Instituto Geologico y Minero de España, 94:29–36

Feng XQ, Chen ZQ, Woods A, Wu SQ, Fang YH, Luo M, Xu YL (2017) Anisian (Middle Triassic) marine ichnocoenoses from the eastern and western margins of the Kamdian Continent, Yunnan Province, SW China: implications for the Triassic biotic recovery. Glob Planet Chang 157:194–213

Feng XQ, Chen ZQ, Bottjer DJ, Fraiser ML, Xu YL, Luo M (2018) Additional records of ichnogenus *Rhizocorallium* from the Lower and Middle Triassic, South China: implications for biotic recovery after the end-Permian mass extinction. Geol Soc Am Bull 130:1197–1215

Fenton CL, Fenton MA (1937) *Archaeonassa*, Cambrian snail trails and burrows. Am Midl Nat 18:454–456

Fernández DE, Pazos PJ, Aguirre-Urreta MB (2010) *Protovirgularia dichotoma- Protovirgularia rugosa*: an example of a compound trace fossil from the lower Cretaceous (Agrio Formation) of the Neuquén Basin, Argentina. Ichnos 17:40–47

Fillion D, Pickerill RK (1990) Ichnology of the upper Cambrian to Lower Ordovician Bell Island and Wabana groups of eastern Newfoundland. Palaeontogr Can 7:1–119

Firtion F (1958) Sur la présence d'ichnites dans le Portlandien de l'Ile d'Oléron (Charente maritime). Annales Universitatis Saraviensis (Naturw.) 7:107–112

Fischer-Ooster C (1858) Die fossilen Fucoiden der Schweizer Alpen, nebst Erörterungen über deren geologisches Alter. Huber, Bern

Fitch A (1850) A historical, topographical and agricultural survey of the County of Washington. Part 2-5. Transactions of the New York Agricultural Society, 9:753–944

Frey RW, Bromley RG (1985) Ichnology of American chalks: the Selma Group (Upper Cretaceous), western Alabama. Can J Earth Sci 22:801–828

Frey RW, Howard JD (1985) Trace fossils from the Panther Member, Star Point Formation (Upper Cretaceous), Coal Creek Can yon, Utah. J Paleontol 59:370–404

Frey RW, Howard JD, Pryor WA (1978) *Ophiomorpha*: its morphologic, taxonomic and environmental significance. Palaeogeography, Palaeoclimatology, Palaeoecology, 23:199–229

Frey RW, Pemberton SG, Fagerstrom JA (1984) Morphological, ethological, and environmental significance of the ichnogenera *Scoyenia* and *Ancorichnus*. J Paleontol 58:511–528

Fu S (1991) Funktion, Verhalten und Einteilung fucoider und lophocteniider Lebensspuren. Courier Forschungs-Institut Senckenberg 135:1–79

Fürsich FT (1973) A revision of the trace fossils *Spongeliomorpha*, *Ophiomorpha* and *Thalassinoides*. Neues Jb Geol Paläontologische Mitteilungen, 719–735

Fürsich FT (1974a) Ichnogenus *Rhizocorallium*. Paläontologische Zeitschrift 48:16–28

Fürsich FT (1974b) On *Diplocraterion* Torell 1870 and the significance of morphological features in vertical, spreiten-bearing, U-shaped trace fossils. J Paleontol 48(5):952–962

Fürsich FT, Palmer TJ (1975) Open crustacean burrows associated with hardgrounds in the Jurassic of the Cotswolds. Proc Geol Assoc 86:171–181

Fürsich FT, Uchman A, Alberti M, Pandey DK (2018) Trace fossils of an amalgamated storm-bed succession from the Jurassic of the Kachchh Basin, India: the significance of time-averaging in ichnology. J Palaeogeogr 7:14–31

Gabelli L, De (1900) Sopra un'interesssante impronta medusoidae. II Pensiero Aristotelico della Scienza Moderna, 1(2):74–78

Gaillard C (2011) A giant limulid trackway (*Kouphichnium lithographicum*) from the lithographic limestones of Cerin (Late Kimmeridgian, France): ethological and environmental implications. Swiss J Geosci 104:57–72

Getty PR, McCarthy TD, Hsieh S, Bush AM (2016) A new reconstruction of continental *Treptichnus* based on exceptionally preserved material from the Jurassic of Massachusetts. J Paleontol 90(2):269–278

Geyer G, Uchman A (1995) Ichnofossil assemblages from the Nama Group (Neoproterozoic-Lower Cambrian) in Namibia and the Proterozoic-Cambrian boundary problem revisited. In: Geyer G, Landing E (eds) Morocco'95. The Lower–Middle Cambrian standard of western Gondwana, Beringria special issue 2, pp 175–202, International Geological Correlation Program 366, Würzburg.

Gibert JM de (1996) A new decapod burrow system from the NW Mediterranean Pliocene. Revista Española de Paleontología 11(2):251–254

Gibert JM de, Jeong K, Martinell J (1999) Ethologic and ontogenic significance of the Pliocene trace fossil *Sinusichnus sinuosus* from the northwestern Mediterranean. Lethaia 32:31–40

Gibert JM de, Benner JS (2002) The trace fossil *Gyrochorte*: ethology and paleoecology. Revista Española de Paleontología 17:1–12

Gibert JM de, Domènech R (2008) Trazas fósiles de nuculoideos (*Protovirgularia*) del Mioceno marino de la Cuenca del Vallès-Penedès. Revista Española de Paleontología 23:129–138

Głuszek A (1995) Invertebrate trace fossils in the continental deposits of an Upper Carboniferous coal-bearing succession. Upper Silesia, Poland. Studia Geol Pol 108:171–202

Głuszek A (1998) Trace fossils from Late Carboniferous storm deposits, Upper Silesia Coal Basin, Poland. Acta Palaeontol Pol 43:517–546

Godbold A, Schoepfer SD, Shen SZ, Henderson CM (2017) Precarious ephemeral refugia during the earliest Triassic. Geology 45:607–610

Goldring R (1996) The sedimentological significance of concentrically laminated burrows from Lower Cretaceous Ca-bentonites, Oxfordshire. J Geol Soc Lond 53:255–263

Goldring R, Pollard JE, Taylor AM (1991) *Anconichnus horizontalis*: a pervasive ichnofabric-forming trace fossil in post-Paleozoic off-shore siliciclastic facies. Palaios 6:250–263

Goldring R, Gruszczynski M, Gatt PA (2002) A bow-form burrow and its sedimentological and paleoecological significance. Palaios 17:622–630

Gong Y (1987) Trace fossils of Early-Middle Devonian clastic formations in southwestern Hunan and their relation to sedimentary environments. Collect Papers Lit Palaogeogr 4:98–116. (in Chinese, with English Abstract)

Greb SF, Chesnut DR Jr (1992) Transgressive channel filling in the Breathitt Formation (Upper Carboniferous), Eastern Kentucky Coal Field, USA. Sediment Geol 5:209–221

Greiner H (1972) Arthropod trace fossils in the lower Devonian Jacquet River Formation of New Brunswick. Can J Earth Sci 9:1772–1777

Grossgeim VA (1961) Nekatoriye noviye gieroglifi iz nizhniemelovikh otl'ozheniy severo-zapaadnokho Kavkaza. Trudy Krasodarskoho Filial'a Vsesoyuznoho Neflegazovoho Nauchno-Issledovatielskoho Instituta, 6:202–206

Haldeman SS (1840) Supplement to Number one of "A monograph of the Limniades, or Freshwater Univalve Shells of North America," containing descriptions of apparently new animals in different classes, and the names and characters of the subgenera in Paludina and Anculosa. J. Dobson, Philadelphia (3)

Hall J (1847) Paleontology of New York, Volume I. C. Van Benthuysen, Albany, 338 pp.

Hammersburg SR, Hasiotis ST, Robison RA (2018) Ichnotaxonomy of the Cambrian Spence Shale Member of the Langston Formation, Wellsville Mountains, Northern Utah, USA. Paleontol Contributions 20:1–66

Han Y, Pickerill RK (1994) Taxonomic reassessment of *Protovirgularia* M'Coy 1850 with new examples from the Paleozoic of New Brunswick, eastern Canada. Ichnos 3:203–212

Häntzschel W (1975) Trace fossils and problematica. W1–W269. In: Teichert C (ed) Treatise on invertebrate paleontology. Part W, Miscellanea supplement 1. Geological Society of America/University of Kansas Press, Boulder/Lawrence

Heer O (1865) Die Unwell der Schweiz. F. Schulthess. Zurich. 622

Heer O (1877) Florafossilis Helvetiae. Die vorweltliche flora der Schweiz. J. Wurster and Co., Zurich, 182

Hembree DI, Hasiotis ST (2008) Miocene vertebrate and invertebrate burrows defining compound paleosols in the Pawnee Creek Formation, Colorado, U.S.A. Palaeogeogr Palaeoclimatol Palaeoecol 270:349–365
Hertweck G (1970) The animal community of a muddy environment and the development of biofacies as effected by the life cycle of the characteristic species. In: Crimes TP, Harper JC (eds) Trace fossils, Geological Journal special issue 3, pp 235–242, Seel House Press, Liverpool.
Hitchcock E (1858) Ichnology of New England. A report on the Sandstone of the Connecticut Valley, especially its fossil footmarks. Commonwealth of Massachusetts, William White, Boston xii+220 pp., 60 pls
Hofmann HJ, Patel IM (1989) Trace fossils from the type "Etcheminian Series" (Lower Cambrian Ratcliffe Brook Formation), Saint John area, New Brunswick, Canada. Geol Mag 126:139–157
Hofmann R, Goudemand N, Wasmer M, Bucher H, Hautmann M (2011) New trace fossil evidence for an early recovery signal in the aftermath of the end-Permian mass extinction. Palaeogeogr Palaeoclimatol Palaeoecol 310:216–226
Hofmann R, Mángano MG, Elicki O, Shinaq R (2012) Paleoecologic and biostratigraphic significance of trace fossils from shallow- to marginal-marine environments from the Middle Cambrian (Stage 5) of Jordan. J Paleontol 86:931–955
Horbury AD, Qing H (2004) 'Pseudobreccias' revealed as calcrete mottling and bioturbation in the late Dinantian of the southern Lake District, UK. Sedimentology 51:19–38
Howard JD, Frey RW (1984) Characteristic trace fossils in nearshore to offshore sequences, Upper Cretaceous of east-central Utah. Can J Earth Sci 21:200–219
Hundt R (1913) Eine Ergänzung zu 'Organische Reste aus dem Untersilur des Hüttchenberges bei Wünschendorf an der Elster'. Zentralbl. Mineral. Geol. Paläontol. 1913:180–181
ICZN (International Commission on Zoological Nomenclature) (2018) International Code of Zoological Nomenclature. http://www.iczn.org/iczn/index.jsp
Izumi K (2012) Formation process of the trace fossil *Phymatoderma granulata* in the Lower Jurassic black shale (Posidonia Shale, southern Germany) and its paleoecological implications. Palaeogeogr Palaeoclimatol Palaeoecol 353–355:116–122
Izumi K (2014) Utility of geochemical analysis of trace fossils: case studies using *Phycosiphon incertum* from the Lower Jurassic shallow-marine (Higashinagano Formation, southwest Japan) and Pliocene deep-marine deposits (Shiramazu Formation, central Japan). Ichnos 21:62–72
Jaglarz P, Uchman A (2010) A hypersaline ichnoassemblage from the Middle Triassic carbonate ramp of the Tatricum domain in the Tatra Mountains, southern Poland. Palaeogeogr Palaeoclimatol Palaeoecol 292:71–81
James UP (1879) Description of new species of fossils and remarks on some others from the lower and Upper Silurian rocks of Ohio. Paleontologist 3:17–24
Jensen S (1997) Trace fossils from the Lower Cambrian Mickwitzia Sandstone, south-central Sweden. Fossils Strata 42:1–110
Jensen S (1999) A Lower Cambrian shallow-water occurrence of the branching 'deep-water' type trace fossil *Dendrorhaphe* from the Lontova Formation, eastern Latvia. Paläontol Z 73:183–197
Jensen S (2003) The Proterozoic and earliest trace fossil record: patterns, problems and perspectives. Integr Comp Biol 43:219–228
Jensen S, Gehling JG, Droser ML (1998) Ediacara-type fossils in Cambrian sediments. Nature 393(6685):567–569
Jensen S, Saylor BZ, Gehling JG, Germs GJB (2000) Complex trace fossils from the terminal Proterozoic of Namibia. Geology 28:143–146
Johnson EW, Briggs DEG, Suthren RJ, Wright JL, Tunnclif SP (1994) Nonmarine arthropod traces from the subaerial Ordovician Borrowdale Volcanic Group, English Lake District. Geol Mag 131:395–406
Jones B, Oldershaw AE, Narbonne GM (1979) Nature and origin of rubbly limestone in the Upper Silurian Read Bay Formation of Arctic Canada. Sediment. Geol. 24:227–252
Kappel O (2003) Ichnofossilien im Campanium des SE-Münsterlanders. Münster Forsch Geologie und Paläontologie 96:1–195
Kazmierczak J, Pszczółkowski A (1969) Burrows of Enteropneusta in Muschelkalk (Middle Triassic) of the Holy Cross Mountains, Poland. Acta Palaeontol Pol 14(2):299–317

Keighley DG, Pickerill RK (1994) The ichnogenus *Beaconites* and its distinction from *Ancorichnus* and *Taenidium*. Palaeontology 37:305–337

Keighley DG, Pickerill RK (1995) The ichnotaxa *Palaeophycus* and *Planolites*: historical perspectives and recommendations. Ichnos 3:301–309

Keighley DG, Pickerill RK (1996) Small *Cruziana*, *Rusophycus*, and related ichnotaxa from eastern Canada: the nomenclatural debate and systematic ichnology. Ichnos 4(4):261–285

Keighley DG, Pickerill RK (1997) Systematic ichnology of the Mabou and Cumberland groups (Carboniferous) of western Cape Breton Island, eastern Canada, 1: burrows, pits, trails, and coprolites. Atl Geol 33:181–215

Kennedy WJ, Garrison RE (1975) Morphology and genesis of nodular chalks and hardgrounds in the Upper Cretaceous of southern England. Sedimentology 22:311–386

Kern JP, Warme JE (1974) Trace fossils and bathymetry of the Upper Cretaceous Point Loma Formation, San Diego, California. Geol Soc Am Bull 85:893–900

Kim JY, Pickerill RK, Wilson RA (2000) *Palaeophycus bolbitermiius* isp. nov. from the Lower Silurian Upsalquitch Formation of New Brunswick, eastern Canada. Atl Geosci 36:131–137

King OA, Stimson MR, Lucas SG (2019) The ichnogenus *Kouphichnium* and related xiphosuran traces from the Steven C. Minkin Paleozoic Footprint Site (Union Chapel Mine), Alabama, USA: ichnotaxonomic and paleoenvironmental implications. Ichnos 26:266–302

Knaust D (2007) Invertebrate trace fossils and ichnodiversity in shallow-marine carbonates of the German Middle Triassic (Muschelkalk). In: Bromley R, Buatois LA, Mángano MG, Genise J, Melchor R (eds) Sediment-organism interactions: a multifaceted ichnology, Society for Sedimentary Geology special publication 88. SEPM (Society for Sedimentary Geology), Tulsa, pp 223–240

Knaust D (2008) *Balanoglossites* Mägdefrau, 1932 from the Middle Triassic of Germany: part of a complex trace fossil probably produced by boring and burrowing polychaetes. Paläontol Z 82:347–372

Knaust D (2010) The end-Permian mass extinction and its aftermath on an equatorial carbonate platform: insights from ichnology. Terra Nova 22:195–202

Knaust D (2013) The ichnogenus *Rhizocorallium*: classification, trace makers, palaeoenvironments and evolution. Earth Sci Rev 126:1–47

Knaust D (2015) Siphonichnidae (new ichnofamily) attributed to the burrowing activity of bivalves: ichnotaxonomy, behaviour and palaeoenvironmental implications. Earth Sci Rev 150:497–519

Knaust D (2017) Atlas of trace fossils in well core: appearance, taxonomy and interpretation. 209p, Springer, Dordrecht.

Knaust D (2018) The ichnogenus *Teichichnus* Seilacher, 1955. Earth Sci Rev 177:386–403

Knaust D (2019) The enigmatic trace fossil *Tisoa* de Serres, 1840. Earth Sci Rev 188:123–147

Knaust D (2021) *Balanoglossites*-burrowed firmgrounds—the most common ichnofabric on earth? Earth Sci Rev 220:103747

Knaust D (2022) Who were the tracemakers of *Protovirgularia*—Molluscs, arthropods, or annelids? Gondwana Res 111:95–102

Knaust D (2023) The multifaceted ichnogenus *Protovirgularia* M'Coy, 1850: taxonomy, producers and environments. Earth Sci Rev 244:104511

Knaust D, Costamagna LG (2012) Ichnology and sedimentology of the Triassic carbonates of north-west Sardinia, Italy. Sedimentology 59:1190–1207

Knaust D, Desrochers A (2020) The trace fossil *Sinusichnus* from the Upper Ordovician of Anticosti Island, eastern Canada. Lethaia 53:145–153

Knaust D, Neumann C (2016) *Asteriacites* von Schlotheim, 1820—the oldest valid ichnogenus name—and other asterozoan-produced trace fossils. Earth Sci Rev 157:111–120

Knaust D, Curran HA, Dronov AV (2012) Shallow-marine carbonates. In: Knaust D, Bromley RG (eds) Trace fossils as indicators of sedimentary environments, Developments in sedimentology 64. Elsevier Science, Oxford, pp 705–750

Knaust D, Uchman A, Hagdorn H (2016) The probable isopod burrow *Sinusichnus seilacheri* isp. n. from the Middle Triassic of Germany: an example of behavioral convergence. Ichnos 23:138–146

Knaust D, Thomas RDK, Curran HA (2018) *Skolithos linearis* Haldeman, 1840 at its early Cambrian type locality, Chickies Rock, Pennsylvania: analysis and designation of a neotype. Earth Sci Rev 185:15–31

Kozur HW, Krainer K, Mostler H (1996) Ichnology and sedimentology of the Early Permian deep-water deposits from the Lercara-Roccapalumba area (western Sicily, Italy). Facies 34:123–150

Książkiewicz M (1968) O niekt6rych problematykach z flissu Karpat Polskich (Czesc): Polskiego TowarzystwaG eologicznego W. Krakowie, 38:3–17

Książkiewicz M (1970) Observations on the ichnofauna of the Polish Carpathians. In: Crimes TP, Harper C (eds) Trace fossils, Geological Journal special issue 3, pp 283–322, Seel House Press, Liverpool.

Książkiewicz M (1977) Trace fossils in the flysch of the Polish Carpathians. Palaeontol Pol 36:1–208

Kulkarni KG, Borkar VD, Petare TJ (2008) Ichnofossils from the Fort Member (Middle Jurassic), Jaisalmer Formation, Rajasthan. J Geol Soc India 71:731–738

Kuwazuru J, Nakadawa S (2018) Discovery of trace fossil *Protovirgularia* from the Shimanto Supergroup in Yakushima Island, Kagoshima Prefecture, Japan. Kagoshima Prefectural Mus Res Rep 37:73–88. (In Japanese)

Li Y, Lu Z, Wang D (1997) Continental trace fossils and sedimentary environments in Liaohe Basin. Petroleum Industry Press, Beijing. (in Chinese, with English Abstract)

Lima JHD, Minter NJ, Netto RG (2017) Insights from functional morphology and neoichnology for determining tracemakers: a case study of the reconstruction of an ancient glacial arthropod-dominated fauna. Lethaia 50:576–590

Linck O (1949) Lebens-Spuren aus dem Schilfsandstein (Mittl. Keuper, km 2) NWWürttembergs und ihre Bedeutung für die Bildungsgeschichte der Stufe. Jahreshefte Vereins Vaterländische Naturkunde Württemberg 97–101:1–100

Lockley MG, Rindsberg AK, Zeiler RM (1987) The paleoenvironmental significance of the near-shore *Curvolithus* ichnofacies. Palaios 2:255–262

López Cabrera MIL, Mángano MG, Buatois LA, Olivero EB, Maples CG (2019) Bivalves on the move: the interplay of extrinsic and intrinsic factors on the morphology of the trace fossil *Protovirgularia*. Palaios 34:349–363

Luo M, Shi GR (2017) First record of the trace fossil *Protovirgularia* from the Middle Permian of southeastern Gondwana (southern Sydney Basin, Australia). Alcheringa 41:335–349

Luo M, George AD, Chen ZQ (2016) Sedimentology and ichnology of two Lower Triassic sections in South China: implications for the biotic recovery following the end-Permian mass extinction. Glob Planet Chang 144:198–212

Luo M, Shi GR, Hu S, Benton MJ, Chen Z-Q, Huang J, Zhang Q, Zhou C, Wen W (2019) Early middle Triassic trace fossils from the Luoping Biota, southwestern China: evidence of recovery from mass extinction. Palaeogeogr Palaeoclimatol Palaeoecol 515:6–22

M'Coy F (1850) On some new genera and species of Silurian Radiata in the collection of the University of Cambridge. Ann. Mag. Nat. Hist. Second Ser. 6:270–290

MacLeay WS (1839) Note on the Annelida. In: Murchison, R. I.: The Silurian System. Part II. Organic Remains. J. Murray, London, 699–701

Mägdefrau K (1932) Über einige Bohrgänge aus dem Unteren Muschelkalk von Jena. Paläontol Z 14:150–160

Mägdefrau K (1934) Über *Phycodes circinatum* Reinh. Richter aus dem thüringischen Ordovicium. N Jb Geol Paläont 72:259–282

Mángano MG, Buatois LA (2004) Reconstructing Early Phanerozoic intertidal ecosystems: ichnology of the Cambrian Campanario Formation in northwest Argentina. Fossils Strata 51:17–38

Mángano MG, Buatois LA (2011) Timing of infaunalization in shallow-marine early Paleozoic communities in Gondwanan settings: discriminating evolutionary and paleogeographic controls. Palaeontologia Electronica 14.2.9A

Mángano MG, Buatois LA (2014) Decoupling of body-plan diversification and ecological structuring during the Ediacaran–Cambrian transition: evolutionary and geobiological feedbacks. Proc R Soc B Biol Sci 281:20140038

Mángano MG, Buatois LA (2016) The Cambrian explosion. In: Mángano MG, Buatois LA (eds) The trace-fossil record of major evolutionary changes, volume 1, Topics in geobiology 39. Springer, Dordrecht, pp 72–126

Mángano MG, Buatois LA, Aceñolaza GF (1996) Trace fossils and sedimentary facies from a Late Cambrian-Early Ordovician tide-dominated shelf (Santa Rosita Formation, northwest Argentina): implications for ichnofacies models of shallow marine successions. Ichnos 5:53–88

Mángano MG, Buatois LA, Maples CG, Lanier WP (1997) *Tonganoxichnus* a new insect trace from the Upper Carboniferous of eastern Kansas. Lethaia 30:113–125

Mángano MG, Buatois LA, West R, Maples CG (1998) Contrasting behavioral and feeding strategies recorded by tidal-flat bivalve trace fossils from the Upper Carboniferous of eastern Kansas. Palaios 13:335–351

Mángano MG, Buatois LA, West RR, Maples CG (1999) The origin and paleoecologic significance of the trace fossil *Asteriacites* in the Pennsylvanian of Kansas and Missouri. Lethaia 32:17–30

Mángano MG, Buatois LA, Maples CG, West RR (2000) A new ichnospecies of *Nereites* from Carboniferous tidal-flat facies of eastern Kansas, USA: implications for the *Nereites-Neonereites* debate. J Paleontol 74:149–157

Mángano MG, Buatois LA, West RR, Maples CG (2002) Ichnology of a Pennsylvanian equatorial tidal flat: the Stull Shale Member at Waverly, eastern Kansas. Bull Kansas Geol Surv 245:1–139

Maples CG, Archer AW (1987) Redescription of Early Pennsylvanian trace-fossil holotypes from the nonmarine Hindostan Whetstone Beds of Indiana. J Paleontol 61:890–897

Maples CG, West RR (1989) *Lockeia*, not *Pelecypodichnus*. J Paleontol 63(5):694–696

Martinsson A (1965) Aspects of a Middle Cambrian Thanatotope on Öland. Geologiska Föreningen i Stockholm Förhandlingar 87:181–230

Mayer, G., 1954. Neue Beobachtungen an Lebensspuren aus dem Unteren Hauptmuschelkalk (Trochitenkalk) von Wiesloch. Neues Jahrb. Geol. Palaontol. Abh. 99, 223–229 pls 14–18.

McCann T (1993) A *Nereites* ichnofacies from the Ordovician-Silurian Welsh Basin. Ichnos 3:39–56

McIlroy D, Brasier MD (2017) Ichnological evidence for the Cambrian explosion in the Ediacaran to Cambrian Succession of Tanafjord, Finnmark, Northern Norway, Geological Society of London, special publications 448. Geological Society, London, pp 351–368

Mekki F, Bouchemla I, Adaci M, Talmat S, Ferré B, Benyouce M (2023) A diverse trace-fossil assemblage from the Middle Jurassic (Bajocian) Teniet El Klakh Formation (western Saharan Atlas, Algeria). Proc Geol Assoc 134:573–589

Melchor RN, Bromley RG, Bedatou E (2010) *Spongeliomorpha* in nonmarine settings: an ichnotaxonomic approach. Earth Environ Sci Trans R Soc Edinburgh 100:429–436

Melchor RN, Genise JF, Buatois LA, Umazano A (2012) Fluvial environments. In: Knaust D, Bromley RG (eds) Trace fossils as indicators of sedimentary environments. Elsevier, Amsterdam, pp 329–378

Metz R (1998) Nematode trails from the late Triassic of Pennsylvania. Ichnos 5:303–308

Metz R (2020) First record of the trace fossil *Protovirgularia* in the Passaic Formation (late Triassic), Newark Supergroup, near Milford, New Jersey. Ichnos 27:428–432

Mikuláš R (1990) The ophiuroid *Taeniaster* as a tracemaker of *Asteriacites*, Ordovician of Czechoslovakia. Ichnos 1:133–137

Mikuláš R (2001) The trace fossil *Rejkovicichnus necrofilus* Mikuláš et. al., 1996 (Middle Cambrian, Czech Republic): an early example of "gardening". Neues Jahrbuch für Geologie und Paläontologie Monatshefte, Heft 1(2001):56–64

Mikuláš R, Meššfis S, Ivanov A, Lukševičs E, Zupiòd I, Stinkulis I (2013) A rich ichnofossil assemblage from the Frasnian (Upper Devonian) deposits at Andoma Hill, Onega Lake, Russia. Bull Geosci 88:389–400

Miller GD (1986) The sediments and trace fossils of the Rough Rock Group on Cracken Edge, Derbyshire. Mercian Geol 10:189–202

Miller SA (1889) North American geology and palaeontology for the use of amateurs, students and scientists. Western Methodist Book Concern, Cincinnati, Ohio, 664

Miller WI (1993) Trace fossil zonation in Cretaceous turbidite facies, northern California. Ichnos 3:11–28
Miller WI (2011) A stroll in the forest of the fucoids: status of *Melatercichnus burkei* Miller, 1991, the doctrine of ichnotaxonomic conservatism and the behavioral ecology of trace fossil variation. Palaeogeogr Palaeoclimatol Palaeoecol 307:109–116
Miller SA, Dyer CB (1878) Contributions to palaeontology. J Cincinnati Soc Nat Hist 1:24–39
Minter NJ, Braddy SJ, Davis RB (2007) Between a rock and a hard place: arthropod trackways and ichnotaxonomy. Lethaia 40:365–375
Monaco P, Caracuel JE, Giannetti A (2009) *Thalassinoides* and *Ophiomorpha* as cross-facies trace fossils of crustaceans from shallow-to-deep-water environments: Mesozoic and Tertiary examples from Italy and Spain. In: Garassino A, Feldmann RM, Teruzzi G (eds) 3rd symposium on Mesozoic and Cenozoic decapod crustaceans—Museo di Storia Naturale di Milano, 23–25 May 2007. Memorie della Società Italiana di Scienze Naturali e del Museo Civico di Storia Naturale di Milano 35, pp 79–82
Moreau JD, Fara E, Gand G, Lafaurie G, Baret L (2014) Gigantism among Late Jurassic limulids: new ichnological evidence from the Causses Basin (Lozère, France) and comments on body-size evolution among horseshoe crabs. Geobios 47:237–253
Morton RC, Myers RA, Gingras MK, Zonneveld J-P (2023) Aquaria-based observations of the ophiuroid *Ophiolepis superba* and the trackways it produces. Palaios 38:98–108
Mude SN (2012) Paleoenvironmental significance of ichnofossils from the Kand Formation of the Cambay Basin, Gujarat, India. Proc Int Acad Ecol Environ Sci 2:12–20
Muñoz DF, Mángano MG, Buatois LA (2019) Unravelling Phanerozoic evolution of radial to rosette trace fossils. Lethaia 52:350–369
Myrow P (1995) *Thalassinoides* and the enigma of Early Paleozoic open-framework burrow systems. Palaios 10:58–74
Nara M (1995) *Rosselia socialis*: a dwelling structure of a probable terebellid polychaete. Lethaia 28:171–178
Nara M (2002) Crowded *Rosselia socialis* in Pleistocene inner shelf deposits: benthic paleoecology during rapid sea-level rise. Palaios 17:268–276
Nara M, Haga M (2007) The youngest record of trace fossil *Rosselia socialis*: occurrence in the Holocene shallow marine deposits of Japan. Paleontol Res 11:21–27
Nara M, Ikari Y (2011) "Deep-sea bivalvian highways": an ethological interpretation of branched *Protovirgularia* of the Palaeogene Muroto-Hanto Group, southwestern Japan. Palaeogeogr Palaeoclimatol Palaeoecol 305:250–255
Narbonne GM (1984) Trace fossils in Upper Silurian tidal flat to basin slope carbonates of Arctic Canada. J Paleontol 58(2):398–415
Naruse H, Nifuku K (2008) Three-dimensional morphology of the ichnofossil *Phycosiphon incertum* and its implication for paleoslope inclination. Palaios 23:270–279
Neto de Carvalho C (2001) Implicações geométricas na etologia do produtor de *Dictyodora tenuis* (M' Coy) no Arenigiano de Barrancos (Portugal): custos e benefícios de ser um pioneiro. Seminarios de Paleontologia de Zaragoza 5:369–378
Neto de Carvalho C, Bayet-Goll A (2023) *Thalassinoides horizontalis* Myrow, 1995 from the Middle-Upper Ordovician shallow marine siliciclastics of Iran (Lashkerak Formation). Comptes Rendus Palevol 22:569–583
Nicholson HA (1873) Contributions to the study of the errant annelides of the older Palaeozoic rocks. R Soc Lond Proc 21:288–290 (also Geological Magazine 10:309–310)
Nielsen J, Görmüş M, Uysal K, Kanbur S (2012) Ichnology of the Miocene Güneyce Formation (Southwest Turkey): oxygenation and sedimentation dynamics. Turk J Earth Sci 21(3):391–405
Nopsca F (1923) Die Familien der Reptilien. Fortschritte der Geologie und Palaontologie, 2:1–210
Ohta S (1984) Star-shaped feeding traces produced by echiurian worms on the deep-sea floor of the Bay of Bengal. Deep-Sea Res 31:1415–1432
Orłowski S (1989) Trace fossils in the Lower Cambrian sequence Świętokrzyskie Mountains, Central Poland. Acta Palaeontol Pol 34:211–231

Orr PJ (1995) A deep-marine ichnofaunal assemblage from Llandovery strata of the Welsh Basin, West Wales, UK. Geol. Mag. 132:267–285

Orr P, Benton M, Trewin N (1996) Deep marine trace fossil assemblages from the lower Carboniferous of Menorca, Balearic Islands, western Mediterranean. Geol J 31:235–258

Osgood RG (1970) Trace fossils of the Cincinnati area. Palaeontogr Am 6:277–444

Pacześna J (1996) The Vendian and Cambrian ichnocoenoses from the Polish part of the East-European Platform. Polish Geological Institute, Warszawa 77:30

Pazos PJ, Heredia AM, Fernández DE, Gutiérrez C, Comerio M (2015) The ichnogenus *Dictyodora* from late Silurian deposits of central-western Argentina: ichnotaxonomy, ethology and ichnostratigrapical perspectives from Gondwana. Palaeogeogr Palaeoclimatol Palaeoecol 439:27–37

Pedley M (1992) Bio-retexturing: early diagenetic fabric modifications in outer-ramp settings–a case study from the Oligo-Miocene of the Central Mediterranean. Sediment Geol 79:173–188

Pemberton SG (1992) Applications of Ichnology to Petroleum Exploration: A Core Workshop: Core Workshop 17: Tulsa, Oklahoma, Society for Sedimentary Geology, 429

Pemberton SG, Frey RW (1982) Trace fossil nomenclature and the *Planolites–Palaeophycus* dilemma. J Paleontol 56(4):843–881

Pemberton SG, MacEachern JA, Ranger MJ (1992) Ichnology and event stratigraphy: the use of trace fossils in recognizing tempestites. In: Pemberton SG (ed) Applications of ichnology to petroleum exploration, Society of Economic Paleontologists and Mineralogists Core Workshop, 17. Tulsa, Oklahoma, pp 15–118

Pemberton SG, MacEachern JA, Dashtgard SE, Bann KL, Gingras MK, Zonneveld J-P (2012) Shorefaces. In: Knaust D, Bromley RG (eds) Trace fossils as indicators of sedimentary environments, Developments in sedimentology 64. Elsevier Science, Oxford, pp 563–604

Petter AL, Steel RJ (2006) Hyperpycnal flow variability and slope organization on an Eocene shelf margin, Central Basin, Spitsbergen. AAPG Bull 90:1451–1472

Pickerill RK (1981) Trace fossils in a Lower Palaeozoic submarine canyon sequence, the Siegas Formation of northwestern New Brunswick, Canada. Atl Geosci 17:36–58

Pickerill RK, Peel JS (1991) *Gordia nodosa* isp. nov. and other trace fossils from the Cass Fjord Formation (Cambrian) of North Greenland. Gronlands Geologiske Undersogelse, Rapport 150:15–28

Pickerill RK, Fyffe LR, Forbes WH (1987) Late Ordovlcian-Early Silurian trace fossils from the Matapedia Group, Tobique River, western New Brunswick. Atl Geosci 23:77–88

Pickerill RK, Norman Y, Han Y (1995) New occurrence of *Phycodes templus* Han and Pickerill, 1994 with a taxonomic reassessment of *Yangziichnus yichangensis* Yang, 1984. Atlantic Geology, 31(1):1–5

Pollard JE (1985) *Isopodichnus*, related arthropod traces and notostracans from Triassic fluvial sediments. Earth Environ Sci Trans R Soc Edinburgh 76:273–285

Pollard JE (1999) *Arenicolites carbonarius* in the Manchester region: ichnofabrics of a deltaic opportunist. In: Pollard J, Taylor A (eds) Fifth international ichnofabric workshop IIW5, Manchester University, 12–14 July 1999, Abstract volume

Pruss SB, Bottjer DJ (2004) Early Triassic trace fossils of the western United States and their implications for prolonged environmental stress from the end-Permian mass extinction. Palaios 19:551–564

Qi Y, Meng Y, Dai M, Li D (2014) Biogenic leopard patch structures from the Zhushadong Formation (Cambrian Series 2), Dengfeng area, western Henan. Geol. Sci. Technol. Informat. 33:1–8 (In Chinese)

Quenstedt FA (1876) Petrefaktenkunde Deutschlands. 4. Die Asteriden und Encriniden nebst Cysti und Blastoideen. Fues's, Leipzig (742 pp., pls. 90–114)

Quenstedt FA (1879) Petrefactenkunde Deutschlands. Korallen (Sternkorallen). In: Abtheilung, Erste, Band, Dritter, Heft, Sechstes (Eds.), Atlas. vol. 6. Fues's Verlag (R. Reisland), Leipzig, 337–496

Rajkonwar C, Tiwari RP, Patel SJ (2013) *Arenicolites helixus* isp. nov. and associated ichnofossils from the Bhuban Formation, Surma Group (Lower-Middle Miocene) of Aizawl, Mizoram, India. Himal Geol 34:18–37

Reineck HE, Gutmann WF, Hertweck G (1967) Das Schlickgebiet südlich Helgoland als Beispiel rezenter Schelfablagerungen. Senckenb Lethaea 48:219–275

Reineck H-E, Dörjes J, Gadow S, Hertweck G (1968) Sedimentologie, Faunenzonierung und Faziesabfolge vor der Ostküste der inneren Deutschen Bucht. Senckenb Lethaea 49:261–309

Retallack GJ (2001) *Scoyenia* burrows from Ordovician palaeosols of the Juniata Formation in Pennsylvania. Palaeontology 44:209–235

Richter R (1850) Aus der thiiringischen Grauwacke. Z Dtsch Ges Geowiss 2:198–206

Richter R (1853) Thüringische Graptolithen. Z Dtsch Ges Geowiss 5:439–464

Richter R (1941) Marken und Spuren im Hunsrück-Schiefer. 3. Fährten als Zeugnisse des Lebens auf dem Meeres-Grunde. Senckenbergiana 23:218–260

Rieth A (1932) Neue Funde spongeliomorpher Fucoiden aus dem Jura Schwabens. Geologie, Palaontologie Abhandlungen, N.F., Jena, 19:257–294

Rindsberg AK (1994) Ichnology of the Upper Mississippian Hartselle Sandstone of Alabama, with notes on other Carboniferous formations. Bull Geol Surv Alabama 158:1–107

Rindsberg AK, Kopaska-Merkel DC (2005) *Treptichnus* and *Arenicolites* from the Steven C. Minkin Paleozoic Footprint Site (Langsettian, Alabama, USA). In: Buta RJ, Rindsberg AK, Kopaska-Merkel DC (eds) Pennsylvanian footprints in the Black Warrior Basin of Alabama, Alabama Paleontological Society monograph 1. Alabama Paleontological Society, Birmingham, pp 121–141

Rindsberg AK, Martin AJ (2003) *Arthrophycus* in the Silurian of Alabama (USA) and the problem of compound trace fossils. Palaeogeogr Palaeoclimatol Palaeoecol 192:187–219

Rodríguez-Tovar FJ, Pérez-Valera F (2008) Trace fossil *Rhizocorallium* from the Middle Triassic of the Betic Cordillera, Southern Spain: characterization and environmental implications. Palaios 23:78–86

Rodríguez-Tovar FJ, Buatois LA, Piñuela L, Mángano MG, García-Ramos JC (2012) Palaeoenvironmental and functional interpretation of *Rhizocorallium jenense spinosus* (ichnosubsp. nov.) from the lower Jurassic of Asturias, northern Spain. Palaeogeogr Palaeoclimatol Palaeoecol 339–341:114–120

Salter JW (1857) On annelide-burrows and surface-markings from the Cambrian rocks of the Longmynd. Q. J. Geol. Soc. 13:199–206 pl. 5

Saporta G de (1872-1873). Paleontologie francaise ou description des fossiles de la France. 2 ser. Vegetaux. Plantes Jurassiques. G. Masson, Paris, 1:506

Savrda CE, Krawinkel H, McCarthy FMG, McHugh CMG, Olson HC, Mountain G (2001) Ichnofabrics of a Pleistocene slope succession, New Jersey margin: relations to climate and sea-level dynamics. Palaeogeogr Palaeoclimatol Palaeoecol 171:41–61

Schatz ER, Mángano MG, Aitken AE, Buatois LA (2013) Response of benthos to stress factors in Holocene Arctic fjord settings: Maktak, Coronation, and North Pangnirtung Fjords, Baffin Island, Canada. Palaeogeogr Palaeoclimatol Palaeoecol 386:652–668

Schlirf M (2000) Upper Jurassic trace fossils from the Boulonnais (northern France). Geol Palaeontol 34:145–213

Schlirf M (2012) *Heliophycus seilacheri* n. isp. and *Biformites insolitus* Linck, 1949 (trace fossils) from the Late Triassic of the Germanic Basin: their taxonomy and palaeoecological relevance. Neues Jahrb Geol Palaontol Abh 263(3):185–198

Schlirf M, Uchman A (2005) Revision of the ichnogenus *Sabellarifex* Richter, 1921 and its relationship to *Skolithos* Haldeman, 1840 and *Polykladichnus* Fürsich, 1981. J Syst Palaeontol 3:115–131

Schlirf M, Uchman A, Kümmel M (2001) Upper Triassic (Keuper) non-marine trace fossils from the HaBberge area (Franconia, south-eastern Germany). Paläontol Z 75(1):71–96

Schmid EE (1876) Der Muschelkalk des östlichen Thüringen. Fromann, Jena (20)

Seilacher A (1953) Studien zur Palichnologie. II. Die fossilien Ruhespuren (Cubichnia). N Jb Geol Paläont 98:87–124

Seilacher A (1955) Spuren und Fazies im Unterkambrium. In: Schindewolf OH, Seilacher A (eds) Beiträge zur Kenntnis des Kambriums in der Salt Range (Pakistan), Akad. Wiss. Lit. Mainz Abh. Math. Nat. Kl. 1955, pp 373–339

Seilacher A (1957) An-aktualistisches Wattenmeer? Paläontol Z 31:198–206
Seilacher A (1960) Lebensspuren als Leitfossilien. Geol Rundsch 49:41–50
Seilacher A (1964) Biogenic sediment structures. In: Imbrie J, Newell N (eds) Approaches to paleoecology. Wiley, New York, pp 296–316
Seilacher A (1977) Pattern analysis of *Paleodictyon* and related trace fossils. In: Crimes TP, Harper JC (eds) Trace fossils 2, Geological Journal, special issue 9. Seel House Press, Liverpool, pp 289–334
Seilacher A (1985) Trilobite palaeobiology and substrate relationships. Earth Environ Sci Trans R Soc Edinburgh 76(2-3):231–237
Seilacher A (1990) Aberrations in bivalve evolution related to photo- and chemosymbiosis. Hist Biol 3(4):289–311
Seilacher A (1992) An updated *Cruziana* stratigraphy of Gondwanan Paleozoic sandstones, 1565–1581. In: Salem MJ (ed) The geology of Lybia 8. Elsevier, Amsterdam
Seilacher A (2007) Trace fossil analysis. Springer, Berlin
Seilacher A, Seilacher E (1994) Bivalvian trace fossils: a lesson from actuopaleontology. Courier Forschungsinstitut Senckenberg 169:5–15
Shibata M, Varricchio DJ (2020) Horseshoe crab trace fossils from the Upper Cretaceous two medicine formation of Montana, USA, and a brief review of the xiphosurid ichnological record. J Paleontol 94(5):887–905
Shinn EA (1968) Burrowing in recent lime sediments of Florida and the Bahamas. J Paleontol 42:879–894
Shu WC, Tong JN, Tian L, Benton MJ, Chu DL, Yu JX, Guo WW (2018) Limuloid trackways from Permian-Triassic continental successions of North China. Palaeogeogr Palaeoclimatol Palaeoecol 508:71–90
Smith JJ, Hasiotis ST, Kraus MJ et al (2008) *Naktodemasis bowni*: new ichnogenus and ichnospecies for adhesive meniscate burrows (AMB), and paleoenvironmental implications, Paleogene Willwood Formation, Bighorn Basin, Wyoming. J Paleontol 82:267–278
Smith JJ, Platt BF, Ludvigson GA, Thomasson JR (2011) Ant-nest ichnofossils in honeycomb calcretes, Neogene Ogallala Formation, High Plains region of western Kansas, U.S.A. Palaeogeogr Palaeoclimatol Palaeoecol 308(3):383–394
Stanley TM, Feldmann RM (1998) Significance of nearshore trace-fossil assemblages of the Cambro-Ordovician deadwood formation and Aladdin Sandstone, South Dakota. Ann Carnegie Museum 67:1–51
Stanley DCA, Pickerill RK (1994) *Planolites constriannulatus* isp. nov. from the Late Ordovician Georgian Bay Formation of southern Ontario, eastern Canada. Ichnos 3:119–123
Stanley DCA, Pickerill RK (1995) *Arenituba*, a new name for the trace fossil ichnogenus *Micatuba* Chamberlain, 1971. J Paleontol 69:612–614
Todaro S, Hollis C, Di Stefano P (2016) Spongy-like porosity in peritidal carbonates: an interaction of cyclic sea-level oscillations, fresh water supply and sediment texture. Sediment Geol 333:70–83
Torell O (1870) Petrificata Suecana Formationis Cambricae. Lunds Universitets Årsskrift 6 (Afdelningen 2) 8:1–14
Uchman A (1995) Taxonomy and palaeoecology of flysch trace fossils: the Marnoso-arenacea formation and associated facies (Miocene, Northern Apennines, Italy). Beringeria 15:1–115
Uchman A (1998) Taxonomy and ethology of flysch trace fossils: revision of the Marian Książkiewicz collection and studies of complementary material. Ann Soc Geol Pol 68:105–218
Uchman A (1999) Ichnology of the Rhenodanubian flysch (Lower Cretaceous-Eocene) in Austria and Germany. Beringeria 25:65–171
Uchman A (2005) *Treptichnus*-like traces made by insect larvae (Diptera: Chironomidae, Tipulidae). In: Buta RJ, Rindsberg AK, Kopaska-Merkel DC (eds) Pennsylvanian footprints in the Black Warrior Basin of Alabama, Alabama Paleontological Society monograph 1. Alabama Paleontological Society, Birmingham, pp 143–146
Uchman A, Krenmayr HG (1995) Trace fossils from lower Miocene (Ottnangian) Molasse deposits of Upper Austria. Paläontol Z 69:503–524

Uchman A, Martyshyn A (2020) Taxis behaviour of burrowing organisms recorded in an Ediacaran trace fossil from Ukraine. Palaeogeogr Palaeoclimatol Palaeoecol 538:109441

Uchman A, Rattazzi B (2019) The trace fossil *Circulichnis* as a record of feeding exploration: new data from deep-sea Oligocene–Miocene deposits of northern Italy. Comptes Rendus Palevol 18:1–12

Uchman A, Bromley RG, Leszczyński S (1998) Ichnogenus *Treptichnus* in Eocene flysch, Carpathians, Poland: taxonomy and preservation. Ichnos 5:269–275

Uchman A, Janbu NE, Nemec W (2004) Trace fossils in the Cretaceous-Eocene flysch of the Sinop-Boyabat Basin, Central Pontides, Turkey. Ann Soc Geol Pol 74:197–235

Vialov OS (1968) On star-shaped problematica. (In Russian, English title). Ezheegodnik Vsesoyuznogo Paleontologicheskivego Obshchestva, 18:326–340

Vialov OS (1971) Rare Mesozoic problematica from the Pamir and the Caucasus. Paleontologicheskii Sbornik 7:85–93

Volk M (1961) *Protovirgularia nereitarum* (Reinhard Richter), eine Lebensspur aus de m Devon Thüringens. Senckenb Lethaea 42:69–75

von Eck H (1872) Rüdersdorf und Umgebung. Eine geognostische Monographie. In: Abhandlungen zur geologischen Specialkarte von Preußen und den thüringischen Staaten, 1, VII, 1–183, 1 plate, 1 map

von Schlotheim F (1820) Die Petrefactenkunde auf ihrem jetzigen Standpunkte durch die Beschreibung seiner Sammlung versteinerter und fossiler Überreste des Thier- und Pflanzenreichs der Vorwelt. Becker, Gotha

von Sternberg GK (1833) Versuch einer geognostisch-botanischen Darstellung der Flora der Vorwelt. IV Heft. C.E. Brenck, Regensburg, 48

Wagner R (1897) Beitrag zur genaueren Kenntniss des Muschelkalks bei Jena. Abhandlungen der Königlich Preussischen Geologischen Landesanstalt, Neue Folge 27:1–105

Wang Y, Lin J-P, Zhao YL, Orr PJ (2009) Palaeoecology of the trace fossil *Gordia* and its interaction with nonmineralizing taxa from the early Middle Cambrian Kaili Biota, Guizhou province, South China. Palaeogeogr Palaeoclimatol Palaeoecol 277:141–148

Weiss E (1884) Vorlegung des Dictyophytum liebeanum Gein. aus der Gegend von Gera. 1884. Gesellschaft Naturforschender Freunde Berlin, Sitzungsberichte, 17

Weller S (1899) Kinderhook faunal studies 1: The fauna of the vermicular sandstone at Northview, Webster County, Missouri. St. Louis Academy of Science, Transactions, 9:9–51

Werner F (2002) Bioturbation structures in marine Holocene sediments of Kiel Bay (western Baltic). Meyniana 54:41–72

Wesolowski LJN, Buatois LA, Mángano MG, Ponce JJ, Carmona NB (2018) Trace fossils, sedimentary facies and parasequence architecture from the Lower Cretaceous Mulichinco Formation of Argentina: the role of fair-weather waves in shoreface deposits. Sediment Geol 367:146–163

Wetzel A (2002) Modern *Nereites* in the South China Sea—ecological association with redox conditions in the sediment. Palaios 17:507–515

Wetzel A (2008) Recent bioturbation in the deep South China Sea: a uniformitarian ichnologic approach. Palaios 23:601–615

Wetzel A (2010) Deep-sea ichnology: observations in modern sediments to interpret fossil counterparts. Acta Geol Pol 60:125–138

Wetzel A, Bromley RG (1994) *Phycosiphon incertum* revisited: *Anconichnus horizontalis* is its junior subjective synonym. J Paleontol 68(6):1396–1402

Wetzel A, Bromley RG (1996) Re-evaluation of the ichnogenus *Helminthopsis*—a new look at the type material. Palaeontology 39:1–19

Wetzel A, Uchman A (1997) Ichnology of deep-sea fan overbank deposits of the Ganei Slates (Eocene, Switzerland)—a classical flysch trace fossil locality studied first by Oswald Heer. Ichnos 5:139–162

Wilson MA, Rigby JK (2000) *Asteriacites lumbricalis* von Schlotheim1820: ophiuroid trace fossils from the Lower Triassic Thaynes Formation, central Utah. Ichnos 7:43–49

Winn K (2006) Bioturbation structures in marine Holocene sediments of the Great Belt (western Baltic). Meyniana 58:157–178

Worsley D, Mørk A (2001) The environmental significance of the trace fossil *Rhizocorallium jenense* in the Lower Triassic of western Spitsbergen. Polar Res 20:37–48

Yang SP (1984) Silurian trace fossils from the Yangzi Gorges and the significance to depositional environments. Acta Palaeontol Sin 23(6):705–715

Yang SP, Hu Y (1992) Silurian trace fossils from the western part of West Qinling. Geoscience 6:384–391. (3 plates)

Yang SP, Song Z (1985) Middle-Upper Triassic trace fossils from Zhada, Ngari, Southwest Xizang (Tibet), and its geologic significance. Xizang (Tibet). Geology 1:1–14. (pl. I–IV.)

Yochelson EL, Fedonkin MA (1997) The type specimens (Middle Cambrian) of the trace fossil *Archaeonassa* Fenton and Fenton. Can J Earth Sci 34:1210–1219

Zenker JC (1836) Historisch-Topographisches Taschenbuch von Jena und seiner Umgebung besonders in Naturwissenschaftlicher u. medizinischer Beziehung. Friedrich Fromann, Jena

Zhang LJ, Knaust D, Zhao Z (2016) Palaeoenvironmental and ecological interpretation of the trace fossil *Rhizocorallium* based on contained iron framoboids (Upper Devonian, South China). Palaeogeogr Palaeoclimatol Palaeoecol 446:144–151

Zhang LJ, Buatois LA, Gabriela Mángano M, Qi YA, Zhang X, Sun S, Tai C (2017) Early Triassic estuarine depauperate *Cruziana* Ichnofacies from the Sichuan area of South China and its implications for the biotic recovery in brackish-water settings after the end-Permian mass extinction. Palaeogeogr Palaeoclimatol Palaeoecol 485:351–360

Zhang LJ, Buatois LA, Mángano MG, Gong YM, Feng QL, Qi YA, Luo M, Zhang X (2018) Uppermost Permian trace fossils along a shelf to slope transect in South China and their implications for oceanic redox evolution and extinction pattern. Glob Planet Chang 167:74–86

Zhang LJ, Buatois LA, Mángano MG, Qi YA, Tai C (2019) Early Triassic trace fossils from South China marginal-marine settings: implications for biotic recovery following the end-Permian mass extinction. Geol Soc Am Bull 131:235–251

Zhang LJ, Fan RY, Dang ZY et al (2020) The youngest known *Dictyodora* from the Late Permian (Lopingian) deep sea in West Qinling, central China. Palaeogeogr Palaeoclimatol Palaeoecol 558:109948

Zhao XM, Tong JN (2010) Two episodic changes of trace fossils through the Permian–Triassic transition in the Meishan cores, Zhejiang Province. Sci China Earth Sci 53:1885–1893

Zheng QF, Cao CQ, Zhang MY (2013) Sedimentary features of the Permian–Triassic boundary sequence of the Meishan section in Changxing County, Zhejiang Province. Sci China Earth Sci 56:956–969

Zonneveld JP, Pemberton SG, Saunders TDA, Pickerill RK (2002) Large robust *Cruziana* from the Middle Triassic of northeastern British Columbia: ethologic, biostratigraphic and paleobiologic significance. Palaios 17:435–448

Chapter 4
Ecological Significances of the Ichnofauna

4.1 Introduction of Ichno-Ecological Proxies

Trace fossils have proven to be a valuable tool for evaluating the recovery of benthic ecosystems following the Permian-Triassic (P-Tr) mass extinction, as well as other mass extinction events. There are several proxies, such as ichnodiversity, ichnodisparity, complexity, burrow size and tiering level, that have been employed to unravel the processes of biotic recovery. In this chapter, we first describe some of these ichno-ecological proxies such as ichnodiversity and ichnodisparity and also introduce another two relatively new ecological proxies—ecospace utilization and ecosystem engineering—and the analysis of the ichnotaxa from China was undertaken by using these proxies. Secondly, we analyze variations in these proxies across the Permian-Triassic mass extinction interval and the Early Triassic, spanning environments from brackish to carbonate ramp, platform, lower shoreface, and shelf-slope-basin settings. This analysis aims to explore potential environmental influences on the behaviors and ecology of trace-making organisms during the P-Tr mass extinction and its aftermath.

Ichnodiversity has been widely used to evaluate general trends in ichnotaxonomic richness in specific sedimentary environments or through geological time, particularly among intervals before and after mass extinctions (Seilacher 1974; Ekdale 1985; Crimes 1994; Bromley 1996; Orr 2001; Uchman 2004; Buatois et al. 2005; Buatois and Mángano 2013). Previous studies have extensively discussed the meaning and significance of ichnodiversity (Buatois and Mángano 2011, 2013; Knaust et al. 2014; Buatois et al. 2016; Luo et al. 2020). In practice, ichnodiversity refers to the number of ichnotaxa present, typically at the ichnogeneric rank, as trace fossil taxonomy is more reliably established at the ichnogenus level.

Provide the results of semiquantitative parameters of the trace fossils, i.e., bioturbation, burrow size, behavioral categories, complexity, and tiering

X. Feng et al., *Ichnology of the Permian-Triassic Mass Extinction in China*,
https://doi.org/10.1007/978-981-96-6639-3_4

Consequently, ichnodiversity is better understood as a measure of richness rather than true diversity (Buatois and Mángano 2013).

In contrast to ichnodiversity, the concept of ichnodisparity has been introduced more recently (Buatois and Mángano 2013; Buatois et al. 2017), with several studies highlighting its significance in understanding evolutionary radiations (Mángano and Buatois 2014; Buatois et al. 2016). Ichnodisparity seeks to capture variations in trace fossil morphology at a broader level than that of ichnogenera, assessed through the identification of architectural design categories. While ichnodiversity measures ichnotaxonomic richness, ichnodisparity quantifies the variability of morphological plans in trace fossils (Buatois et al. 2017). This metric may enable the evaluation of major morphological changes in trace fossils, driven by innovations in body plans and locomotory systems. Buatois et al. (2017) proposed 79 types of architectural design according to bioturbation structures and bioerosion structures.

The analysis of modes of life and the characterization of ecospace utilization are commonly applied to body fossils. Bambach et al. (2007) and Bush et al. (2007) developed a framework for identifying modes of life based on categorization across three parameters: tiering, motility, and feeding mode. Categories within these parameters define the total potential ecospace, and the modes of life of individual taxa can be mapped within this framework to determine occupied ecospace at any given time. Trace fossils offer an excellent model system for analyzing benthic ecospace occupation using these principles (Minter et al. 2017; Buatois et al. 2020). The extent of occupied ecospace is quantified by the number of modes of life, categorized according to three parameters: (1) tiering, subdivided into surficial, semi-infaunal, shallow infaunal, intermediate infaunal, and deep infaunal tiers; (2) motility, subdivided into motile, facultatively motile, and non-motile types; and (3) feeding mode, subdivided into suspension feeders, non-specialized deposit feeders, specialized deposit feeders, predators, chemosymbiosis, farming, and trapping.

Tiering refers to the vertical position of an animal within the sediment when in its regular life position. For a trace fossil to form, an animal must interact with a substrate. Trace fossils provide direct evidence of the life positions of animals, enabling precise identification of infaunal tiering levels (Ausich and Bottjer 1982). The system used here defines a shallow infaunal tier for depths up to 6 cm below the substrate-water interface, an intermediate infaunal tier for depths of 6–12 cm, and a deep infaunal tier for depths greater than 12 cm. This framework distinguishes between an intermediate tier and a truly deep tier of infaunal animals. Accordingly, each trace fossil is classified as surficial, semi-infaunal (0–0.5 cm), shallow (0.5–6 cm), intermediate (6–12 cm), or deep (>12 cm) (Mángano and Buatois 2014; Bottjer and Ausich 1986).

Motility is categorized into motile, facultatively motile, and non-motile types (Minter et al. 2017). Most trace-making animals exhibit some degree of motility. Generally, organisms capable of producing temporary bioturbation structures that reflect continuous movement are considered motile. Animals that are typically stationary but capable of movement are classified as facultatively motile.

The mode of feeding is divided into five categories: suspension feeding, non-specialized deposit feeding, specialized deposit feeding, predation, chemosymbiosis, farming, and trapping (Minter et al. 2017). Suspension feeders capture food particles from the water column, while deposit feeders ingest food particles from the substrate. Deposit feeders are further classified based on trace fossil morphology: non-specialized deposit feeders produce non-patterned and overcrossing trails, whereas specialized deposit feeders create exploration and non-overcrossing locomotory burrows. Predators are trace-makers inferred to have captured prey. Chemosymbiosis involves endosymbiosis with chemoautotrophic bacteria, as seen in downward-radiating burrows like *Chondrites*. Farming refers to the cultivation of bacteria or fungi for feeding, often on the internal surfaces of burrows or chambers. Trapping involves the passive capture of migrating meiofauna or microorganisms, exemplified by complex, patterned, meandering, spiral, radiating, or networked trace fossils known as graphoglyptids (Minter et al. 2017).

For the measures of ecospace utilization, a three-dimensional coordinate axis is a nice way to explain. One can assume each of the three parameters (tiering, motility and feeding mode) represents x, y, z axis. One data of ecospace utilization in a three-dimensional coordinate system must consist of three values in x, y, z axis (x, y, z). In practice, we first confirm the values of the x, y, z axis and then plot them in the three-dimensional cubes.

Ecosystem engineering is classified using three proxies: (1) tiering, (2) mode of sediment interaction, and (3) mode of sediment modification. The tiering categories have been well-documented in studies of ecospace utilization. The ways in which trace-making animals interact with the substrate (mode of sediment interaction) include intrusion, compression, backfilling, and excavation. Intrusion refers to the displacement of sediments as the animal moves through, with the sediment closing behind it (Minter et al. 2017). Compression involves the movement and compaction of sediments around the trace-making animal as it passes through. Backfilling is the active backward movement of sediments around the animal. Excavation refers to the active loosening and relocation of sediments. Each ichnogenus was assigned one of these four sediment interaction modes. A detailed list of sediment interaction modes for most ichnogenera can be found in previous studies (Minter et al. 2017).

The mode of sediment modification is assessed by categorizing trace fossils based on how their makers affected and reworked sediment (Buatois et al. 2020). This framework is adapted from the conceptual models of marine benthic ecology. Sediment modification can be classified into four types: biodiffusion, gallery biodiffusion, conveyor, and regenerator. Biodiffusion involves the short-distance movement of sediment particles, while gallery biodiffusion refers to the rapid redistribution of sediment particles from one part of the sediment profile to another. The concept of conveyors encompasses both upward and downward sediment transport, as well as lateral movement. This category includes animals that actively transport sediment particles across and within tiers. Regenerators are animals that actively move sediment from deeper layers to the surface, where it may be redistributed by physical sedimentary processes such as currents (Feng et al. 2022). Each ichnogenus was assigned one of the above sediment interaction and modification modes based on previous descriptions and classifications.

Similar to the measures of ecospace utilization, a three-dimensional coordinate axis is a nice way to explain the measures of ecosystem engineering. One can assume each of the three parameters (tiering, mode of sediment interaction, mode of sediment modification) represents x, y, z axis. One data of ecosystem engineering in a three-dimensional coordinate system must consist of three values in x, y, z axis (x, y, z). In practice, we first confirm the values of the x, y, z axis and then plot them in the three-dimensional cubes. Although both ecospace utilization and ecosystem engineering metrics take tiering into consideration, as mentioned above, an ecospace utilization or ecosystem engineering metric must consist of x, y, and z values, and if tiering represents the z axis, the same z values also do not ensure they have same x and y values. Actually, the other two parameters (x, y) of ecospace utilization or ecosystem engineering measure very different things, and thus, they should also be independent of each other.

Finally, to reconstruct and compare the four ichnological metrics across five environmental settings during the P-Tr mass extinction and the Early Triassic, we organized the collected data into nine half-substage time intervals. These subdivisions are primarily based on biostratigraphical data, supplemented by chemostratigraphical data where available. The intervals include the early Changhsingian, late Changhsingian, early Griesbachian, late Griesbachian, Dienerian, early Smithian, late Smithian, early Spathian, and late Spathian.

4.2 Ecological Characteristics of Ichnofauna in the Aftermath of the P-Tr Mass Extinction in Brackish Setting

4.2.1 Ichnodiversity and Ichnodisparity

In brackish environmental settings, no strata were exposed in China; thus we could not evaluate the richness of ichnoassemblage in the Late Permian, as well as the ecological variations across the P-Tr mass extinction. However, the horizons that yield abundant trace fossils were exposed in the Houzhougongmiao and Longmendong sections from western China. Trace fossils were collected from the siltstone and mudstone beds of the Sunjiagou Formation from the Houzhougongmiao section and from the Dongchuan and Feixianguan formations of the Longmendong section. There are thirteen ichnogenera in the late Griesbachian, they are *Arenicolites*, *Asteriacites*, *Biformites*, *Didymaulichnus*, *Diplichnites*, *Kouphichnium*, *Laevicyclus*, *Monomorphichnus*, *Planolites*, *Protovirgularia*, *Ptychoplasma*, *Treptichnus*, and *Undichna* (Table 4.1). The richness of ichnogenera displayed a sharp decrease through the late Griesbachian to Dienerian (Fig. 4.1). There were only six ichnogenera in the Dienerian. The diversity of trace fossils peaked in the early Smithian, with fourteen ichnogenera. They are *Arenicolites*, *Asteriacites*, *Biformites*, *Cylindrichnus*, *Diplocraterion*, *Diplopodichnus*, *Helminthopsis*, *Lockeia*, *Palaeophycus*, *Planolites*,

Table 4.1 Ichnogenera and ecological proxies in half-substage time bins, collected from the studied sections in a brackish environment

Ichnogenus	Architectural design	Tier	Motility	Feeding	Interaction	Modification
Late Griesbachian						
Arenicolites	Vertical single U- and Y-shaped burrows	Intermediate	Facultative	Suspension feeder	Compression	Gallery biodiffusive
Asteriacites	Pentameral-shaped imprints and burrows	Surficial	Free	Non-specialized deposit/ predator	Compression	Biodiffusive
Biformites	Trackways and scratch imprints	Surficial	Free	Non-specialized deposit/ predator	Compression	Biodiffusive
Didymaulichnus	Bilobate trails and paired grooves	Semi-infaunal	Free	Non-specialized deposit/ predator	Excavation	Regenerator
Diplichnites	Trackways and scratch imprints	Surficial	Free	Non-specialized deposit/ predator	Compression	Biodiffusive
Kouphichnium	Trackways and scratch imprints	Surficial	Free	Non-specialized deposit/ predator	Compression	Biodiffusive
Laevicyclus	Vertical unbranched burrows	Semi-infaunal	Facultative	Suspension/predator	Compression	Biodiffusive
Monomorphichnus	Trackways and scratch imprints	Surficial	Free	Non-specialized deposit	Compression	Biodiffusive
Planolites	Simple, actively-filled (massive) horizontal to oblique structures	Shallow	Free	Non-specialized deposit	Backfill	Conveyor
Protovirgularia	Chevronate trails	Shallow	Free	Non-specialized deposit	Intrusion	Biodiffusive
Ptychoplasma	Isolated and serial oval to almond-shaped burrows	Semi-infaunal	Free	Non-specialized deposit	Compression	Biodiffusive
Treptichnus	Horizontal burrows with horizontal to vertical branches	Semi-infaunal	Free	Specialized deposit	Compression	Gallery biodiffusive
Undichna	Simple horizontal trails	Surficial	Free	Predator	Compression	Biodiffusive

(continued)

Table 4.1 (continued)

Ichnogenus	Architectural design	Tier	Motility	Feeding	Interaction	Modification
Dienerian						
Cylindrichnus	Vertical concentrically filled burrows	Semi-infaunal	Free	Non-specialized deposit	Backfill	Conveyor
Diplocraterion	Vertical single U- and Y-shaped burrows	Deep	Facultative	Suspension	Compression	Gallery biodiffusive
Palaeophycus	Passively-filled horizontal to oblique burrows	Shallow	Facultative	Suspension/predator	Compression	Gallery biodiffusive
Planolites	Simple, actively-filled (massive) horizontal to oblique structures	Shallow	Free	Non-specialized deposit	Backfill	Conveyor
Siphonichnus	Vertical unbranched burrows	Deep	Facultative	Suspension/predator	Compression	Gallery biodiffusive
Teichichnus	Horizontal burrows with simple vertically oriented spreiten	Intermediate	Free	Non-specialized deposit	Backfill	Conveyor
Early Smithian						
Arenicolites	Vertical single U- and Y-shaped burrows	Intermediate	Facultative	Suspension	Compression	Gallery biodiffusive
Asteriacites	Pentameral-shaped imprints and burrows	Surficial	Free	Non-specialized deposit/predator	Compression	Biodiffusive
Biformites	Trackways and scratch imprints	Surficial	Free	Non-specialized deposit/predator	Compression	Biodiffusive
Cylindrichnus	Vertical concentrically filled burrows	Semi-infaunal	Free	Non-specialized deposit	Backfill	Conveyor
Diplocraterion	Vertical single U- and Y-shaped burrows	Deep	Facultative	Suspension	Compression	Gallery biodiffusive
Diplopodichnus	Bilobate trails and paired grooves	Surficial	Free	Non-specialized deposit/predator	Compression	Biodiffusive
Helminthopsis	Simple horizontal trails	Semi-infaunal	Free	Non-specialized deposit	Backfill	Conveyor

Lockeia	Isolated and serial oval to almond-shaped burrows	Semi-infaunal	Free	Non-specialized deposit	Compression	Biodiffusive
Palaeophycus	Passively-filled horizontal to oblique burrows	Shallow	Facultative	Suspension/predator	Compression	Gallery biodiffusive
Planolites	Simple, actively-filled (massive) horizontal to oblique structures	Shallow	Free	Non-specialized deposit	Backfill	Conveyor
Ptychoplasma	Isolated and serial oval to almond-shaped burrows	Semi-infaunal	Free	Non-specialized deposit	Compression	Biodiffusive
Rhizocorallium	Burrows with horizontal spreiten	Shallow	Facultative	Non-specialized deposit	Excavation	Gallery biodiffusive
Siphonichnus	Vertical unbranched burrows	Deep	Facultative	Suspension/predator	Compression	Gallery biodiffusive
Teichichnus	Horizontal burrows with simple vertically oriented spreiten	Intermediate	Free	Non-specialized deposit	Backfill	Conveyor
Late Smithian						
Arenicolites	Vertical single U- and Y-shaped burrows	Intermediate	Facultative	Suspension	Compression	Gallery biodiffusive
Asteriacites	Pentameral-shaped imprints and burrows	Surficial	Free	Non-specialized deposit/ predator	Compression	Biodiffusive
Cylindrichnus	Vertical concentrically filled burrows	Semi-infaunal	Free	Non-specialized deposit	Backfill	Conveyor
Diplocraterion	Vertical single U- and Y-shaped burrows	Deep	Facultative	Suspension	Compression	Gallery biodiffusive
Diplopodichnus	Bilobate trails and paired grooves	Surficial	Free	Non-specialized deposit/ predator	Compression	Biodiffusive
Lockeia	Isolated and serial oval to almond-shaped burrows	Semi-infaunal	Free	Non-specialized deposit	Compression	Biodiffusive

(continued)

Table 4.1 (continued)

Ichnogenus	Architectural design	Tier	Motility	Feeding	Interaction	Modification
Palaeophycus	Passively filled horizontal to oblique burrows	Shallow	Facultative	Suspension/predator	Compression	Gallery biodiffusive
Planolites	Simple, actively filled (massive) horizontal to oblique structures	Shallow	Free	Non-specialized deposit	Backfill	Conveyor
Ptychoplasma	Isolated and serial oval to almond-shaped burrows	Semi-infaunal	Free	Non-specialized deposit	Compression	Biodiffusive
Rhizocorallium	Burrows with horizontal spreiten	Shallow	Facultative	Non-specialized deposit	Excavation	Gallery biodiffusive
Siphonichnus	Vertical unbranched burrows	Deep	Facultative	Suspension/predator	Compression	Gallery biodiffusive
Teichichnus	Horizontal burrows with simple vertically oriented spreiten	Intermediate	Free	Non-specialized deposit	Backfill	Conveyor
Early Spathian						
Helminthopsis	Simple horizontal trails	Semi-infaunal	Free	Non-specialized deposit	Backfill	Conveyor
Palaeophycus	Passively filled horizontal to oblique burrows	Shallow	Facultative	Suspension/predator	Compression	Gallery biodiffusive
Phycodes	Horizontal burrows with horizontal to vertical branches	Semi-infaunal	Free	Specialized deposit feeder	Backfill	Conveyor
Taenidium	Simple, actively filled (meniscate) horizontal to oblique structures	Shallow	Free	Non-specialized deposit/ predator	Backfill	Conveyor
Thalassinoides	Maze and boxwork burrows	Deep	Free	Non-specialized deposit/ predator	Excavation	Regenerator
Late Spathian						
Asteriacites	Pentameral-shaped imprints and burrows	Surficial	Free	Non-specialized deposit/ predator	Compression	Biodiffusive
Helminthopsis	Simple horizontal trails	Semi-infaunal	Free	Non-specialized deposit	Backfill	Conveyor

Planolites	Simple, actively filled (massive) horizontal to oblique structures	Shallow	Free	Non-specialized deposit	Backfill	Conveyor
Ptychoplasma	Isolated and serial oval to almond-shaped burrows	Semi-infaunal	Free	Non-specialized deposit	Compression	Biodiffusive
Rhizocorallium	Burrows with horizontal spreiten	Shallow	Facultative	Non-specialized deposit	Excavation	Gallery biodiffusive
Thalassinoides	Maze and boxwork burrows	Deep	Free	Non-specialized deposit/ predator	Excavation	Regenerator

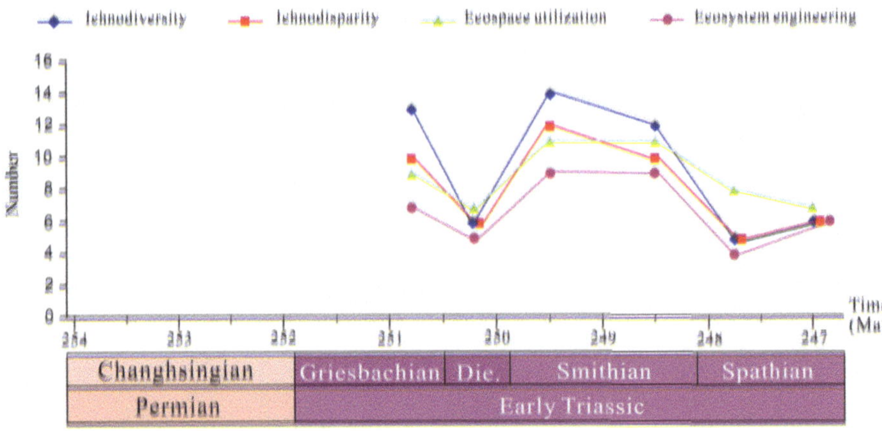

Fig. 4.1 Colonization of brackish environmental setting, plot of changes of the four metrics in the Early Triassic

Ptychoplasma, *Rhizocorallium*, *Siphonichnus*, and *Teichichnus* (Table 4.1). The ichnodiversity was twelve in the late Smithian, followed by a sharp drop during the Spathian (Fig. 4.1). The richness of ichnogenera was six and five in the early and late Spathian, respectively.

Ichnodisparity showed a very similar tendency with ichnodiversity, which also decreased through the late Griesbachian to Dienerian, then started to increase, and peaked in the early Smithian (Fig. 4.1). The richness of architectural designs declined from the Smithian to Spathian. Most architectural designs were retained during the whole Early Triassic. The ichnodisparity across the P-Tr mass extinction is listed in Table 4.1 in detail.

4.2.2 Ecospace Utilization and Ecosystem Engineering

Nine out of 90 potential infaunal ecospace categories were occupied in the late Griesbachian, based on combined data from half-substage time bins in brackish environmental settings (Fig. 4.1). The number of modes of life is seven in the Dienerian, eleven in the early and late Smithian, eight in the early Spathian, and seven in the late Spathian (Fig. 4.1). Generally, ecospace utilization therefore decreased from the late Griesbachian to Dienerian, followed by a sharp increase to the Smithian, then continued to fall during the Spathian (Fig. 4.1). More specifically, ecological categories in the surficial and semi-infaunal tier levels vanished through late Griesbachian to Dienerian, but in the shallow to deep tiers, many new ecospace categories appeared. In the semi-infaunal tiering level, five infaunal ecospace categories were occupied in the late Griesbachian; they were freely motile, nonspecialized deposit feeders and specialized deposit feeders, predators, facultatively motile, suspension feeders and predators (Fig. 4.2a). These ecospaces most

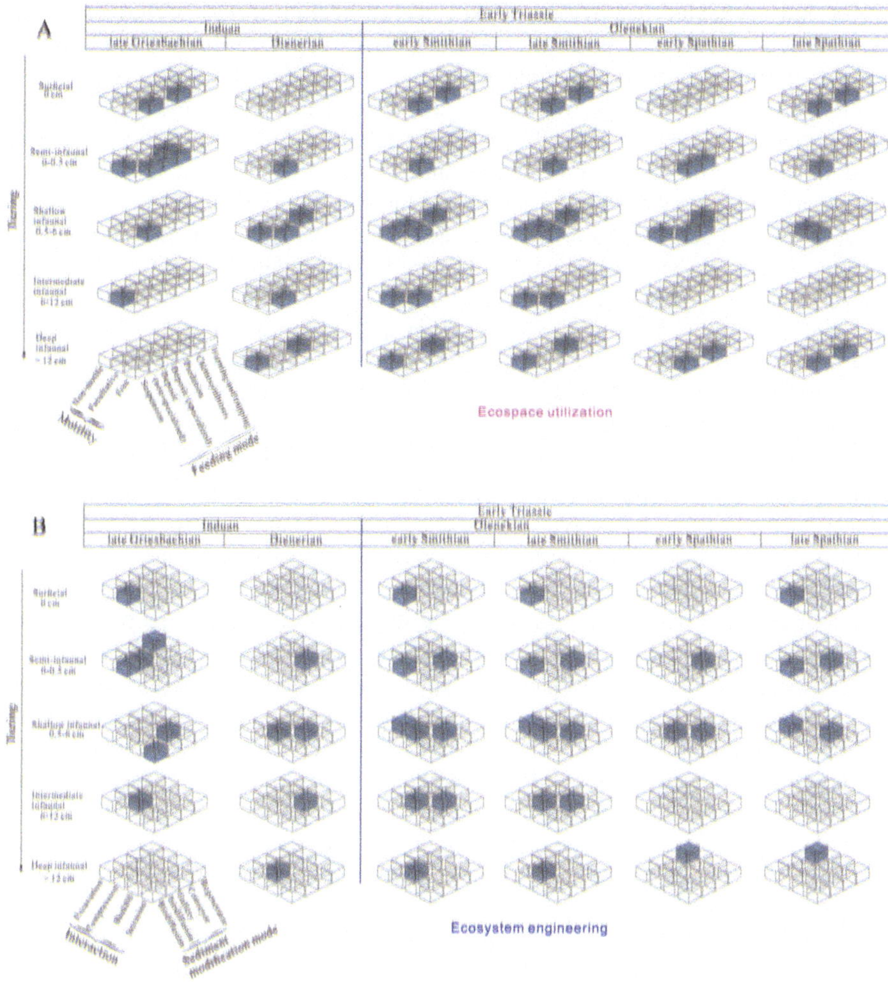

Fig. 4.2 Colonization of brackish environmental setting through the Early Triassic. (**a**) Cubes of ecospace utilization. (**b**) Cubes of ecosystem engineering

disappeared in the Dienerian, with only freely motile, nonspecialized deposit feeders persisting, and this condition was kept during the whole Early Triassic. In the shallow-infaunal tier, both facultatively motile suspension feeders and predators newly appeared in the Dienerian and held thereafter (Fig. 4.2a). There was only one infaunal ecospace category in the intermediate tiering level in the late Griesbachian, and it increased to two in the Smithian and disappeared in the Spathian. In the deep infaunal tier, no ecospace was occupied in the Griesbachian, and two were newly occupied in the Dienerian, and they persisted throughout the Early Triassic (Fig. 4.2a).

The most abundant categories of ecospace utilization in the late Griesbachian were surficial-tiering, freely motile, nonspecialized deposit feeders and predators. Shallow-tiering, freely motile, nonspecialized deposit feeders was also common (Table 4.2). Ecospace utilization categories of the shallow tier rebounded in the early Smithian, while those in the deep tier rebounded earlier in the Dienerian. Facultatively motile suspension feeders and predators were very common in both shallow and deep tiers during the Smithian (Table 4.2).

Of all possible combinations of substrate interaction and sediment modification mechanisms, seven were observed in the late Griesbachian, five in the Dienerian, nine in the early and late Smithian, four in the early Spathian, and six in the late Spathian, according to combined data from half-substage time bins in brackish environmental settings (Fig. 4.1). Generally, the number of ecosystem engineering increased from Induan to Olenekian and declined from the late Smithian to Spathian (Fig. 4.1). Ecosystem engineering styles in all tiers are persistent during the Early Triassic. More specifically, in the semi-infaunal tier, three types of ecosystem engineering occurred during the late Griesbachian, but these disappeared in the Dienerian, replaced by a new type. In the late Griesbachian, biodiffusors and gallery biodiffusors interacted with sediment through compression, while regenerators interacted through excavation. In the Dienerian, sediment conveyors interacted with sediment through backfilling (Fig. 4.2b). During the Smithian and Spathian, biodiffusors interacted with sediment through compression. At the shallow-tiering level, the number of ecosystem engineering types remained consistent throughout the Early Triassic, but the specific types varied between the Induan and Olenekian. During the Smithian-Spathian, gallery biodiffusors interacted with sediment through compression and excavation, while sediment conveyors interacted through backfilling. In the intermediate tier, the impact on the substrate showed a slight increase from the Induan to the Olenekian, though the types of ecosystem engineering remained unchanged. In the deep tier, no ecosystem engineering occurred during the late Griesbachian. Only one type of ecosystem engineering was observed throughout the Early Triassic in this tier. Gallery biodiffusors interacted with sediment through compression from the Dienerian to the Smithian, while regenerators interacted through excavation during the Spathian (Fig. 4.2b).

The occurrence of ecosystem engineering modes was sporadic during the Griesbachian and Dienerian, except for the biodiffusors that interacted with sediment through compressions in the surficial tier, which owned 9 occurrences in the late Griesbachian (Table 4.3). Most ecosystem engineering modes became quite abundant during the Smithian, especially those that occupied shallow and deep tiers (Table 4.3). The occurrence of ecosystem engineering styles was highest in the deep tier during the Dienerian, which had 9 occurrences (Table 4.3). Shallow-tier gallery biodiffusors that interacted with sediment through compression and deep-tier gallery biodiffusors that interacted with sediment through compression were the most abundant during the Smithian. In semi-infaunal tier, sediment conveyors that interacted with sediment through backfilling were also common in the early Smithian (Table 4.3).

Table 4.2 Abundance of ichnogenera in each mode of life (ecospace utilization) in a brackish environmental setting in half-substage time bins

No.	Ecospace utilization	Late Griesbachian	Dienerian	Early Smithian	Late Smithian	Early Spathian	Late Spathian
1	Surficial + Free + Deposit (non-specialized)	9	0	4	2	0	0
2	Surficial + Free + Predator	9	0	4	2	0	0
3	Semi-infaunal + Free + Deposit (non-specialized)	2	2	7	5	1	2
4	Semi-infaunal + Free + Predator	1	0	0	0	0	0
5	Semi-infaunal + Facultative + Suspension	1	0	0	0	0	0
6	Semi-infaunal + Facultative + Predator	1	0	0	0	0	0
7	Semi-infaunal + Free + Deposit (specialized)	1	0	0	0	1	0
8	Semi-infaunal + Free + Farming	0	0	0	0	0	0
9	Shallow + Facultative + Suspension	1	4	7	6	3	0
10	Shallow + Facultative + Predator	0	4	5	3	3	0
11	Shallow + Free + Deposit (non-specialized)	6	4	3	3	2	3
12	Shallow + Free + Deposit (specialized)	1	0	0	0	0	0
13	Shallow + Free + Predator	0	0	0	0	2	1
14	Shallow + Facultative + Deposit (non-specialized)	0	0	2	2	0	2
15	Shallow + Free + Farming	0	0	0	0	0	0

(continued)

Table 4.2 (continued)

No.	Ecospace utilization	Late Griesbachian	Dienerian	Early Smithian	Late Smithian	Early Spathian	Late Spathian
16	Intermediate + Facultative + Suspension	0	0	0	0	0	0
17	Intermediate + Free + Deposit (non-specialized)	0	3	3	2	0	0
18	Intermediate + Free + Deposit (specialized)	0	0	0	0	0	0
19	Intermediate + Free + Predator	0	0	0	0	0	0
20	Deep + Facultative + Suspension	0	9	6	6	0	0
21	Deep + Facultative + Predator	0	4	4	2	0	0
22	Deep + Facultative + Deposit (specialized)	0	0	0	0	0	0
23	Deep + Free + Deposit (non-specialized)	0	0	0	0	0	0
24	Deep + Free + Predator	0	0	0	0	0	0
25	Deep + Facultative + Farming	0	0	0	0	0	0
26	Deep + Facultative + Chemosymbiosis	0	0	0	0	0	0

4.3 Ecological Characteristics of Ichnofauna Across the P-Tr Mass Extinction and Its Aftermaths in Carbonate Ramp Setting

4.3.1 Ichnodiversity and Ichnodisparity

There were nine ichnogenera in the late Changhsingian from South China, *Balanoglossites*, *Chondrites*, *Palaeophycus*, *Phycosiphon*, *Planolites*, *Siphonichnus*, *Taenidium*, and *Thalassinoides*, and *Zoophycos* (Table 4.4). Most trace fossils were from the uppermost Changxing Formation and lowermost Yinkeng Formation of the GSSP Meishan section, Zhejiang Province, South China. Similar to trace fossils in carbonate platform setting, only simple, horizontal traces such as *Planolites* and *Helminthoidichnites* distributed in the Griesbachian-Dienerian. The number of ichnogenera increased notably from the Dienerian to the early Smithian, continued to rise, peaking during the late Smithian, and then experienced a slight decline in the

Table 4.3 Abundance of ichnogenera in each impact upon sediment (ecosystem engineering) in all environments in half-substage time bins

No.	Ecosystem engineering	Late Griesbachian	Dienerian	Early Smithian	Late Smithian	Early Spathian	Late Spathian
1	Surficial + Compression + Biodiffusion	9	0	4	2	0	1
2	Semi-infaunal + Compression + Gallery Biodiffusion	0	0	0	0	0	0
3	Semi-infaunal + Backfill + Conveyor	0	2	5	2	2	1
4	Semi-infaunal + Excavation + Regenerator	1	0	0	0	0	0
5	Semi-infaunal + Compression + Biodiffusion	2	0	2	3	0	1
6	Shallow + Compression + Gallery Biodiffusion	3	4	7	5	3	0
7	Shallow + Backfill + Conveyor	2	4	3	3	1	2
8	Shallow + Intrusion + Biodiffusion	2	0	0	0	0	0
9	Shallow + Excavation + Gallery Biodiffusion	0	0	3	2	1	2
10	Intermediate + Compression + Gallery Biodiffusion	0	0	0	0	0	0
11	Intermediate + Backfill + Conveyor	0	3	3	2	0	0
12	Intermediate + Excavation + Regenerator	0	0	0	0	0	0
13	Deep + Compression + Gallery Biodiffusive	0	9	6	6	0	0

(continued)

Table 4.3 (continued)

No.	Ecosystem engineering	Late Griesbachian	Dienerian	Early Smithian	Late Smithian	Early Spathian	Late Spathian
14	Deep + Backfill + Conveyor	0	0	0	0	0	0
15	Deep + Excavation + Regenerator	0	0	0	0	0	0
16	Deep + Excavation + Conveyor	0	0	0	0	0	0

early Spathian before rebounding in the late Spathian (Fig. 4.3). The Early Triassic trace fossils were collected from the limestone beds of the Helongshan and Nanlinghu formations of the Chaohu and Susong sections located in lower Yangtze region, also from the Daye and Jialingjiang formations of the Shatian and Daxiakou sections in middle Yangtze region, and from muddy limestone beds of the Daye and Anshun formations from the Dangwu and Gaimao sections in the upper Yangtze region. The ichnogenera in the Smithian were *Arenicolites*, *Archaeonassa*, *Arenituba*, *Chondrites*, *Cochlichnus*, *Gordia*, *Gyrochorte*, *Helminthoidichnites*, *Helminthopsis*, *Kouphichnium*, *Laevicyclus*, *Lockeia*, *Palaeophycus*, *Planolites*, *Phycodes*, *Phycosiphon*, *Psammichnites*, *Sinusichnus*, *Skolithos*, *Teichichnus*, *Thalassinoides*, and *Treptichnus*. The ichnogenera distributed in the Spathian were *Arenicolites*, *Archaeonassa*, *Arenituba*, *Circulichnis*, *Diplocraterion*, *Gordia*, *Gyrochorte*, *Helminthopsis*, *Helminthoidichnites*, *Laevicyclus*, *Palaeophycus*, *Planolites*, *Psammichnites*, *Rhizocorallium*, *Sinusichnus*, *Skolithos*, *Spongeliomorpha*, *Thalassinoides*, and *Treptichnus* (Table 4.4). Ichnodisparity showed a very similar tendency with ichnodiversity but with a lower value, which also decreased across the P-Tr mass extinction and was rare during the Griesbachian-Dienerian and sharply increased in the Smithian, followed by a slight decline in the early Spathian, and a rise in the late Spathian (Fig. 4.3).

4.3.2 Ecospace Utilization and Ecosystem Engineering

In the late Changhsingian, the infaunal ecospace categories (modes of life) were sufficiently occupied, with a total of eleven categories utilized, primarily in the shallow and deep tiers, based on combined data from half-substage time bins in a carbonate ramp environment (Fig. 4.3). However, the number of modes of life declined significantly during the Griesbachian-Dienerian periods. A rebound in ecospace utilization occurred in the early Smithian, with thirteen ecospace categories being occupied (Fig. 4.3). The late Smithian marked the peak of infaunal ecospace colonization, with nineteen ecospace cubes utilized. In contrast, the early Spathian saw fourteen ecospace categories occupied, while the late Spathian showed a slight

Table 4.4 Ichnogenera and ecological proxies in half-substage time bins, collected from the studied sections in the carbonate ramp environment

Ichnogenus	Architectural design	Tier	Motility	Feeding	Interaction	Modification
Late Changhsingian						
Balanoglossites	Vertical single U- and Y-shaped burrows	Deep	Free	Non-specialized deposit/ predator	Excavation	Regenerator
Chondrites	Burrows with shaft or bunch with downwards radiating probes	Deep	Facultative	Chemosymbiosis	Compression	Gallery biodiffusive
Palaeophycus	Passively-filled horizontal to oblique burrows	Shallow	Facultative	Suspension/predator	Compression	Gallery biodiffusive
Phycosiphon	Burrows with horizontal spreiten	Intermediate	Free	Specialized deposit feeder	Backfill	Conveyor
Planolites	Simple, actively-filled (massive) horizontal to oblique structures	Shallow	Free	Non-specialized deposit	Backfill	Conveyor
Siphonichnus	Vertical unbranched burrows	Deep	Facultative	Suspension/predator	Compression	Gallery biodiffusive
Taenidium	Simple actively filled (meniscate) horizontal to oblique structures	Shallow	Free	Non-specialized deposit/ predator	Backfill	Conveyor
Thalassinoides	Maze and boxwork burrows	Deep	Free	Non-specialized deposit/ predator	Excavation	Regenerator
Zoophycos	Burrows with helicoidal spreiten	Deep	Facultative	Farming	Excavation	Conveyor
Early Griesbachian						
Planolites	Simple, actively filled (massive) horizontal to oblique structures	Shallow	Free	Non-specialized deposit	Backfill	Conveyor
Helminthoidichnites	Simple horizontal trails	Semi-infaunal	Free	Non-specialized deposit	Backfill	Conveyor
Late Griesbachian						

(continued)

Table 4.4 (continued)

Ichnogenus	Architectural design	Tier	Motility	Feeding	Interaction	Modification
Planolites	Simple, actively filled (massive) horizontal to oblique structures	Shallow	Free	Non-specialized deposit	Backfill	Conveyor
Helminthoidichnites	Simple horizontal trails	Semi-infaunal	Free	Non-specialized deposit	Backfill	Conveyor
Dienerian						
Planolites	Simple, actively filled (massive) horizontal to oblique structures	Shallow	Free	Non-specialized deposit	Backfill	Conveyor
Palaeophycus	Passively filled horizontal to oblique burrows	Shallow	Facultative	Suspension/predator	Compression	Gallery biodiffusive
Early Smithian						
Archaeonassa	Simple horizontal trails	Semi-infaunal	Free	Non-specialized deposit	Compression	Biodiffusive
Arenicolites	Vertical single U- and Y-shaped burrows	Intermediate	Facultative	Suspension	Compression	Gallery biodiffusive
Chondrites	Burrows with shaft or bunch with downwards radiating probes	Deep	Facultative	Chemosymbiosis	Compression	Gallery biodiffusive
Cochlichnus	Simple horizontal trails	Semi-infaunal	Free	Non-specialized deposit	Compression	Biodiffusive
Gyrochorte	Burrows with complex vertically oriented spreiten	Shallow	Free	Non-specialized deposit	Backfill	Conveyor
Helminthoidichnites	Simple horizontal trails	Semi-infaunal	Free	Non-specialized deposit	Backfill	Conveyor
Helminthopsis	Simple horizontal trails	Semi-infaunal	Free	Non-specialized deposit	Backfill	Conveyor
Kouphichnium	Trackways and scratch imprints	Surficial	Free	Non-specialized deposit/predator	Compression	Biodiffusive
Laevicyclus	Vertical unbranched burrows	Semi-infaunal	Facultative	Suspension/predator	Compression	Biodiffusive

Palaeophycus	Passively filled horizontal to oblique burrows	Shallow	Facultative	Suspension/predator	Compression	Gallery biodiffusive
Planolites	Simple, actively filled (massive) horizontal to oblique structures	Shallow	Free	Non-specialized deposit	Backfill	Conveyor
Phycodes	Horizontal burrows with horizontal to vertical branches	Semi-infaunal	Free	Specialized deposit feeder	Backfill	Conveyor
Thalassinoides	Maze and boxwork burrows	Deep	Free	Non-specialized deposit/predator	Excavation	Regenerator
Treptichnus	Horizontal burrows with horizontal to vertical branches	Semi-infaunal	Free	Specialized deposit feeder	Compression	Gallery biodiffusive
Late Smithian						
Arenicolites	Vertical single U- and Y-shaped burrows	Intermediate	Facultative	Suspension	Compression	Gallery biodiffusive
Archaeonassa	Simple horizontal trails	Semi-infaunal	Free	Non-specialized deposit	Compression	Biodiffusive
Arenituba	Radial to rosetted structures	Shallow	Free	Specialized deposit feeder	Compression	Gallery biodiffusive
Chondrites	Burrows with shaft or bunch with downwards radiating probes	Deep	Facultative	Chemosymbiosis	Compression	Gallery biodiffusive
Gordia	Simple horizontal trails	Semi-infaunal	Free	Non-specialized deposit	Backfill	Conveyor
Gyrochorte	Burrows with complex vertically oriented spreiten	Shallow	Free	Non-specialized deposit	Backfill	Conveyor
Helminthoidichnites	Simple horizontal trails	Semi-infaunal	Free	Non-specialized deposit	Backfill	Conveyor
Kouphichnium	Trackways and scratch imprints	Surficial	Free	Non-specialized deposit/predator	Compression	Biodiffusive

(continued)

Table 4.4 (continued)

Ichnogenus	Architectural design	Tier	Motility	Feeding	Interaction	Modification
Laevicyclus	Vertical unbranched burrows	Semi-infaunal	Facultative	Suspension/predator	Compression	Biodiffusive
Lockeia	Isolated and serial oval to almond-shaped burrows	Semi-infaunal	Free	Non-specialized deposit	Compression	Biodiffusive
Palaeophycus	Passively filled horizontal to oblique burrows	Shallow	Facultative	Suspension/predator	Compression	Gallery biodiffusive
Planolites	Simple, actively filled (massive) horizontal to oblique structures	Shallow	Free	Non-specialized deposit	Backfill	Conveyor
Phycodes	Horizontal burrows with horizontal to vertical branches	Semi-infaunal	Free	Specialized deposit feeder	Backfill	Conveyor
Phycosiphon	Burrows with horizontal spreiten	Intermediate	Free	Specialized deposit feeder	Backfill	Conveyor
Psammichnites	Complex actively filled horizontal structures	Shallow	Free	Specialized deposit feeder	Backfill	Conveyor
Sinusichnus	Maze and boxwork burrows	Intermediate	Free	Non-specialized deposit/predator	Excavation	Regenerator
Skolithos	Vertical unbranched burrows	Deep	Facultative	Suspension/predator	Compression	Gallery biodiffusive
Teichichnus	Horizontal burrows with simple vertically oriented spreiten	Intermediate	Free	Non-specialized deposit	Backfill	Conveyor
Thalassinoides	Maze and boxwork burrows	Deep	Free	Non-specialized deposit/predator	Excavation	Regenerator
Treptichnus	Horizontal burrows with horizontal to vertical branches	Semi-infaunal	Free	Specialized deposit feeder	Compression	Gallery biodiffusive
Early Spathian						
Arenicolites	Vertical single U- and Y-shaped burrows	Intermediate	Facultative	Suspension	Compression	Gallery biodiffusive
Arenituba	Radial to rosetted structures	Shallow	Free	Specialized deposit feeder	Compression	Gallery biodiffusive

Archaeonassa	Simple horizontal trails	Semi-infaunal	Free	Non-specialized deposit	Compression	Biodiffusive
Diplocraterion	Vertical single U- and Y-shaped burrows	Deep	Facultative	Suspension	Compression	Gallery biodiffusive
Gordia	Simple horizontal trails	Semi-infaunal	Free	Non-specialized deposit	Backfill	Conveyor
Gyrochorte	Burrows with complex vertically oriented spreiten	Shallow	Free	Non-specialized deposit	Backfill	Conveyor
Helminthoidichnites	Simple horizontal trails	Semi-infaunal	Free	Non-specialized deposit	Backfill	Conveyor
Laevicyclus	Vertical unbranched burrows	Semi-infaunal	Facultative	Suspension/predator	Compression	Biodiffusive
Palaeophycus	Passively filled horizontal to oblique burrows	Shallow	Facultative	Suspension/predator	Compression	Gallery biodiffusive
Planolites	Simple, actively filled (massive) horizontal to oblique structures	Shallow	Free	Non-specialized deposit	Backfill	Conveyor
Psammichnites	Complex actively filled horizontal structures	Shallow	Free	Specialized deposit feeder	Backfill	Conveyor
Rhizocorallium	Burrows with horizontal spreiten	Shallow	Facultative	Non-specialized deposit	Excavation	Gallery biodiffusive
Skolithos	Vertical unbranched burrows	Deep	Facultative	Suspension/predator	Compression	Gallery biodiffusive
Spongeliomorpha	Maze and boxwork burrows	Deep	Free	Non-specialized deposit/ predator	Excavation	Regenerator
Thalassinoides	Maze and boxwork burrows	Deep	Free	Non-specialized deposit/ predator	Excavation	Regenerator
Treptichnus	Horizontal burrows with horizontal to vertical branches	Semi-infaunal	Free	Specialized deposit feeder	Compression	Gallery biodiffusive

(continued)

Table 4.4 (continued)

Ichnogenus	Architectural design	Tier	Motility	Feeding	Interaction	Modification
Late Spathian						
Arenituba	Radial to rosetted structures	Shallow	Free	Specialized deposit feeder	Compression	Gallery biodiffusive
Arenicolites	Vertical single U- and Y-shaped burrows	Intermediate	Facultative	Suspension	Compression	Gallery biodiffusive
Circulichnis	Simple horizontal trails	Semi-infaunal	Free	Non-specialized deposit	Backfill	Conveyor
Diplocraterion	Vertical single U- and Y-shaped burrows	Deep	Facultative	Suspension	Compression	Gallery biodiffusive
Gordia	Simple horizontal trails	Semi-infaunal	Free	Non-specialized deposit	Backfill	Conveyor
Gyrochorte	Burrows with complex vertically oriented spreiten	Shallow	Free	Non-specialized deposit	Backfill	Conveyor
Helminthopsis	Simple horizontal trails	Semi-infaunal	Free	Non-specialized deposit	Backfill	Conveyor
Helminthoidichnites	Simple horizontal trails	Semi-infaunal	Free	Non-specialized deposit	Backfill	Conveyor
Laevicyclus	Vertical unbranched burrows	Semi-infaunal	Facultative	Suspension/predator	Compression	Biodiffusive
Palaeophycus	Passively filled horizontal to oblique burrows	Shallow	Facultative	Suspension/predator	Compression	Gallery biodiffusive
Planolites	Simple, actively filled (massive) horizontal to oblique structures	Shallow	Free	Non-specialized deposit	Backfill	Conveyor
Psammichnites	Complex actively filled horizontal structures	Shallow	Free	Specialized deposit feeder	Backfill	Conveyor
Rhizocorallium	Burrows with horizontal spreiten	Shallow	Facultative	Non-specialized deposit	Excavation	Gallery biodiffusive

Sinusichnus	Maze and boxwork burrows	Intermediate	Free	Non-specialized deposit/predator	Excavation	Regenerator
Skolithos	Vertical unbranched burrows	Deep	Facultative	Suspension/predator	Compression	Gallery biodiffusive
Spongeliomorpha	Maze and boxwork burrows	Deep	Free	Non-specialized deposit/predator	Excavation	Regenerator
Thalassinoides	Maze and boxwork burrows	Deep	Free	Non-specialized deposit/predator	Excavation	Regenerator
Treptichnus	Horizontal burrows with horizontal to vertical branches	Semi-infaunal	Free	Specialized deposit feeder	Compression	Gallery biodiffusive

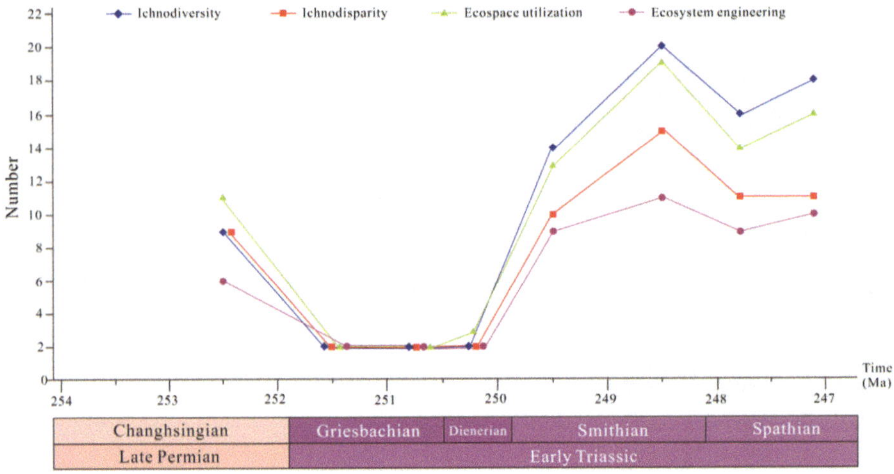

Fig. 4.3 Colonization of carbonate ramp environmental setting through the Late Permian to Early Triassic

increase to sixteen (Fig. 4.3). Overall, ecospace utilization decreased from the late Changhsingian to the early Griesbachian and remained at very low levels throughout the Griesbachian-Dienerian periods. It then rebounded significantly in the early Smithian, continued to rise to a peak in the late Smithian, experienced a slight decline in the early Spathian, and finally saw a modest increase in the late Spathian (Fig. 4.3).

More specifically, in surficial and semi-infaunal tiers, rare ecological space was utilized before and after the mass extinction, as well as during the whole Induan; only free motile, nonspecialized deposit feeders emerged in semi-infaunal tier after the mass extinction, the number of ecospace categories rebounded during the Smithian and Spathian. During the Smithian, freely motile, nonspecialized deposit feeders and predators occupied the surficial tier, while facultatively motile suspension feeders and predators, along with freely motile nonspecialized and specialized deposit feeders, utilized the shallow tier (Fig. 4.4a). In the shallow tier, four ecospace categories were occupied in the late Changhsingian, three of which vanished across the P-Tr mass extinction, and only freely motile, nonspecialized deposit feeders persisted. Ecospace utilization rebounded in the Smithian and Spathian. In the intermediate tier, only one ecospace cube existed pre-extinction, which disappeared across the P-Tr mass extinction; and the following Induan, the number of ecospace utilization rebounded and exceeded the pre-extinction level in the Smithian and Spathian (Fig. 4.4a). Freely motile, nonspecialized and specialized deposit feeders and predators, as well as facultatively motile, suspension feeders, occupied the intermediate tier during the late Smithian and Spathian. The ecospace of the deep tier was largely occupied in the late Changhsingian, with six ecospace categories. They were represented by facultatively motile, suspension feeders, predators, chemosymbiosis of meiofauna or bacteria, and farming meiofauna, as well as freely

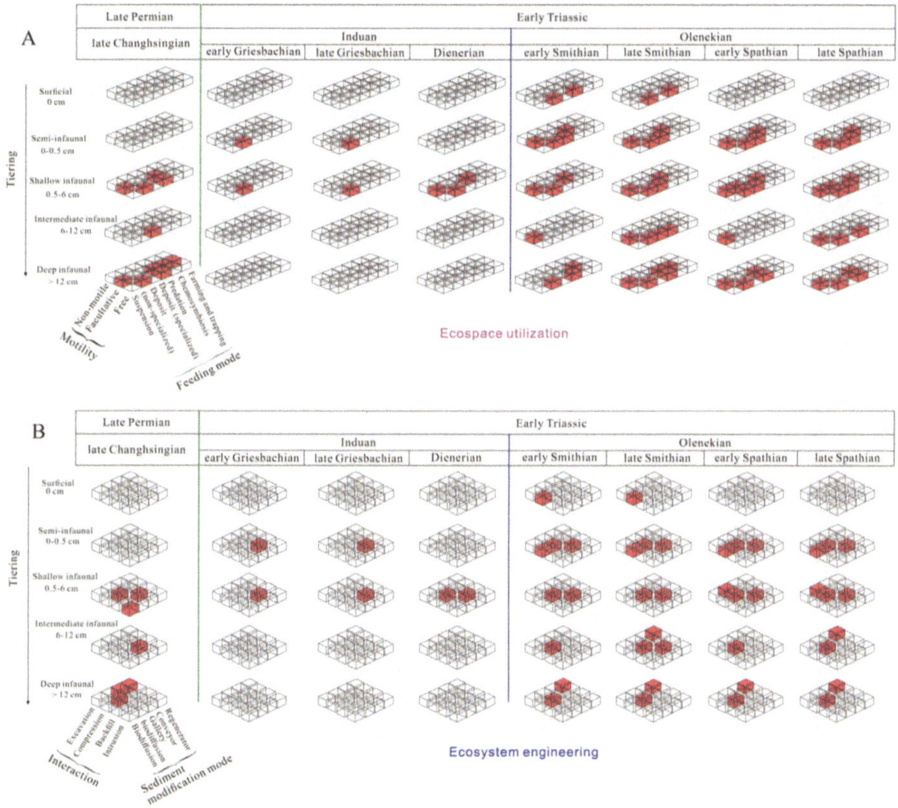

Fig. 4.4 Colonization of carbonate ramp environmental setting through the Late Permian to Early Triassic. (**a**) Cubes of ecospace utilization. (**b**) Cubes of ecosystem engineering

motile, nonspecialized deposit feeders and predators; however, all of which disappeared across the extinction and the aftermath. Ecospace in deep tier was re-utilized in the Smithian and rebounded but never exceeded the pre-extinction level in the late Smithan to Spathian. The type of ecospace categories in the late Early Triassic (Smithian-Spathian) was largely the same as those in the pre-extinction (Fig. 4.4a).

Although categories of ecospace utilization in the shallow and deep tiers were abundant in the late Changhsingian, their abundance was rather low, with only 2–3 occurrences (Table 4.5). The occurrence of categories of ecospace utilization in the Griesbachian and Dienerian was also rather sporadic. Freely motile, nonspecialized deposit feeders in shallow tier were the most common in the Dienerian, with 7 occurrences (Table 4.5). In contrast, the occurrence of categories of ecospace utilization became abundant during the Smithian and Spathian. For example, in the early Smithian, there were 23 occurrences in semi-infaunal tier, 31 in the shallow tier, 9 in the deep tier, among which freely motile, nonspecialized deposit feeders in semi-infaunal tier were the most abundant, with 12 occurrences, the second most were freely motile, nonspecialized deposit feeders, facultatively motile, suspension

Table 4.5 Abundance of ichnogenera in each mode of life (ecospace utilization) in carbonate ramp environmental setting in half-substage time bins

No.	Ecospace utilization	Late Changhsingian	Early Griesbachian	Late Griesbachian	Dienerian	Early Smithian	Late Smithian	Early Spathian	Late Spathian
1	Surficial + Free + Deposit (non-specialized)	0	0	0	0	2	2	0	0
2	Surficial + Free + Predator	0	0	0	0	2	2	0	0
3	Semi-infaunal + Free + Deposit (non-specialized)	0	0	3	0	12	12	13	10
4	Semi-infaunal + Free + Predator	0	0	0	0	0	0	0	0
5	Semi-infaunal + Facultative + Suspension	0	0	0	0	2	1	1	1
6	Semi-infaunal + Facultative + Predator	0	0	0	0	2	1	1	1
7	Semi-infaunal + Free + Deposit (specialized)	0	0	0	0	7	9	1	3
8	Semi-infaunal + Free + Farming	0	0	0	0	0	0	0	0
9	Shallow + Facultative + Suspension	0	0	0	3	10	11	11	6
10	Shallow + Facultative + Predator	0	0	0	3	10	11	11	6
11	Shallow + Free + Deposit (non-specialized)	3	4	2	7	10	10	19	3
12	Shallow + Free + Deposit (specialized)	0	0	0	0	1	3	6	2
13	Shallow + Free + Predator	1	0	0	0	0	2	0	0
14	Shallow + Facultative + Deposit (non-specialized)	0	0	0	0	0	0	6	6
15	Shallow + Free + Farming	0	0	0	0	0	0	0	0
16	Intermediate + Facultative + Suspension	0	0	0	0	2	4	6	1

17	Intermediate + Free + Deposit (non-specialized)	1	0	0	0	1	2	0	2
18	Intermediate + Free + Deposit (specialized)	1	0	0	0	0	2	0	0
19	Intermediate + Free + Predator	1	0	0	0	1	0	0	2
20	Deep + Facultative + Suspension	1	0	0	0	0	2	14	1
21	Deep + Facultative + Predator	1	0	0	0	0	2	8	1
22	Deep + Facultative + Deposit (specialized)	0	0	0	0	0	0	0	0
23	Deep + Free + Deposit (non-specialized)	2	0	0	0	4	5	13	6
24	Deep + Free + Predator	2	0	0	0	4	5	13	6
25	Deep + Facultative + Farming	1	0	0	0	0	0	0	0
26	Deep + Facultative + Chemosymbiosis	2	0	0	0	1	1	0	0

feeders and predators in shallow tier, each had 10 occurrences (Table 4.5). The occurrence of categories of ecospace utilization during the late Smithian was very similar to those in the early Smithian. During the early Spathian, many categories of ecospace utilization became abundant; freely motile, nonspecialized deposit feeders in semi-infaunal tier had 13 occurrences, they increased to 19 in the shallow tier and 13 in the deep tier, both facultatively motile, suspension feeders and predators owned 11 occurrences in the shallow tier, and they increased to 14 in the deep tier (Table 4.5). In all, the semi-, shallow, and deep tiers were well colonized by those freely motile, nonspecialized deposit feeders and facultatively motile, suspension feeders and predators during the late Early Triassic.

The number of mechanisms of substrate interaction and sediment modification (ecosystem engineering) was 7 in the late Changhsingian, 2 in the Griesbachian, 2 in the Dienerian, 9 in the early Smithian, 11 in the late Smithian, 9 in the early Spathian, and 10 in the late Spathian (Fig. 4.3). Generally, the number of ecosystem engineering sharply declined across the P-Tr boundary and kept at a low level in the Induan, then remarkably increased from the Dienerian to early Smithian and continued to rise to form a plateau in the Smithian-Spathian (Fig. 4.3). Ecosystem engineering styles existed in both shallow and deep tiers in the late Changhsingian, only in semi- and shallow tiers in the Griesbachian-Dienerian, and extended to most tiers during the Smithian-Spathian (Fig. 4.3). The types of ecosystem engineering were also consistent during Smithian-Spathian, which largely exceeded their pre-extinction counterpart.

More specifically, in the semi-infaunal tier, no ecosystem engineering occurred before the extinction. In the earliest Triassic, only one type of ecosystem engineering emerged: sediment conveyors, which interacted with sediment through backfilling. According to the Smithian and Spathian, additional types of ecosystem engineering appeared, including biodiffusors and gallery biodiffusors, which interacted with sediment through compression (Fig. 4.4b). In the shallow tier, three mechanisms of substrate interaction and sediment modification were present in the late Changhsingian: sediment conveyors (backfilling), biodiffusors (intrusion), and gallery biodiffusors (compression). Ecosystem engineering rebounded during the Smithian-Spathian, with the addition of gallery biodiffusors interacting through compression and excavation, alongside sediment conveyors. In the intermediate tier, only one type of ecosystem engineering existed before the extinction, but it disappeared entirely during the extinction and remained absent throughout the Induan. However, two to three types of ecosystem engineering re-emerged in the Smithian-Spathian. In the deep infaunal tier, three types of ecosystem engineering were present before the extinction: regenerators and sediment conveyors (excavation) and gallery biodiffusors (compression). All were eliminated during the P-Tr mass extinction and remained absent throughout the Induan. By the Smithian and Spathian, gallery biodiffusors (compression) and regenerators (excavation) had rebounded (Fig. 4.4b).

Similar to the pattern of ecospace utilization, although categories of ecosystem engineering in the shallow and deep tiers were relatively diverse, their abundance was rather low, with only 1–3 occurrences during the late Changhsingian (Table 4.6).

Table 4.6 Abundance of ichnogenera in each impact upon sediment (ecosystem engineering) in carbonate ramp environmental setting in half-substage time bins

No.	Ecosystem engineering	Late Changhsingian	Early Griesbachian	Late Griesbachian	Dienerian	Early Smithian	Late Smithian	Early Spathian	Late Spathian
1	Surficial + Compression + Biodiffusion	0	0	0	0	2	2	0	0
2	Semi-infaunal + Compression + Gallery Biodiffusion	0	0	0	0	6	5	1	3
3	Semi-infaunal + Backfill + Conveyor	0	0	3	0	8	12	12	10
4	Semi-infaunal + Excavation + Regenerator	0	0	0	0	0	0	0	0
5	Semi-infaunal + Compression + Biodiffusion	0	0	0	0	7	5	2	1
6	Shallow + Compression + Gallery Biodiffusion	1	0	0	3	10	12	14	7
7	Shallow + Backfill + Conveyor	3	4	2	7	11	10	22	4
8	Shallow + Intrusion + Biodiffusion	0	0	0	0	0	0	0	0
9	Shallow + Excavation + Gallery Biodiffusion	0	0	0	0	0	0	7	7
10	Intermediate + Compression + Gallery Biodiffusion	0	0	0	0	2	4	6	1
11	Intermediate + Backfill + Conveyor	1	0	0	0	0	4	0	0
12	Intermediate + Excavation + Regenerator	1	0	0	0	1	0	0	2

(continued)

Table 4.6 (continued)

No.	Ecosystem engineering	Late Changhsingian	Early Griesbachian	Late Griesbachian	Dienerian	Early Smithian	Late Smithian	Early Spathian	Late Spathian
13	Deep + Compression + Gallery Biodiffusive	3	0	0	0	1	3	14	1
14	Deep + Backfill + Conveyor	0	0	0	0	0	0	0	0
15	Deep + Excavation + Regenerator	2	0	0	0	4	5	13	6
16	Deep + Excavation + Conveyor	1	0	0	0	0	0	0	0

These were also the same during the Griesbachian. The occurrence of ecosystem engineering styles in the shallow tier was highest during the Dienerian, with 10 occurrences in total (Table 4.6). Sediment conveyors that interacted with sediment through backfilling were the most abundant, which had 7 occurrences. The occurrence of categories of ecosystem engineering became abundant during the Smithian and Spathian. In the early Smithian, ecosystem engineering occurrences were distributed as follows: 21 in the semi-infaunal tier, 22 in the shallow tier, and 5 in the deep tier. Among these, the most common types were gallery biodiffusors, which interacted with sediment through compression, and sediment conveyors, which interacted through backfilling. Each of these types accounted for 11 occurrences in the shallow tier (Table 4.6). In the late Smithian, ecosystem engineering occurrences were distributed as follows: 22 in the semi-infaunal tier, 22 in the shallow tier, 8 in the intermediate tier, and 8 in the deep tier. Among these, gallery biodiffusors interacting with sediment through compression were most common in the shallow tier, with 12 occurrences, while sediment conveyors interacting through backfilling were dominant in the semi-infaunal tier (12 occurrences) and also frequent in the shallow tier (10 occurrences) (Table 4.6). During the early Spathian, occurrences of ecosystem engineering reached their highest levels across all tiers: 15 in the semi-infaunal tier, 43 in the shallow tier, and 27 in the deep tier. Sediment conveyors interacting through backfilling were the most abundant, with 12 occurrences in the semi-infaunal tier and 22 in the shallow tier. Gallery biodiffusors interacting through compression were prominent in both the shallow and deep tiers, with 14 occurrences each, while regenerators interacting through excavation were frequent in the deep tier, with 13 occurrences. The patterns in the late Spathian were similar to those in the early Spathian, though with fewer occurrences in each tier.

4.4 Ecological Characteristics of Ichnofauna Across the Permian-Triassic Mass Extinction and Its Aftermaths in Carbonate Platform Setting

4.4.1 Ichnodiversity and Ichnodisparity

Four late Changhsingian ichnogenera were collected from the uppermost Changxing Formation of the Xiandaohu section, Hubei Province, South China. They were *Balanoglossites*, *Planolites*, *Thalassinoides*, and *Zoophycos*. Only simple, horizontal *Planolites* burrows were found in the strata of Griesbachian-Dienerian in age (Table 4.7). The richness of ichnogenera increased sharply in the early Smithian, plateaued throughout the Smithian to early Spathian, and then declined sharply in the late Spathian (Fig. 4.5). The Early Triassic trace fossils were collected from the limestone beds of the Daye Formation from the Lichuan section and from the muddy limestone beds of the Yelang and Jialingjiang Formations from the Xiejiacao section. The ichnogenera in the Smithian are *Bifungites*, *Circulichnis*, *Gyrochorte*,

Table 4.7 Ichnogenera and ecological proxies in half-substage time bins, collected from the studied sections in the carbonate platform environment

Ichnogenus	Architectural design	Tier	Motility	Feeding	Interaction	Modification
Late Changhsingian						
Balanoglossites	Vertical single U- and Y-shaped burrows	Deep	Free	Non-specialized deposit/ predator	Excavation	Regenerator
Planolites	Simple, actively filled (massive) horizontal to oblique structures	Shallow	Free	Non-specialized deposit	Backfill	Conveyor
Thalassinoides	Maze and boxwork burrows	Deep	Free	Non-specialized deposit/ predator	Excavation	Regenerator
Zoophycos	Burrows with helicoidal spreiten	Deep	Facultative	Farming	Excavation	Conveyor
Early Griesbachian						
Planolites	Simple, actively filled (massive) horizontal to oblique structures	Shallow	Free	Non-specialized deposit	Backfill	Conveyor
Late Griesbachian						
Planolites	Simple, actively-filled (massive) horizontal to oblique structures	Shallow	Free	Non-specialized deposit	Backfill	Conveyor
Dienerian						
Bifungites	Dumbbell and arrow-shaped burrows	Shallow	Facultative	Suspension	Compression	Gallery biodiffusive
Phycodes	Horizontal burrows with horizontal to vertical branches	Semi-infaunal	Free	Specialized deposit feeder	Backfill	Conveyor
Planolites	Simple, actively filled (massive) horizontal to oblique structures	Shallow	Free	Non-specialized deposit	Backfill	Conveyor
Early Smithian						
Bifungites	Dumbbell and arrow-shaped burrows	Shallow	Facultative	Suspension	Compression	Gallery biodiffusive
Circulichnis	Simple horizontal trails	Semi-infaunal	Free	Non-specialized deposit	Backfill	Conveyor

Gyrochorte	Burrows with complex vertically oriented spreiten	Shallow	Free	Non-specialized deposit	Backfill	Conveyor
Helminthopsis	Simple horizontal trails	Semi-infaunal	Free	Non-specialized deposit	Backfill	Conveyor
Helminthoidichnites	Simple horizontal trails	Semi-infaunal	Free	Non-specialized deposit	Backfill	Conveyor
Palaeophycus	Passively filled horizontal to oblique burrows	Shallow	Facultative	Suspension/predator	Compression	Gallery biodiffusive
Phycodes	Horizontal burrows with horizontal to vertical branches	Semi-infaunal	Free	Specialized deposit feeder	Backfill	Conveyor
Planolites	Simple, actively filled (massive) horizontal to oblique structures	Shallow	Free	Non-specialized deposit	Backfill	Conveyor
Sinusichnus	Maze and boxwork burrows	Intermediate	Free	Non-specialized deposit/predator	Excavation	Regenerator
Psammichnites	Complex actively filled horizontal structures	Shallow	Free	Specialized deposit feeder	Backfill	Conveyor
Skolithos	Vertical unbranched burrows	Deep	Facultative	Suspension/predator	Compression	Gallery biodiffusive
Thalassinoides	Maze and boxwork burrows	Deep	Free	Non-specialized deposit/predator	Excavation	Regenerator
Treptichnus	Horizontal burrows with horizontal to vertical branches	Semi-infaunal	Free	Specialized deposit feeder	Compression	Gallery biodiffusive
Late Smithian						
Arenicolites	Vertical single U- and Y-shaped burrows	Intermediate	Facultative	Suspension	Compression	Gallery biodiffusive
Arenituba	Radial to rosetted structures	Shallow	Free	Specialized deposit feeder	Compression	Gallery biodiffusive

(continued)

Table 4.7 (continued)

Ichnogenus	Architectural design	Tier	Motility	Feeding	Interaction	Modification
Circulichnis	Simple horizontal trails	Semi-infaunal	Free	Non-specialized deposit	Backfill	Conveyor
Cochlichnus	Simple horizontal trails	Semi-infaunal	Free	Non-specialized deposit	Compression	Biodiffusive
Gordia	Simple horizontal trails	Semi-infaunal	Free	Non-specialized deposit	Backfill	Conveyor
Gyrochorte	Burrows with complex vertically oriented spreiten	Shallow	Free	Non-specialized deposit	Backfill	Conveyor
Helminthopsis	Simple horizontal trails	Semi-infaunal	Free	Non-specialized deposit	Backfill	Conveyor
Laevicyclus	Vertical unbranched burrows	Semi-infaunal	Facultative	Suspension/predator	Compression	Biodiffusive
Palaeophycus	Passively filled horizontal to oblique burrows	Shallow	Facultative	Suspension/predator	Compression	Gallery biodiffusive
Planolites	Simple, actively filled (massive) horizontal to oblique structures	Shallow	Free	Non-specialized deposit	Backfill	Conveyor
Psammichnites	Complex actively filled horizontal structures	Shallow	Free	Specialized deposit feeder	Backfill	Conveyor
Phycodes	Horizontal burrows with horizontal to vertical branches	Semi-infaunal	Free	Specialized deposit feeder	Backfill	Conveyor
Thalassinoides	Maze and boxwork burrows	Deep	Free	Non-specialized deposit/predator	Excavation	Regenerator
Treptichnus	Horizontal burrows with horizontal to vertical branches	Semi-infaunal	Free	Specialized deposit feeder	Compression	Gallery biodiffusive
Early Spathian						
Asteriacites	Pentameral-shaped imprints and burrows	Surficial	Free	Non-specialized deposit/predator	Compression	Biodiffusive

Balanoglossites	Vertical single U- and Y-shaped burrows	Deep	Free	Non-specialized deposit/ predator	Excavation	Regenerator
Biformites	Trackways and scratch imprints	Surficial	Free	Non-specialized deposit/ predator	Compression	Biodiffusive
Diplocraterion	Vertical single U- and Y-shaped burrows	Deep	Facultative	Suspension	Compression	Gallery biodiffusive
Gyrochorte	Burrows with complex vertically oriented spreiten	Shallow	Free	Non-specialized deposit	Backfill	Conveyor
Helminthopsis	Simple horizontal trails	Semi-infaunal	Free	Non-specialized deposit	Backfill	Conveyor
Laevicyclus	Vertical unbranched burrows	Semi-infaunal	Facultative	Suspension/predator	Compression	Biodiffusive
Lockeia	Isolated and serial oval to almond-shaped burrows	Semi-infaunal	Free	Non-specialized deposit	Compression	Biodiffusive
Palaeophycus	Passively filled horizontal to oblique burrows	Shallow	Facultative	Suspension/predator	Compression	Gallery biodiffusive
Phycodes	Horizontal burrows with horizontal to vertical branches	Semi-infaunal	Free	Specialized deposit feeder	Backfill	Conveyor
Planolites	Simple, actively filled (massive) horizontal to oblique structures	Shallow	Free	Non-specialized deposit	Backfill	Conveyor
Rhizocorallium	Burrows with horizontal spreiten	Shallow	Facultative	Non-specialized deposit	Excavation	Gallery biodiffusive
Sinusichnus	Maze and boxwork burrows	Intermediate	Free	Non-specialized deposit/ predator	Excavation	Regenerator
Skolithos	Vertical unbranched burrows	Deep	Facultative	Suspension/predator	Compression	Gallery biodiffusive

(continued)

Table 4.7 (continued)

Ichnogenus	Architectural design	Tier	Motility	Feeding	Interaction	Modification
Thalassinoides	Maze and boxwork burrows	Deep	Free	Non-specialized deposit/ predator	Excavation	Regenerator
Treptichnus	Horizontal burrows with horizontal to vertical branches	Semi-infaunal	Free	Specialized deposit feeder	Compression	Gallery biodiffusive
Late Spathian						
Palaeophycus	Passively filled horizontal to oblique burrows	Shallow	Facultative	Suspension/predator	Compression	Gallery biodiffusive
Planolites	Simple, actively filled (massive) horizontal to oblique structures	Shallow	Free	Non-specialized deposit	Backfill	Conveyor
Thalassinoides	Maze and boxwork burrows	Deep	Free	Non-specialized deposit/ predator	Excavation	Regenerator

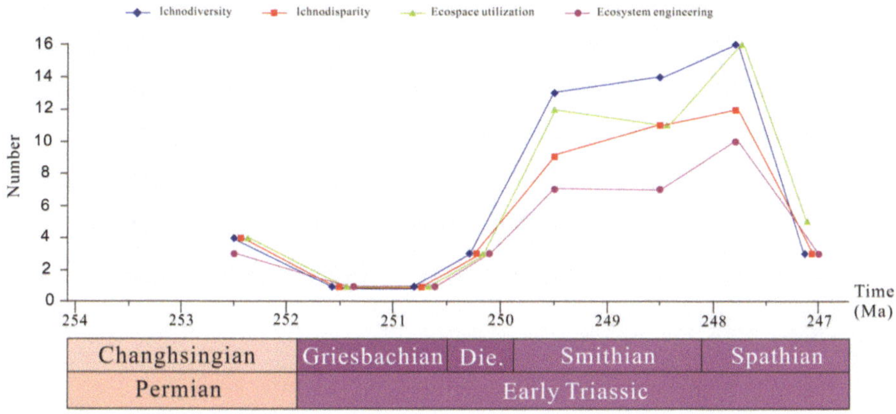

Fig. 4.5 Colonization of carbonate platform environmental setting through the Late Permian to Early Triassic

Helminthopsis, *Helminthoidichnites*, *Palaeophycus*, *Phycodes*, *Planolites*, *Sinusichnus*, *Psammichnites*, *Skolithos*, *Thalassinoides*, and *Treptichnus*. The ichnogenera in the Spathian are *Asteriacites*, *Balanoglossites*, *Biformites*, *Diplocraterion*, *Gyrochorte*, *Helminthopsis*, *Laevicyclus*, *Lockeia*, *Palaeophycus*, *Phycodes*, *Planolites*, *Rhizocorallium*, *Sinusichnus*, *Skolithos*, *Thalassinoides*, and *Treptichnus* (Table 4.7). Ichnodisparity showed a very similar tendency with ichnodiversity, which also decreased across the P-Tr mass extinction and was rare during the Griesbachian-Dienerian, and sharply increased in the Smithian, followed by a steep decline in the late Spathian (Fig. 4.5).

4.4.2 Ecospace Utilization and Ecosystem Engineering

Only four infaunal ecospace categories (modes of life) were occupied in the late Changhsingian, based on combined data from half-substage time bins in a carbonate platform environment (Fig. 4.5). The number of modes of life remained very low during the Griesbachian-Dienerian periods. The number of ecospace categories is twelve in the early Smithian, eleven in the late Smithian, sixteen in the early Spathian, and five in the late Spathian (Fig. 4.5). Ecospace utilization therefore generally decreased from the late Changhsingian to early Griesbachian and kept at a very low level during the Griesbachian-Dienerian periods, then rebounded with a sharp increase in the early Smithian, and continued to increase to the early Spathian, followed by a sharp decrease in the late Spathian (Fig. 4.5). More specifically, ecological categories of free motile, deposit feeders and predators, and facultatively motile, suspension feeders and predators in the semi-, shallow, intermediate, and several other infaunal ecospace categories began to occur in the Smithian-Spathian (Fig. 4.6a). Of the deep tier, three modes of life occurred in the late Changhsingian,

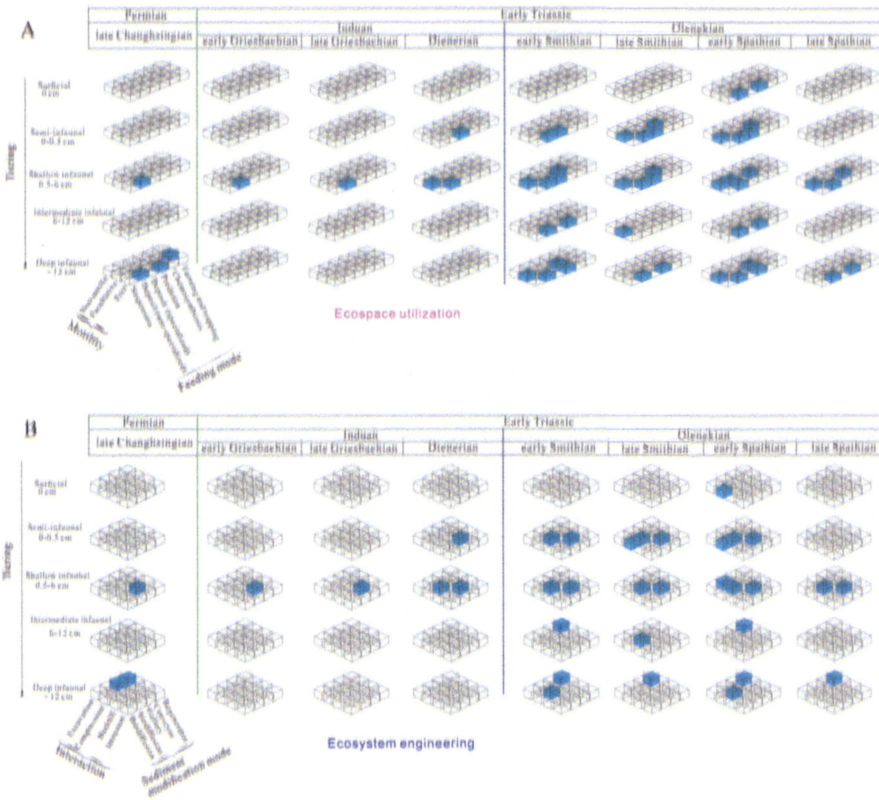

Fig. 4.6 Colonization of carbonate platform environmental setting through the Late Permian to Early Triassic. (**a**) Cubes of ecospace utilization. (**b**) Cubes of ecosystem engineering

and they vanished across the P-Tr mass extinction, as well as the whole Induan (Fig. 4.6a). The ecospace of the deep tier was occupied once again in the Smithian; ecological categories were dominated by freely motile, nonspecialized deposit feeders and predators, as well as facultatively motile, suspension feeders and predators (Fig. 4.6a).

The most abundant categories of ecospace utilization were deep-tiering, freely motile, nonspecialized deposit feeders and predators in the late Changhsingian. Other ecospace utilization categories did not exist. All categories of ecospace utilization did not exist or were just sporadically distributed in each infaunal tiering level during the Griesbachian-Dienerian (Table 4.8). Facultatively motile, suspension feeders of the shallow infaunal tier were most common in the early Smithian, with six occurrences (Table 4.8). Freely motile, nonspecialized deposit feeders in the shallow tier were the most abundant ecospace utilization category in the late Smithian, with nine occurrences. The second most abundant category was also freely motile, nonspecialized deposit feeders, but in the semi-infaunal tier, with seven occurrences (Table 4.8). By the late Spathian, freely motile, nonspecialized

Table 4.8 Abundance of ichnogenera in each mode of life (ecospace utilization) in carbonate platform environmental setting in half-substage time bins

No.	Ecospace utilization	Late Changhsingian	Early Griesbachian	Late Griesbachian	Dienerian	Early Smithian	Late Smithian	Early Spathian	Late Spathian
1	Surficial + Free + Deposit (non-specialized)	0	0	0	0	0	0	3	0
2	Surficial + Free + Predator	0	0	0	0	0	0	3	0
3	Semi-infaunal + Free + Deposit (non-specialized)	0	0	0	0	2	7	6	0
4	Semi-infaunal + Free + Predator	0	0	0	0	0	0	0	0
5	Semi-infaunal + Facultative + Suspension	0	0	0	0	0	2	1	0
6	Semi-infaunal + Facultative + Predator	0	0	0	0	0	2	1	0
7	Semi-infaunal + Free + Deposit (specialized)	0	0	0	2	2	4	4	0
8	Semi-infaunal + Free + Farming	0	0	0	0	0	0	0	0
9	Shallow + Facultative + Suspension	0	0	0	2	6	2	5	0
10	Shallow + Facultative + Predator	0	0	0	0	3	2	5	0
11	Shallow + Free + Deposit (non-specialized)	1	2	2	3	2	9	8	1
12	Shallow + Free + Deposit (specialized)	0	0	0	0	2	3	0	0
13	Shallow + Free + Predator	0	0	0	0	0	0	0	0
14	Shallow + Facultative + Deposit (non-specialized)	0	0	0	0	0	0	3	0

(continued)

Table 4.8 (continued)

No.	Ecospace utilization	Late Changhsingian	Early Griesbachian	Late Griesbachian	Dienerian	Early Smithian	Late Smithian	Early Spathian	Late Spathian
15	Shallow + Free + Farming	0	0	0	0	0	0	0	0
16	Intermediate + Facultative + Suspension	0	0	0	0	0	1	0	0
17	Intermediate + Free + Deposit (non-specialized)	0	0	0	0	3	0	4	0
18	Intermediate + Free + Deposit (specialized)	0	0	0	0	0	0	0	0
19	Intermediate + Free + Predator	0	0	0	0	3	0	4	0
20	Deep + Facultative + Suspension	0	0	0	0	1	0	4	0
21	Deep + Facultative + Predator	0	0	0	0	1	0	2	0
22	Deep + Facultative + Deposit (specialized)	0	0	0	0	0	0	0	0
23	Deep + Free + Deposit (non-specialized)	3	0	0	0	3	3	6	1
24	Deep + Free + Predator	3	0	0	0	3	3	6	1
25	Deep + Facultative + Farming	1	0	0	0	0	0	0	0
26	Deep + Facultative + Chemosymbiosis	0	0	0	0	0	0	0	0

deposit feeders remained the most abundant across the semi-infaunal, shallow, and deep tiers (Table 4.8).

Based on combined data from half-substage time bins in a carbonate platform environment, the number of potential combinations of mechanisms for substrate interaction and sediment modification (ecosystem engineering) varied across periods: 3 occurred in the late Changhsingian, 1 in the Griesbachian, 3 in the Dienerian, 7 in the early Smithian, 7 in the late Smithian, and 10 in the early Spathian (Fig. 4.5). Generally, the number of ecosystem engineering of most infaunal tiers across the P-Tr boundary was very low (Fig. 4.5). Ecosystem engineering remarkably increased from the Dienerian to early Smithian and continued to rise to reach a peak in the early Spathian. Ecosystem engineering styles existed in both shallow and deep tiers in the late Changhsingian, only in shallow tier in the Griesbachian, in semi- and shallow tiers in the Dienerian, and extended to most tiers during the Smithian-Spathian (Fig. 4.5). The types of ecosystem engineering were also consistent during this period.

More specifically, in semi-infaunal tier, no ecosystem engineering occurred before and after the extinction; only one type of ecosystem engineering occurred in the Dienerian, which was sediment conveyors that interacted with sediment through backfilling (Fig. 4.6b). Besides the sediment conveyors, biodiffusors and gallery biodiffusors interacted with sediment through compression in the Smithian and Spathian (Fig. 4.6b). In the shallow tier, sediment conveyors interacted with sediment through backfilling in the late Changhsingian, and they persisted into the early Griesbachian. Besides the sediment conveyors, gallery biodiffusors interacted with sediment through compression and excavation in the Smithian and Spathian (Fig. 4.6b). In the intermediate tier, no ecosystem engineering occurred before and after the extinction, and only one occurred in the Smithian-Spathian. In the deep infaunal tier, regenerators and sediment conveyors interacted with sediment through excavation during the late Changhsingian. However, all ecosystem engineering activities were eliminated during the P-Tr mass extinction and remained absent throughout the entire Induan. New types of ecosystem engineering emerged in the Smithian and Spathian, including gallery biodiffusors interacting with sediment through compression, as well as regenerators and sediment conveyors interacting through excavation (Fig. 4.6b).

The occurrence of ecosystem engineering modes was extremely low in the late Changhsingian and Griesbachian (Table 4.9). The occurrence of ecosystem engineering styles in the shallow tier is highest during the Dienerian. Sediment conveyors that interacted with sediment through backfilling had 4 occurrences (Table 4.9). In the early Smithian, ecosystem engineering modes in the shallow tier were the most abundant, with 10 occurrences. The most common types were gallery biodiffusors interacting with sediment through compression and sediment conveyors interacting through backfilling. In late Smithian, both semi-infaunal and shallow tiers owned the most abundant ecosystem engineering styles; the former had 13 occurrences, while the latter had 11 occurrences (Table 4.9). Sediment conveyors that interacted with sediment through backfilling were the most abundant both in semi- and shallow tiers; each had 7 occurrences. In the early Spathian, ecosystem

Table 4.9 Abundance of ichnogenera in each impact upon sediment (ecosystem engineering) in carbonate platform environmental setting in half-substage time bins

No.	Ecosystem engineering	Late Changhsingian	Early Griesbachian	Late Griesbachian	Dienerian	Early Smithian	Late Smithian	Early Spathian	Late Spathian
1	Surficial + Compression + Biodiffusion	0	0	0	0	0	0	3	0
2	Semi-infaunal + Compression + Gallery Biodiffusion	0	0	0	0	1	3	1	0
3	Semi-infaunal + Backfill + Conveyor	0	0	0	2	3	7	6	0
4	Semi-infaunal + Excavation + Regenerator	0	0	0	0	0	0	0	0
5	Semi-infaunal + Compression + Biodiffusion	0	0	0	0	0	3	4	0
6	Shallow + Compression + Gallery Biodiffusion	0	0	0	2	6	4	5	0
7	Shallow + Backfill + Conveyor	1	3	1	4	4	7	8	1
8	Shallow + Intrusion + Biodiffusion	0	0	0	0	0	0	0	0
9	Shallow + Excavation + Gallery Biodiffusion	0	0	0	0	0	0	3	0
10	Intermediate + Compression + Gallery Biodiffusion	0	0	0	0	0	1	0	0
11	Intermediate + Backfill + Conveyor	0	0	0	0	0	0	0	0
12	Intermediate + Excavation + Regenerator	0	0	0	0	3	0	4	0
13	Deep + Compression + Gallery Biodiffusive	0	0	0	0	1	0	4	0

14	Deep + Backfill + Conveyor	0	0	0	0	0	0	0	0
15	Deep + Excavation + Regenerator	3	0	0	0	3	3	6	1
16	Deep + Excavation + Conveyor	1	0	0	0	0	0	0	0

engineering styles in all tiers became very abundant, 3 occurrences in the surficial tier, 11 in the semi-infaunal tier, 16 in the shallow tier, 4 in the intermediate tier, and 10 in the deep tier, among which sediment conveyors that interacted with sediment through backfilling were the most abundant both in semi- and shallow tiers, while regenerators that interacted with sediment through excavation were the most common in deep tier (Table 4.9).

4.5 Ecological Characteristics of Ichnofauna Across the P-Tr Mass Extinction and Its Aftermaths in Siliciclastic Lower Shoreface Setting

4.5.1 Ichnodiversity and Ichnodisparity

There are seven ichnogenera during the Late Permian; they are *Phycodes*, *Planolites*, *Protovirgularia*, *Rhizocorallium*, *Rosselia*, *Skolithos*, and *Thalassinoides* (Table 4.10). These fossil materials were largely collected from the upper part of the Talung Formation from the Penglaitan section. The richness of ichnogenera displayed a slight increase across the Permian-Triassic mass extinction, and this increase continued to nine ichnogenera in the late Griesbachian, which was also the peak of the ichnodiversity during the P-Tr transition (Fig. 4.7). The number of ichnogenera started to drop in the Dienerian and fell to only four ichnogenera during the Smithian (Olenekian Stage). The Early Triassic trace fossils were collected from sandstone beds of the Xiahuancang Formation at the Zhihema section. This section is part of the southern Qilian Basin, which was situated on the northern margin of the Paleotethys Ocean during the Early Triassic. The ichnogenera during the Griesbachian are *Arenicolites*, *Diplocraterion*, *Lockeia*, *Palaeophycus*, *Planolites*, *Rhizocorallium*, *Skolithos*, *Thalassinoides*, *Bifungites*, *Gyrochorte*, *Phycodes*, *Skolithos*, and *Treptichnus*, among which, vertical burrow forms such as *Skolithos* and *Diplocraterion* are most abundant in the outcrop in the early Griesbachian, while horizontal burrows such as *Gyrochorte*, and *Palaeophycus* are most abundant in the late Griesbachian (Table 4.10). Ichnodisparity showed a very similar tendency with ichnodiversity, which also increased across the P-Tr mass extinction and declined during the Smithian (Fig. 4.7). In addition, the type of architectural designs differed obviously between the Late Permian and the Early Triassic (Fig. 4.7). The ichnodisparity across the P-Tr mass extinction are listed in Table 4.10 in detail.

4.5.2 Ecospace Utilization and Ecosystem Engineering

Eight out of 90 potential infaunal ecospace categories (modes of life) were occupied in the late Changhsingian, based on combined data from half-substage time bins in a lower shoreface environmental setting (Fig. 4.7). The number of occupied modes

Table 4.10 Ichnogenera and ecological proxies in half-substage time bins, collected from the studied sections in the lower shoreface environment

Ichnogenus	Architectural design	Tier	Motility	Feeding	Interaction	Modification
Late Changhsingian						
Phycodes	Horizontal burrows with horizontal to vertical branches	Semi-infaunal	Free	Specialized deposit feeder	Backfill	Conveyor
Planolites	Simple, actively filled (massive) horizontal to oblique structures	Shallow	Free	Non-specialized deposit	Backfill	Conveyor
Protovirgularia	Chevronate trails	Shallow	Free	Non-specialized deposit	Intrusion	Biodiffusive
Rhizocorallium	Burrows with horizontal spreiten	Shallow	Facultative	Non-specialized deposit	Excavation	Gallery biodiffusive
Rosselia	Vertical concentrically filled burrows	Intermediate	Facultative	Suspension	Compression	Gallery biodiffusive
Skolithos	Vertical unbranched burrows	Deep	Facultative	Suspension/predator	Compression	Gallery biodiffusive
Thalassinoides	Maze and boxwork burrows	Deep	Free	Non-specialized deposit/predator	Excavation	Regenerator
Early Griesbachian						
Arenicolites	Vertical single U- and Y-shaped burrows	Intermediate	Facultative	Suspension	Compression	Gallery biodiffusive
Diplocraterion	Vertical single U- and Y-shaped burrows	Deep	Facultative	Suspension	Compression	Gallery biodiffusive
Lockeia	Isolated and serial oval to almond-shaped burrows	Semi-infaunal	Free	Non-specialized deposit	Compression	Biodiffusive
Palaeophycus	Passively filled horizontal to oblique burrows	Shallow	Facultative	Suspension/predator	Compression	Gallery biodiffusive

(continued)

Table 4.10 (continued)

Ichnogenus	Architectural design	Tier	Motility	Feeding	Interaction	Modification
Planolites	Simple, actively filled (massive) horizontal to oblique structures	Shallow	Free	Non-specialized deposit	Backfill	Conveyor
Rhizocorallium	Burrows with horizontal spreiten	Shallow	Facultative	Non-specialized deposit	Excavation	Gallery biodiffusive
Skolithos	Vertical unbranched burrows	Deep	Facultative	Suspension/predator	Compression	Gallery biodiffusive
Thalassinoides	Maze and boxwork burrows	Deep	Free	Non-specialized deposit/ predator	Excavation	Regenerator
Late Griesbachian						
Bifungites	Dumbbell and arrow-shaped burrows	Shallow	Facultative	Suspension	Compression	Gallery biodiffusive
Diplocraterion	Vertical single U- and Y-shaped burrows	Deep	Facultative	Suspension	Compression	Gallery biodiffusive
Gyrochorte	Burrows with complex vertically oriented spreiten	Shallow	Free	Non-specialized deposit	Backfill	Conveyor
Palaeophycus	Passively filled horizontal to oblique burrows	Shallow	Facultative	Suspension/predator	Compression	Gallery biodiffusive
Phycodes	Horizontal burrows with horizontal to vertical branches	Semi-infaunal	Free	Specialized deposit feeder	Backfill	Conveyor
Planolites	Simple, actively filled (massive) horizontal to oblique structures	Shallow	Free	Non-specialized deposit	Backfill	Conveyor
Rhizocorallium	Burrows with horizontal spreiten	Shallow	Facultative	Non-specialized deposit	Excavation	Gallery biodiffusive
Skolithos	Vertical unbranched burrows	Deep	Facultative	Suspension/predator	Compression	Gallery biodiffusive
Treptichnus	Horizontal burrows with horizontal to vertical branches	Semi-infaunal	Free	Specialized deposit feeder	Compression	Gallery biodiffusive
Dienerian						

Cochlichnus	Simple horizontal trails	Semi-infaunal	Free	Non-specialized deposit	Compression	Biodiffusive
Gyrochorte	Burrows with complex vertically oriented spreiten	Shallow	Free	Non-specialized deposit	Backfill	Conveyor
Planolites	Simple, actively filled (massive) horizontal to oblique structures	Shallow	Free	Non-specialized deposit	Backfill	Conveyor
Palaeophycus	Passively filled horizontal to oblique burrows	Shallow	Facultative	Suspension/predator	Compression	Gallery biodiffusive
Rhizocorallium	Burrows with horizontal spreiten	Shallow	Facultative	Non-specialized deposit	Excavation	Gallery biodiffusive
Skolithos	Vertical unbranched burrows	Deep	Facultative	Suspension/predator	Compression	Gallery biodiffusive
Thalassinoides	Maze and boxwork burrows	Deep	Free	Non-specialized deposit/predator	Excavation	Regenerator
Early Smithian						
Palaeophycus	Passively filled horizontal to oblique burrows	Shallow	Facultative	Suspension/predator	Compression	Gallery biodiffusive
Rhizocorallium	Burrows with horizontal spreiten	Shallow	Facultative	Non-specialized deposit	Excavation	Gallery biodiffusive
Skolithos	Vertical unbranched burrows	Deep	Facultative	Suspension/predator	Compression	Gallery biodiffusive
Thalassinoides	Maze and boxwork burrows	Deep	Free	Non-specialized deposit/predator	Excavation	Regenerator

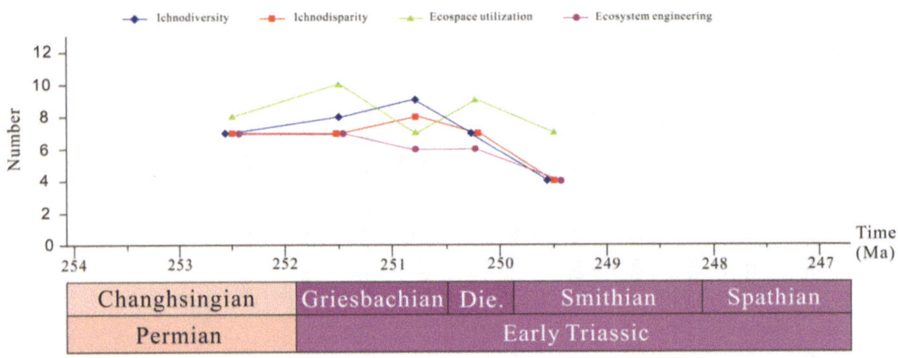

Fig. 4.7 Colonization of siliciclastic lower shoreface environmental setting through the Late Permian to Early Triassic. Plot of changes of the four metrics in siliciclastic lower shoreface settings

of life increased to ten by the early Griesbachian, seven in the late Griesbachian, nine in the Dienerian, and seven in the early Simthian (Fig. 4.7). Ecospace utilization therefore generally increases from the late Changhsingian to early Griesbachian, followed by a decrease to the late Griesbachian, then rebound during the Dienerian, followed by another decline during the early Smithian (Fig. 4.7). More specifically, ecological categories of free motile, deposit feeders in the semi-infaunal tier level persisted across the P-Tr extinction and the aftermath. In the shallow-infaunal tier, both facultatively and freely motile, nonspecialized deposit feeders persisted across the P-Tr extinction (Fig. 4.8a). In this tiering level, there were two new ecological categories emerged in the aftermath of the extinction; they are facultatively motile, suspension feeders and facultatively motile, predators (Fig. 4.8a). In the shallow-tiering level, all ecological categories remained in the whole Induan stage (Fig. 4.8a). All facultatively motile categories were retained, while freely motile, nonspecialized deposit feeders disappeared in the Early Smithian. There was only one infaunal ecospace category in the intermediate tiering level, the facultatively motile, suspension feeders in the late Changhsingian, but it still persisted to the early Griesbachian and disappeared thereafter (Fig. 4.8a). In the deep infaunal tier, four ecospace categories were occupied in the late Changhsingian, and they did not disappear across the P-Tr extinction. Freely motile, nonspecialized deposit feeders and predators disappeared in the late Griesbachian and re-occurred through the Dienerian to Smithian (Fig. 4.8a).

The most abundant categories of ecospace utilization were shallow-tiering, freely motile, nonspecialized deposit feeders, and deep-tiering ecospace categories in the late Changhsingian (Table 4.11). Other ecospace utilization categories were rather sporadic. For shallow infaunal tiers, ecospace utilization decreased in occurrence across the P-Tr extinction, particularly for freely motile, nonspecialized deposit feeders (Table 4.11). In contrast, deep infaunal tiers saw an increase in occurrence during the same period, especially for facultatively motile suspension feeders. Other deep-tiering ecospace categories experienced only a slight decline

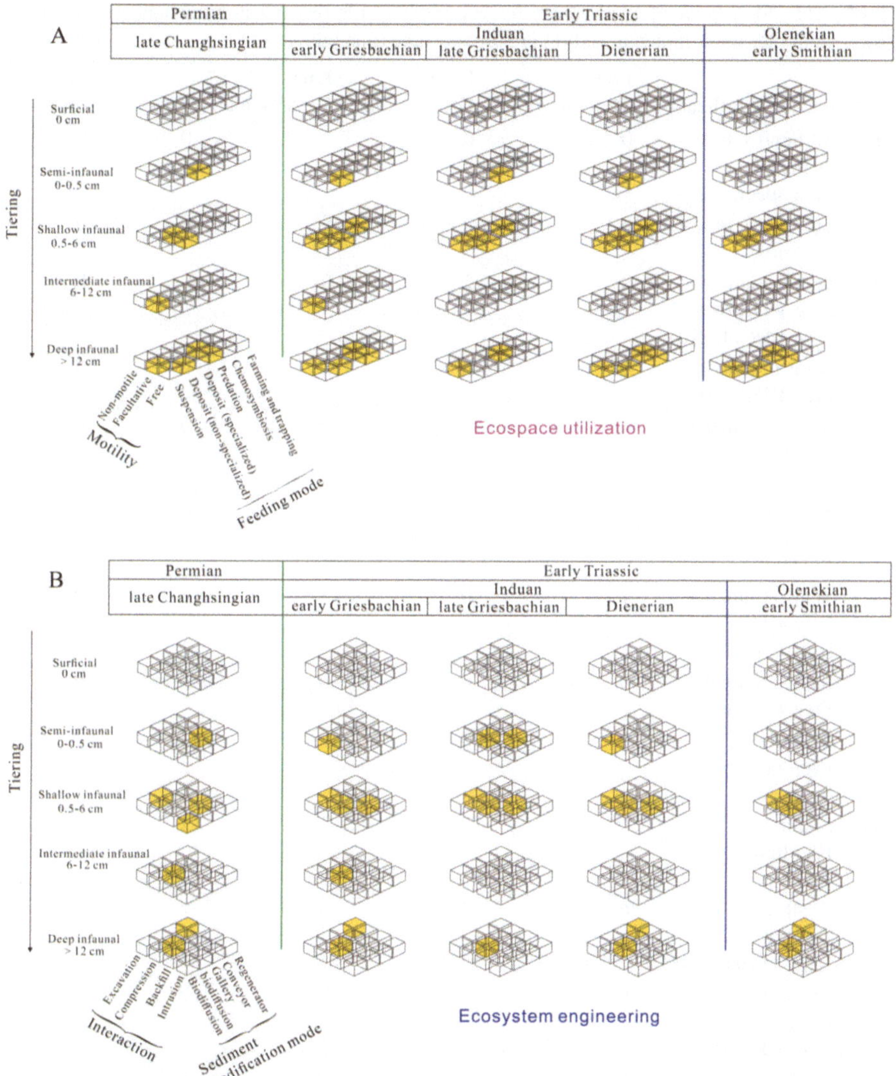

Fig. 4.8 Colonization of siliciclastic lower shoreface environmental setting through the Late Permian to Early Triassic. (**a**) Cubes of ecospace utilization. (**b**) Cubes of ecosystem engineering

across the P-Tr extinction (Table 4.11). Ecological categories of freely motile, non-specialized deposit feeders in the shallow infaunal tier were sporadic during the Griesbachian-Dienerian but rebounded in the Smithian. In contrast, ecological categories of deep infaunal tiers rebounded earlier, in the early Griesbachian, and remained relatively abundant throughout the Early Triassic (Table 4.11).

Based on combined data from half-substage time bins in a lower shoreface environmental setting, the number of potential combinations of mechanisms for

Table 4.11 Abundance of ichnogenera in each mode of life (ecospace utilization) in siliciclastic lower shoreface environmental setting in half-substage time bins

No.	Ecospace utilization	Late Changhsingian	Early Griesbachian	Late Griesbachian	Dienerian	Early Smithian
1	Surficial + Free + Deposit (non-specialized)	0	0	0	0	0
2	Surficial + Free + Predator	0	0	0	0	0
3	Semi-infaunal + Free + Deposit (non-specialized)	0	2	0	0	0
4	Semi-infaunal + Free + Predator	0	0	0	0	0
5	Semi-infaunal + Facultative + Suspension	0	0	0	0	0
6	Semi-infaunal + Facultative + Predator	0	0	0	0	0
7	Semi-infaunal + Free + Deposit (specialized)	1	0	3	0	0
8	Semi-infaunal + Free + Farming	0	0	0	0	0
9	Shallow + Facultative + Suspension	0	0	3	1	3
10	Shallow + Facultative + Predator	0	0	2	1	3
11	Shallow + Free + Deposit (non-specialized)	6	1	4	4	0
12	Shallow + Free + Deposit (specialized)	0	0	0	2	0
13	Shallow + Free + Predator	0	0	0	0	0
14	Shallow + Facultative + Deposit (non-specialized)	2	1	1	1	3
15	Shallow + Free + Farming	0	0	0	0	0

(continued)

Table 4.11 (continued)

No.	Ecospace utilization	Late Changhsingian	Early Griesbachian	Late Griesbachian	Dienerian	Early Smithian
16	Intermediate + Facultative + Suspension	2	3	1	1	2
17	Intermediate + Free + Deposit (non-specialized)	0	0	0	0	2
18	Intermediate + Free + Deposit (specialized)	0	0	0	0	0
19	Intermediate + Free + Predator	0	0	0	0	0
20	Deep + Facultative + Suspension	3	7	4	3	2
21	Deep + Facultative + Predator	3	2	3	3	2
22	Deep + Facultative + Deposit (specialized)	0	0	0	0	0
23	Deep + Free + Deposit (non-specialized)	3	2	0	1	3
24	Deep + Free + Predator	3	2	0	1	3
25	Deep + Facultative + Farming	0	0	0	0	0
26	Deep + Facultative + Chemosymbiosis	0	0	0	0	0

substrate interaction and sediment modification (ecosystem engineering) varied across periods: 7 occurred in the late Changhsingian, 7 in the early Griesbachian, 6 in the late Griesbachian, 6 in the Dienerian, and only 4 in the early Smithian (Fig. 4.7). Overall, there was no significant loss of ecosystem engineering in most infaunal tiers across the P-Tr boundary (Fig. 4.7). However, ecosystem engineering experienced a slight decline from the late Griesbachian to the early Smithian. Ecosystem engineering styles in semi-infaunal, shallow, and deep tiers remained persistent from the mass extinction through the early Smithian. Unfortunately, the lack of exposed localities and trace fossil-bearing successions from the late Smithian to Spathian periods prevents further analysis of ichnological metrics in this environmental habitat.

More specifically, in the semi-infaunal tier, only one type of ecosystem engineering was present in both the late Changhsingian and early Griesbachian, though the forms differed. During the late Changhsingian, sediment conveyors interacted with sediment through backfilling, whereas in the early Griesbachian, biodiffusors interacted with sediment through compression (Fig. 4.8b). In the shallow-tiering level, the number of ecosystem engineering types remained consistent across the Permian-Triassic (P-Tr) boundary and throughout the Early Triassic, but the specific types varied (Fig. 4.8b). In the late Changhsingian, sediment conveyors interacted with sediment via backfilling, gallery biodiffusors through excavation, and biodiffusors through intrusion (Fig. 4.8b). By the early Griesbachian, gallery biodiffusors interacted with sediment through excavation and compression, while sediment conveyors continued to interact through backfilling. These ecosystem engineering types remained unchanged in the shallow tier for the rest of the Early Triassic (Fig. 4.8b). In the intermediate and deep infaunal tiers, the number of impacts on the substrate remained consistent across the P-Tr mass extinction and persisted throughout the Early Triassic. In the deep tier, gallery biodiffusors interacted with sediment through compression, and regenerators interacted with sediment through excavation, both in the late Changhsingian and early Griesbachian (Fig. 4.8b).

The occurrence of ecosystem engineering modes rebounded rapidly during the late Griesbachian and remained persistent throughout the Early Triassic in most tiers (Table 4.12). The highest occurrence of ecosystem engineering styles was observed in the deep tier during the early Griesbachian. In the late Changhsingian, the most abundant modes included shallow-tier sediment conveyors interacting with sediment through backfilling, biodiffusors interacting through intrusion, deep-tier gallery biodiffusors interacting through compression, and regenerators interacting through excavation (Table 4.12). Similarly, in the Early Triassic, the most abundant modes were shallow-tier sediment conveyors interacting through backfilling, shallow-tier gallery biodiffusors interacting through compression, deep-tier gallery biodiffusors interacting through compression, and regenerators interacting through excavation (Table 4.12).

4.6 Ecological Characteristics of Ichnofauna Across the P-Tr Mass Extinction and Its Aftermaths in Deep Marine Setting

4.6.1 Ichnodiversity and Ichnodisparity

The number of ichnogenera in late Changhsingian is much more diverse in deep marine settings than in other environmental settings; eighteen ichnogenera were collected from the studied sections and literature from China (Table 4.13). Most trace fossils were from the deep slope Maomaolong Formation of the Malonggou section in North China, as well as the Talung Formation of the shelf Shangsi and Dongpan sections in South China. Ichnodiversity sharply decreased across the P-Tr mass extinction, and only seven ichnogenera occurred in the early Griesbachian (Fig. 4.9). This result contrasts with former conclusions, which proposed that mass extinction had no or limited impact on benthic organisms in deep marine when compared with the catastrophic collapses in shallow marine and terrestrial

Table 4.12 Abundance of ichnogenera in each impact upon sediment (ecosystem engineering) in siliciclastic lower shoreface environmental setting in half-substage time bins

No.	Ecosystem engineering	Late Changhsingian	Early Griesbachian	Late Griesbachian	Dienerian	Early Smithian
1	Surficial + Compression + Biodiffusion	0	0	0	0	0
2	Semi-infaunal + Compression + Gallery Biodiffusion	0	0	2	0	0
3	Semi-infaunal + Backfill + Conveyor	1	0	1	0	0
4	Semi-infaunal + Excavation + Regenerator	0	0	0	0	0
5	Semi-infaunal + Compression + Biodiffusion	0	2	0	0	0
6	Shallow + Compression + Gallery Biodiffusion	0	1	3	3	3
7	Shallow + Backfill + Conveyor	3	1	4	4	0
8	Shallow + Intrusion + Biodiffusion	3	0	0	0	0
9	Shallow + Excavation + Gallery Biodiffusion	2	1	1	1	3
10	Intermediate + Compression + Gallery Biodiffusion	2	3	1	1	2
11	Intermediate + Backfill + Conveyor	0	0	0	0	1
12	Intermediate + Excavation + Regenerator	0	0	0	0	1
13	Deep + Compression + Gallery Biodiffusive	3	7	4	3	2

(continued)

Table 4.12 (continued)

No.	Ecosystem engineering	Late Changhsingian	Early Griesbachian	Late Griesbachian	Dienerian	Early Smithian
14	Deep + Backfill + Conveyor	0	0	0	0	0
15	Deep + Excavation + Regenerator	3	2	0	1	3
16	Deep + Excavation + Conveyor	0	0	0	0	0

environments (Uchman 2004). Ichnodiversity rebounded in the late Griesbachian, followed by another decline in the Dienerian, then started to rise to form another peak in the early Smithian and continued to decrease to the Spathian (Fig. 4.9). Because very limited deep marine strata and outcrops of the Spathian were exposed in China, the trace fossil data were not enough for our analysis; thus, the lower ichnodiversity could be sample bias (Fig. 4.9).

The ichnogenera in the Late Permian included *Arenicolites*, *Chondrites*, *Cruziana*, *Dictyodora*, *Diplichnites*, *Gyrochorte*, *Halopoa*, *Helminthopsis*, *Helminthoidichnites*, *Lockeia*, *Nereites*, *Ophiomorpha*, *Palaeophycus*, *Planolites*, *Phycodes*, *Protovirgularia*, *Psammichnites*, *Rusophycus*, *Taenidium*, *Thalassinoides*, *Treptichnus*, *Zoophycos*, *Helminthorhaphe*, *Phycosiphon*, *Protopaleodictyon*, and *Skolithos* (Table 4.13). Ichnogenera in the early Griesbachian included *Helminthopsis*, *Palaeophycus*, *Paleodictyon*, *Spirorhaphe*, *Cylindrichnus*, *Planolites*, and *Thalassinoides* (Table 4.13). Ichnogenera in other time bins were listed in Table 4.13 in detail. Ichnodisparity showed a very similar tendency with ichnodiversity, but with lower value, which also sharply decreased across the P-Tr mass extinction and rebounded in the late Griesbachian, followed by a decrease in the Dienerian, and increased in the early Smithian, then declined to the Spathian (Fig. 4.9). The details of ichnodisparity are displayed in Table 4.13.

4.6.2 Ecospace Utilization and Ecosystem Engineering

In the late Changhsingian, the infaunal ecospace categories (modes of life) were sufficiently occupied, with fourteen ecospace categories distributed across the semi-infaunal, shallow, intermediate, and deep tiers (Fig. 4.10). The number of ecospace categories declined to five in the early Griesbachian and rebounded to twelve in the late Griesbachian (Fig. 4.9). Modes of life slightly decreased in the Dienerian and rebounded during the Smithian; thirteen ecospace cubes were utilized during the early and late Smithian (Fig. 4.9). Ecospace utilization therefore sharply decreased from the late Changhsingian to early Griesbachian, followed by a rapid rise during the early Griesbachian, then decreased in the Dienerian, followed by another rise in the early Smithian (Fig. 4.9). Thus, the general tendency of ecospace utilization in

Table 4.13 Ichnogenera and ecological proxies in half-substage time bins, collected from the studied sections in shelf-slope-basin environment

Ichnogenus	Architectural design	Tier	Motility	Feeding	Interaction	Modification
Late Changhsingian						
Arenicolites	Vertical single U- and Y-shaped burrows	Intermediate	Facultative	Suspension	Compression	Gallery biodiffusive
Chondrites	Burrows with shaft or bunch with downwards radiating probes	Deep	Facultative	Chemosymbiosis	Compression	Gallery biodiffusive
Cruziana	Bilobate trails and paired grooves	Semi-infaunal	Free	Non-specialized deposit	Excavation	Regenerator
Dictyodora	Burrows with complex vertically oriented spreiten	Deep	Facultative	Specialized deposit feeder	Backfill	Conveyor
Diplichnites	Trackways and scratch imprints	Surficial	Free	Non-specialized deposit/predator	Compression	Biodiffusive
Gordia	Simple horizontal trails	Semi-infaunal	Free	Non-specialized deposit	Backfill	Conveyor
Gyrochorte	Burrows with complex vertically oriented spreiten	Shallow	Free	Non-specialized deposit	Backfill	Conveyor
Halopoa	Horizontal burrows with simple vertically oriented spreiten	Shallow	Free	Non-specialized deposit	Compression	Gallery biodiffusive
Helminthopsis	Simple horizontal trails	Semi-infaunal	Free	Non-specialized deposit	Backfill	Conveyor
Helminthoidichnites	Simple horizontal trails	Semi-infaunal	Free	Non-specialized deposit	Backfill	Conveyor
Lockeia	Isolated and serial oval to almond-shaped burrows	Semi-infaunal	Free	Non-specialized deposit	Compression	Biodiffusive
Nereites	Complex actively filled horizontal structures	Shallow	Free	Specialized deposit feeder	Backfill	Conveyor

(continued)

Table 4.13 (continued)

Ichnogenus	Architectural design	Tier	Motility	Feeding	Interaction	Modification
Ophiomorpha	Maze and boxwork burrows	Deep	Free	Non-specialized deposit/ predator	Compression	Gallery biodiffusive
Palaeophycus	Passively filled horizontal to oblique burrows	Shallow	Facultative	Suspension/predator	Compression	Gallery biodiffusive
Planolites	Simple, actively filled (massive) horizontal to oblique structures	Shallow	Free	Non-specialized deposit	Backfill	Conveyor
Phycodes	Horizontal burrows with horizontal to vertical branches	Semi-infaunal	Free	Specialized deposit feeder	Backfill	Conveyor
Protovirgularia	Chevronate trails	Shallow	Free	Non-specialized deposit	Intrusion	Biodiffusive
Psammichnites	Complex actively filled horizontal structures	Shallow	Free	Specialized deposit feeder	Backfill	Conveyor
Rusophycus	Bilaterally symmetrical short, scratched impressions and burrows	Surficial	Free	Non-specialized deposit/ predator	Compression	Biodiffusive
Taenidium	Simple actively filled (meniscate) horizontal to oblique structures	Shallow	Free	Non-specialized deposit/ predator	Backfill	Conveyor
Thalassinoides	Maze and boxwork burrows	Deep	Free	Non-specialized deposit/ predator	Excavation	Regenerator
Treptichnus	Horizontal burrows with horizontal to vertical branches	Semi-infaunal	Free	Specialized deposit feeder	Compression	Gallery biodiffusive
Zoophycos	Burrows with helicoidal spreiten	Deep	Facultative	Farming	Excavation	Conveyor
Helminthorhaphe	Guided meandering graphoglyptids	Semi-infaunal	Free	Farming	Compression	Gallery biodiffusive
Phycosiphon	Burrows with horizontal spreiten	Intermediate	Free	Specialized deposit feeder	Backfill	Conveyor
Protopaleodictyon	Regular to irregular network graphoglyptids	Semi-infaunal	Free	Farming	Compression	Gallery biodiffusive
Skolithos	Vertical unbranched burrows	Deep	Facultative	Suspension/predator	Compression	Gallery biodiffusive
Early Griesbachian						

Helminthopsis	Simple horizontal trails	Semi-infaunal	Free	Non-specialized deposit	Backfill	Conveyor
Palaeophycus	Passively filled horizontal to oblique burrows	Shallow	Facultative	Suspension/predator	Compression	Gallery biodiffusive
Paleodictyon	Regular to irregular network graphoglyptids	Semi-infaunal	Free	Farming	Compression	Gallery biodiffusive
Spirorhaphe	Spiral graphoglyptids	Shallow	Free	Farming	Compression	Gallery biodiffusive
Cylindrichnus	Vertical concentrically filled burrows	Semi-infaunal	Free	Non-specialized deposit	Backfill	Conveyor
Planolites	Simple, actively filled (massive) horizontal to oblique structures	Shallow	Free	Non-specialized deposit	Backfill	Conveyor
Thalassinoides	Maze and boxwork burrows	Deep	Free	Non-specialized deposit/predator	Excavation	Regenerator
Late Griesbachian						
Acanthorhaphe	Regular to irregular network graphoglyptids	Semi-infaunal	Free	Farming	Compression	Gallery biodiffusive
Capodistria	Radial to rosetted structures	Shallow	Free	Specialized deposit feeder	Compression	Gallery biodiffusive
Circulichnis	Simple horizontal trails	Semi-infaunal	Free	Non-specialized deposit	Backfill	Conveyor
Cruziana	Bilobate trails and paired grooves	Semi-infaunal	Free	Non-specialized deposit	Excavation	Regenerator
Chondrites	Burrows with shaft or bunch with downwards radiating probes	Deep	Facultative	Chemosymbiosis	Compression	Gallery biodiffusive
Dendrotichnium	Uniramous meandering graphoglyptids	Semi-infaunal	Free	Farming	Compression	Gallery biodiffusive

(continued)

Table 4.13 (continued)

Ichnogenus	Architectural design	Tier	Motility	Feeding	Interaction	Modification
Gordia	Simple horizontal trails	Semi-infaunal	Free	Non-specialized deposit	Backfill	Conveyor
Helminthopsis	Simple horizontal trails	Semi-infaunal	Free	Non-specialized deposit	Backfill	Conveyor
Helminthoidichnites	Simple horizontal trails	Semi-infaunal	Free	Non-specialized deposit	Backfill	Conveyor
Lorenzinia	Radial graphoglyptids	Semi-infaunal	Free	Farming	Compression	Gallery biodiffusive
Palaeophycus	Passively filled horizontal to oblique burrows	Shallow	Facultative	Suspension/predator	Compression	Gallery biodiffusive
Planolites	Simple, actively filled (massive) horizontal to oblique structures	Shallow	Free	Non-specialized deposit	Backfill	Conveyor
Rusophycus	Bilaterally symmetrical short, scratched impressions and burrows	Surficial	Free	Non-specialized deposit/predator	Compression	Biodiffusive
Skolithos	Vertical unbranched burrows	Deep	Facultative	Suspension/predator	Compression	Gallery biodiffusive
Treptichnus	Horizontal burrows with horizontal to vertical branches	Semi-infaunal	Free	Specialized deposit feeder	Compression	Gallery biodiffusive
Ubinia	Uniramous meandering graphoglyptids	Semi-infaunal	Free	Farming	Compression	Gallery biodiffusive
Dienerian						
Gordia	Simple horizontal trails	Semi-infaunal	Free	Non-specialized deposit	Backfill	Conveyor
Helminthoidichnites	Simple horizontal trails	Semi-infaunal	Free	Non-specialized deposit	Backfill	Conveyor
Helminthopsis	Simple horizontal trails	Semi-infaunal	Free	Non-specialized deposit	Backfill	Conveyor
Nereites	Complex actively filled horizontal structures	Shallow	Free	Specialized deposit feeder	Backfill	Conveyor

Palaeophycus	Passively filled horizontal to oblique burrows	Shallow	Facultative	Suspension/predator	Compression	Gallery biodiffusive
Planolites	Simple, actively filled (massive) horizontal to oblique structures	Shallow	Free	Non-specialized deposit	Backfill	Conveyor
Skolithos	Vertical unbranched burrows	Deep	Facultative	Suspension/predator	Compression	Gallery biodiffusive
Taenidium	Simple actively filled (meniscate) horizontal to oblique structures	Shallow	Free	Non-specialized deposit/ predator	Backfill	Conveyor
Early Smithian						
Bergaueria	Vertical plug-shaped burrows	Shallow	Facultative	Suspension	Compression	Gallery biodiffusive
Cosmorhaphe	Guided meandering graphoglyptids	Semi-infaunal	Free	Farming	Compression	Gallery biodiffusive
Glockerichnus	Radial graphoglyptids	Semi-infaunal	Free	Farming	Compression	Gallery biodiffusive
Gordia	Simple horizontal trails	Semi-infaunal	Free	Non-specialized deposit	Backfill	Conveyor
Helminthopsis	Simple horizontal trails	Semi-infaunal	Free	Non-specialized deposit	Backfill	Conveyor
Palaeophycus	Passively filled horizontal to oblique burrows	Shallow	Facultative	Suspension/predator	Compression	Gallery biodiffusive
Paleodictyon	Regular to irregular network graphoglyptids	Semi-infaunal	Free	Farming	Compression	Gallery biodiffusive
Phycodes	Horizontal burrows with horizontal to vertical branches	Semi-infaunal	Free	Specialized deposit feeder	Backfill	Conveyor
Skolithos	Vertical unbranched burrows	Deep	Facultative	Suspension/predator	Compression	Gallery biodiffusive

(continued)

Table 4.13 (continued)

Ichnogenus	Architectural design	Tier	Motility	Feeding	Interaction	Modification
Spirorhaphe	Spiral graphoglyptids	Shallow	Free	Farming	Compression	Gallery biodiffusive
Taenidium	Simple actively filled (meniscate) horizontal to oblique structures	Shallow	Free	Non-specialized deposit/ predator	Backfill	Conveyor
Thalassinoides	Maze and boxwork burrows	Deep	Free	Non-specialized deposit/ predator	Excavation	Regenerator
Zoophycos	Burrows with helicoidal spreiten	Deep	Facultative	Farming	Excavation	Conveyor
Late Smithian						
Gordia	Simple horizontal trails	Semi-infaunal	Free	Non-specialized deposit	Backfill	Conveyor
Helminthopsis	Simple horizontal trails	Semi-infaunal	Free	Non-specialized deposit	Backfill	Conveyor
Palaeophycus	Passively filled horizontal to oblique burrows	Shallow	Facultative	Suspension/predator	Compression	Gallery biodiffusive
Paleodictyon	Regular to irregular network graphoglyptids	Semi-infaunal	Free	Farming	Compression	Gallery biodiffusive
Phycodes	Horizontal burrows with horizontal to vertical branches	Semi-infaunal	Free	Specialized deposit feeder	Backfill	Conveyor
Skolithos	Vertical unbranched burrows	Deep	Facultative	Suspension/predator	Compression	Gallery biodiffusive
Spirorhaphe	Spiral graphoglyptids	Shallow	Free	Farming	Compression	Gallery biodiffusive
Taenidium	Simple actively filled (meniscate) horizontal to oblique structures	Shallow	Free	Non-specialized deposit/ predator	Backfill	Conveyor
Thalassinoides	Maze and boxwork burrows	Deep	Free	Non-specialized deposit/ predator	Excavation	Regenerator
Zoophycos	Burrows with helicoidal spreiten	Deep	Facultative	Farming	Excavation	Conveyor

Early Spathian						
Paleodictyon	Regular to irregular network graphoglyptids	Semi-infaunal	Free	Farming	Compression	Gallery biodiffusive
Palaeophycus	Passively filled horizontal to oblique burrows	Shallow	Facultative	Suspension/predator	Compression	Gallery biodiffusive
Phycodes	Horizontal burrows with horizontal to vertical branches	Semi-infaunal	Free	Specialized deposit feeder	Backfill	Conveyor
Skolithos	Vertical unbranched burrows	Deep	Facultative	Suspension/predator	Compression	Gallery biodiffusive

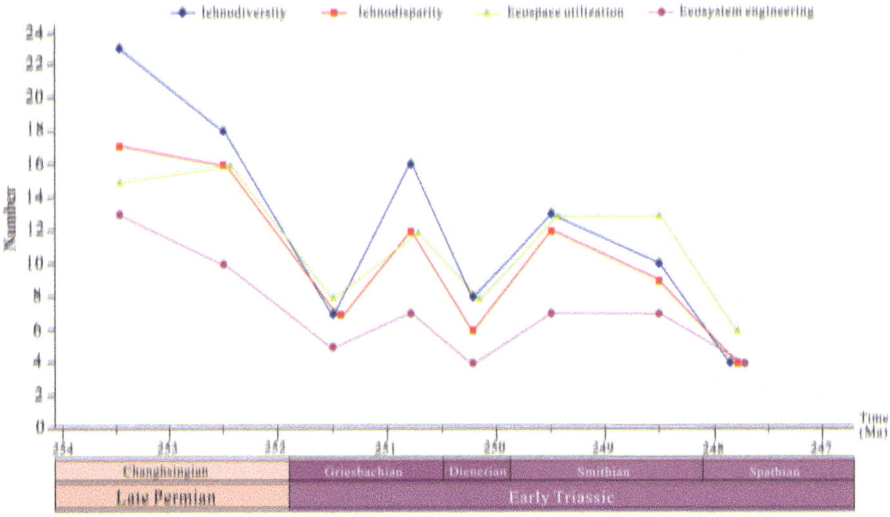

Fig. 4.9 Colonization of shelf-slope-basin environmental setting through the Late Permian to Early Triassic

deep marine environments is more tumultuous than in other environmental settings (Fig. 4.9).

More specifically, in the late Changhsingian, three ecospace categories were utilized in semi-infaunal tier, five in the shallow tier, two in the intermediate tier, and four in the deep tier. Thus the shallow and deep tiers were most utilized (Fig. 4.10a). Freely motile, nonspecialized and specialized deposit feeders, predators, and facultatively motile, suspension feeders and predators occupied the shallow tier (Fig. 4.10a). Freely motile, nonspecialized and predators, and facultatively motile, chemosymbiosis of meiofauna or bacteria and farming meiofauna occupied the deep tier in the late Changhsingian (Fig. 4.10a). Organisms were removed in intermediate and deep tiers across the extinction, and those who utilized the semi- and shallow tiers survived the crisis during the early Griesbachian (Fig. 4.10a). The surficial tier was first occupied by nonspecialized deposit feeders and predators during the late Griesbachian; the deep tier was utilized once again after the extinction, which was colonized by facultatively motile, suspension feeders, predators and chemosymbiosis animals in semi- and shallow tiers, most pre-extinction ecospace categories recovered (Fig. 4.10a). All pre-extinction ecospace categories were retained in the shallow tier during the Dienerian. Most ecospace categories recovered during the Smithian (Fig. 4.10a).

The abundance of categories of ecospace utilization in the late Changhsingian was relatively high when compared with the same categories in other environmental settings. There were 44 occurrences in the shallow tier and 54 occurrences in the deep tier, among which freely motile, nonspecialized deposit feeders were the most abundant, with 23 occurrences in the shallow tier and 16 occurrences in the deep tier

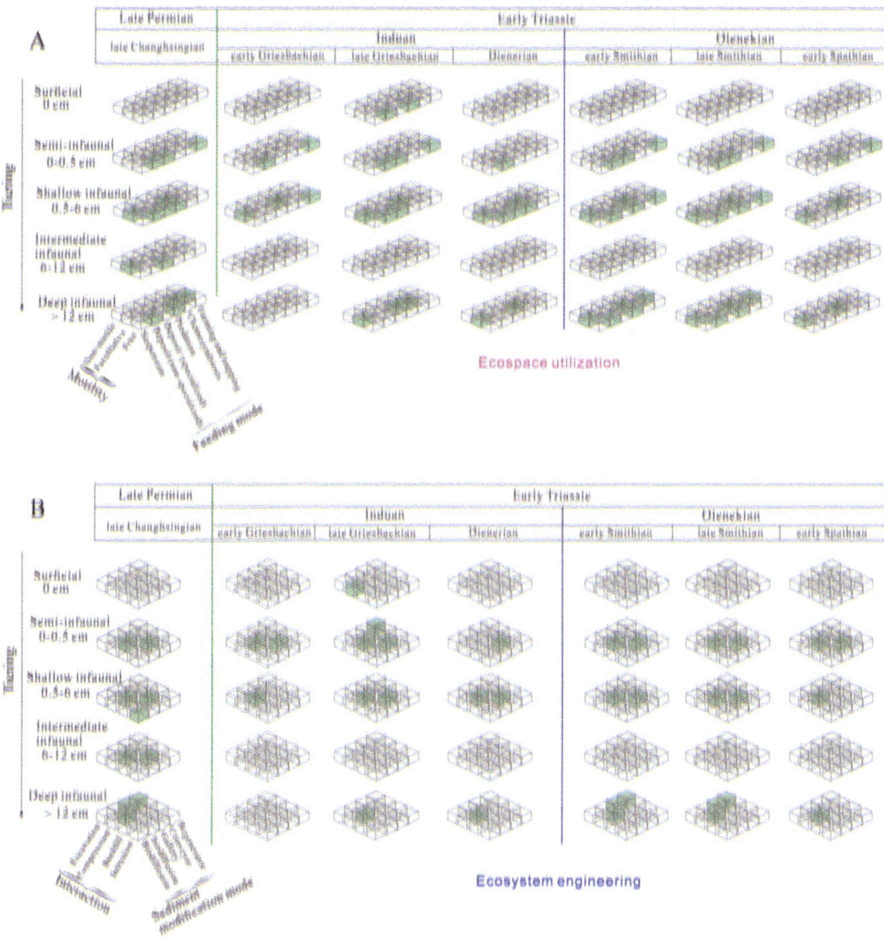

Fig. 4.10 Colonization of shelf-slope-basin environmental setting through the Late Permian to Early Triassic. (**a**) Cubes of ecospace utilization. (**b**) Cubes of ecosystem engineering

(Table 4.14). Most categories of ecospace utilization in the Griesbachian and Dienerian were also rather sporadic (Table 4.14). Freely motile, farming organisms in semi-infaunal tier were the most common in the Dienerian, with 12 occurrences (Table 4.14). When compared with late Changhsingian, although categories of ecospace utilization in the semi-, shallow and deep tiers were abundant in the Smithian-Spathian, their abundance was rather low, with only 4–6 occurrences (Table 4.14).

In the late Changhsingian, there were ten mechanisms of substrate interaction and sediment modification (ecosystem engineering). This number dropped to three in the early Griesbachian, rebounded to seven in the late Griesbachian, declined to four in the Dienerian, and then increased to seven in both the early and late Smithian (Fig. 4.9). Overall, the number of ecosystem engineering mechanisms sharply

Table 4.14 Abundance of ichnogenera in each mode of life (ecospace utilization) in shelf-slope-basin environmental setting in half-substage time bins

No.	Ecospace utilization	Late Changhsingian	Early Griesbachian	Late Griesbachian	Dienerian	Early Smithian	Late Smithian	Early Spathian	Late Spathian
1	Surficial + Free + Deposit (non-specialized)	2	0	1	0	0	0	0	0
2	Surficial + Free + Predator	2	0	1	0	0	0	0	0
3	Semi-infaunal + Free + Deposit (non-specialized)	6	3	8	5	5	4	0	0
4	Semi-infaunal + Free + Predator	0	0	0	0	0	0	0	0
5	Semi-infaunal + Facultative + Suspension	0	0	0	0	0	0	0	0
6	Semi-infaunal + Facultative + Predator	0	0	0	0	0	0	0	0
7	Semi-infaunal + Free + Deposit (specialized)	1	0	1	0	1	3	0	0
8	Semi-infaunal + Free + Farming	5	2	12	0	6	3	2	0
9	Shallow + Facultative + Suspension	8	2	1	3	5	2	0	0
10	Shallow + Facultative + Predator	8	2	1	3	4	2	0	0
11	Shallow + Free + Deposit (non-specialized)	23	1	2	9	4	0	0	0
12	Shallow + Free + Deposit (specialized)	3	0	3	2	0	0	0	0
13	Shallow + Free + Predator	2	0	0	3	1	1	0	0
14	Shallow + Facultative + Deposit (non-specialized)	0	0	0	0	0	0	0	1
15	Shallow + Free + Farming	0	2	0	0	2	2	0	0
16	Intermediate + Facultative + Suspension	1	0	0	0	0	0	0	0

17	Intermediate + Free + Deposit (non-specialized)	0	0	0	0	0	2	0	0
18	Intermediate + Free + Deposit (specialized)	4	0	0	0	0	0	0	0
19	Intermediate + Free + Predator	0	0	0	0	0	2	0	0
20	Deep + Facultative + Suspension	2	0	1	1	2	1	0	0
21	Deep + Facultative + Predator	2	0	1	1	2	1	2	0
22	Deep + Facultative + Deposit (specialized)	0	0	0	0	0	0	0	0
23	Deep + Free + Deposit (non-specialized)	16	2	0	0	1	4	0	0
24	Deep + Free + Predator	16	2	0	0	1	4	0	0
25	Deep + Facultative + Farming	10	0	0	0	2	3	0	0
26	Deep + Facultative + Chemosymbiosis	12	0	1	0	0	0	0	0

declined across the Permian-Triassic (P-Tr) boundary, rebounded quickly in the late Griesbachian, experienced a slight decline in the Dienerian, and then increased again during the Smithian (Fig. 4.9). Ecosystem engineering styles were present in all tiers except the surficial tier during the late Changhsingian. However, after the mass extinction, they persisted only in the semi- and shallow tiers (Fig. 4.9). During the Dienerian to Spathian, ecosystem engineering re-emerged in the semi-, shallow, and deep tiers but did not reappear in the intermediate tier. The types of ecosystem engineering remained consistent during the Smithian-Spathian and were largely similar to their pre-extinction counterparts (Fig. 4.9).

In the semi-infaunal tier, gallery biodiffusors interacting with sediment through compression and sediment conveyors interacting through backfilling were present in the late Changhsingian and persisted into the Early Triassic (Fig. 4.10b). In the shallow tier, three mechanisms of substrate interaction and sediment modification occurred in the late Changhsingian: sediment conveyors interacting through backfilling, biodiffusors interacting through intrusion, and gallery biodiffusors interacting through compression (Fig. 4.10b). However, only sediment conveyors interacting through backfilling persisted into the early Griesbachian. These sediment conveyors reoccurred in the late Griesbachian and remained present throughout the Early Triassic (Fig. 4.10b). In the intermediate tier, two ecosystem engineering categories existed before the extinction: gallery biodiffusors interacting through compression and sediment conveyors interacting through backfilling. Both disappeared during the extinction and did not reappear throughout the Early Triassic (Fig. 4.10b). In the deep infaunal tier, three ecosystem engineering categories were present before the extinction: regenerators and sediment conveyors interacting through excavation and gallery biodiffusors interacting through compression. All of these disappeared in the earliest Triassic, with only gallery biodiffusors interacting through compression re-occurring during the late Griesbachian to Dienerian. All pre-extinction ecosystem engineering categories rebounded in the Smithian and early Spathian (Fig. 4.10b).

Similar to the pattern of ecospace utilization, the occurrence of ecosystem engineering categories in most tiers—particularly in the shallow and deep tiers—was relatively high in the late Changhsingian (Table 4.15). There were 12 occurrences in the semi-infaunal tier, 33 in the shallow tier, 6 in the intermediate tier, and 40 in the deep tier. Among these, sediment conveyors interacting with sediment through backfilling were the most common type in the shallow tier, with 23 occurrences. In the deep tier, gallery biodiffusors interacting through compression were the most abundant, with 16 occurrences, followed by regenerators interacting through excavation, with 14 occurrences (Table 4.15). During the Griesbachian and Dienerian, most categories of ecosystem engineering were relatively sporadic, except for gallery biodiffusors interacting through compression in the semi-infaunal tier, which had 13 occurrences in the late Griesbachian (Table 4.15). Compared to the late Changhsingian, the diversity of ecosystem engineering categories in the semi-, shallow, and deep tiers during the Smithian-Spathian remained notable, but their abundance was significantly lower, with only 2–7 occurrences per category (Table 4.15).

Table 4.15 Abundance of ichnogenera in each impact upon sediment (ecosystem engineering) in shelf-slope-basin environmental setting in half-substage time bins

No.	Ecosystem engineering	Late Changhsingian	Early Griesbachian	Late Griesbachian	Dienerian	Early Smithian	Late Smithian	Early Spathian	Late Spathian
1	Surficial + Compression + Biodiffusion	2	0	1	0	0	0	0	0
2	Semi-infaunal + Compression + Gallery Biodiffusion	5	2	13	0	6	2	1	0
3	Semi-infaunal + Backfill + Conveyor	6	3	7	5	6	7	0	0
4	Semi-infaunal + Excavation + Regenerator	1	0	1	0	0	0	0	0
5	Semi-infaunal + Compression + Biodiffusion	0	0	0	0	0	0	0	0
6	Shallow + Compression + Gallery Biodiffusion	8	4	4	3	7	4	0	0
7	Shallow + Backfill + Conveyor	23	0	3	10	4	7	0	0
8	Shallow + Intrusion + Biodiffusion	2	0	0	0	0	0	0	0
9	Shallow + Excavation + Gallery Biodiffusion	0	0	0	0	0	0	0	0
10	Intermediate + Compression + Gallery Biodiffusion	1	0	0	0	0	0	0	0
11	Intermediate + Backfill + Conveyor	4	0	0	0	0	0	0	0
12	Intermediate + Excavation + Regenerator	1	0	0	0	0	2	0	0

(continued)

Table 4.15 (continued)

No.	Ecosystem engineering	Late Changhsingian	Early Griesbachian	Late Griesbachian	Dienerian	Early Smithian	Late Smithian	Early Spathian	Late Spathian
13	Deep + Compression + Gallery Biodiffusive	16	0	2	1	2	2	0	0
14	Deep + Backfill + Conveyor	0	0	0	0	0	0	0	0
15	Deep + Excavation + Regenerator	14	2	0	0	1	4	0	0
16	Deep + Excavation + Conveyor	10	0	0	0	2	3	0	0

References

Ausich WI, Bottjer DJ (1982) Tiering in suspension feeding communities on soft substrata throughout the Phanerozoic. Science 216:173–174

Bambach RK, Bush AM, Erwin DH (2007) Autecology and the filling of ecospace: key metazoan radiations. Palaeontology 50:1–22

Bottjer DJ, Ausich WI (1986) Phanerozoic development of tiering in soft-substrate suspension-feeding communities. Paleobiology 12:400–420

Bromley RG (1996) Trace fossils: biology, taphonomy and application, 2nd edn. Chapman and Hall, London

Buatois LA, Mángano MG (2011) Ichnology: organism-substrate interaction in space and time. Cambridge University Press, New York, pp 1–358

Buatois LA, Mángano MG (2013) Ichnodiversity and ichnodisparity: significance and caveats. Lethaia 46:281–292

Buatois LA, Gingras MK, MacEachern J, Mángano MG, Zonneveld J-P, Pemberton SG, Netto RG, Martin AJ (2005) Colonization of brackish-water systems through time: evidence from the trace-fossil record. PALAIOS 20:321–347

Buatois LA, Mángano MG, Olea RA, Wilson MA (2016) Decoupled evolution of soft and hard substrate communities during the Cambrian explosion and Great Ordovician Biodiversification event. Proc Natl Acad Sci USA 113:6945–6948

Buatois LA, Wisshak M, Wilson MA, Mángano GM (2017) Categories of architectural designs in trace fossils: a measure of ichnodisparity. Earth Sci Rev 164:102–181

Buatois LA, Mángano MG, Minter NJ, Zhou K, Wisshak M, Wilson MA, Olea RA (2020) Quantifying ecospace utilization and ecosystem engineering during the early Phanerozoic—the role of bioturbation and bioerosion. Sci Adv 6:eabb0618

Bush AM, Bambach RK, Daley GM (2007) Changes in theoretical ecospace utilization in marine fossil assemblages between the mid-Paleozoic and late Cenozoic. Paleobiology 33:76–97

Crimes TP (1994) The period of early evolutionary failure and the dawn of evolutionary success. In: Donovan SK (ed) The palaeobiology of trace fossils. Wiley, Chichester, pp 105–133

Ekdale AA (1985) Paleoecology of the marine endobenthos. Palaeogeogr Palaeoclimatol Palaeoecol 50:63–81

Feng XQ, Chen ZQ, Benton MJ, Su C, Bottjer DJ, Cribb AT, Li Z, Zhao L, Zhu G, Huang Y, Guo Z (2022) Resilience of infaunal ecosystems during the Early Triassic greenhouse Earth. Sci Adv 8:eabo0597

Knaust D, Warchoł M, Kane IA (2014) Ichnodiversity and ichnoabundance: revealing depositional trends in a confined turbidite system. Sedimentology 61:2218–2267

Luo M, Shi GR, Buatois LA, Chen ZQ (2020) Trace fossil as proxy for biotic recovery after the end-Permian: a critical review. Earth Sci Rev 203:103059

Mángano MG, Buatois LA (2014) Decoupling of body-plan diversification and ecological structuring during the Ediacaran–Cambrian transition: evolutionary and geobiological feedbacks. Proc R Soc B 281:e20140038

Minter NJ, Buatois LA, Mángano MG, Davies NS, Gibling MR, Macnaughton RB, Labandeira CC (2017) Early bursts of diversification defined the faunal colonization of land. Nat Ecol Evol 1:1–10

Orr PJ (2001) Colonization of the deep-marine environment during the early Phanerozoic: the ichnofaunal record. Geol J 36:265–278

Seilacher A (1974) Flysch trace fossils: evolution of behavioural diversity in the deep-sea. Neues Jb Geol Paläontol Monat 1974:233–245

Uchman A (2004) Phanerozoic history of deep-sea trace fossils. In: McIlroy D (ed) The application of ichnology to palaeoenvironmental and stratigraphic analysis, Geological society special publications 228. Geological Society, London, pp 125–139

Chapter 5
Implications for the Biotic Recovery Following the P-Tr Mass Extinction

5.1 Behavioral and Ecological Categories of Infauna and Skeletonized Animals Across the P-Tr Mass Extinction in Marine Environmental Settings

During the P-Tr transition, trace fossil metrics exhibit distinct patterns across environmental settings. In brackish environments, all metric trajectories demonstrate a marked decline from the Griesbachian to Dienerian, rise during the early Smithian, and then diminish slightly across the Smithian-Spathian boundary (Figs. 4.1 and 4.2). For carbonate ramp settings, all indices experienced a sharp decline during the P-Tr extinction event, stabilized at reduced levels through the late Griesbachian–Dienerian, reached a peak in the late Smithian, and subsequently saw a minor downturn (Figs. 4.3 and 4.4). In carbonate platform settings, metrics underwent moderate declines during the P-Tr extinction, remained subdued through the late Griesbachian–Dienerian, rose in the early Smithian, and ultimately surpassed pre-extinction values to establish a prominent plateau towards the early Spathian (Figs. 4.5 and 4.6). By contrast, lower shoreface facies sections display an inverse trend to carbonate platforms (Figs. 4.7 and 4.8). Within shelf-slope to basin settings, all metrics displayed rebounds during the early and late Griesbachian, culminating in significant recovery by the early Smithian (Figs. 4.9 and 4.10). These patterns indicate that behavioral variations among burrowing fauna during the P-Tr interval show environment-dependent selectivity. Post-extinction, infaunal behavioral and ecological diversity recovered locally in deeper settings (shelf-slope-basin, lower shoreface) during the early and late Griesbachian, with marine environments achieving full recovery by the early Smithian.

Implications for biotic recovery in the aftermath of the Permian-Triassic mass extinction, and propose recovery models and patterns of marine ecosystems in different environmental habitats

X. Feng et al., *Ichnology of the Permian-Triassic Mass Extinction in China*,
https://doi.org/10.1007/978-981-96-6639-3_5

To reconstruct and compare the four ichnological metrics across five environmental settings during the P-Tr mass extinction, we organized the collection data from each setting into nine half-substage time intervals (Chap. 4), and in this Chapter, we further integrated all data in each of the nine half-substage from the five environmental settings to generate an overall tendency of the four ichnological metrics. The half-substage time bin statistics for behavioral and ecological categories from the studied sections reveal that 17 of all potential ecospace infaunal categories (modes of life) were occupied during the Changhsingian (Fig. 5.1a). Modes of life declined sharply across the P-Tr extinction, surged in the late Griesbachian–Spathian, and ultimately exceeded pre-extinction levels, with the exception of a minor drop during the Dienerian (Fig. 5.1a). Ecospace utilization decreased

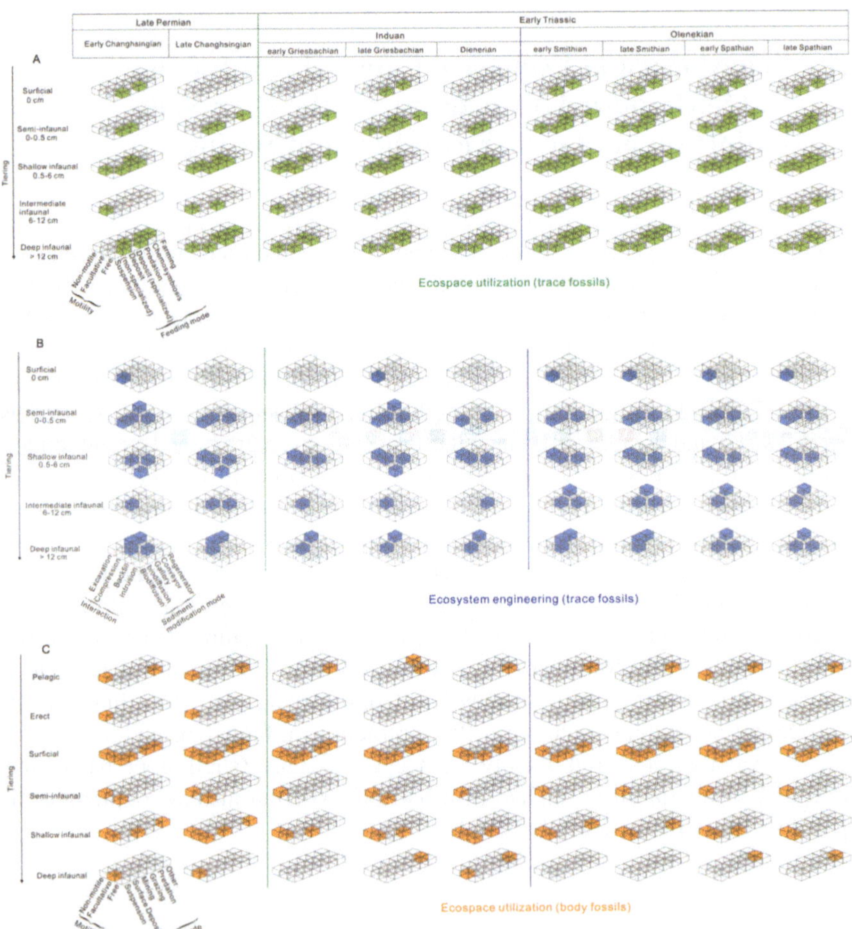

Fig. 5.1 Plot of changes in ecological cubes in all environments through the late Permian to early Triassic. (**a**) Ecospace utilization of infauna based on trace fossils. (**b**) Ecosystem engineering of infauna based on trace fossils. (**c**) Ecospace utilization of skeletonized animals based on body fossils

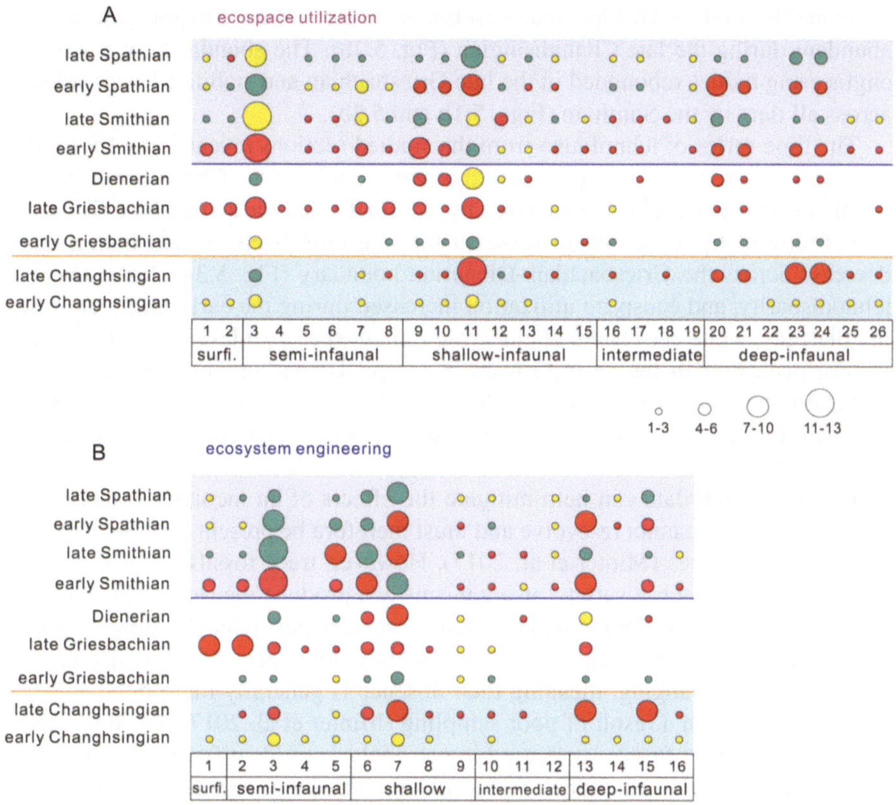

Fig. 5.2 Abundance of ichnogenera in all environments in each mode of life (**a**) and ecosystem engineering (**b**) across the studied interval. Increases (red), decreases (green) and no change (yellow) from the previous time bin. The number codes in each tier in A and B are coincident with those in Tables S5–6

significantly across the P-Tr extinction, particularly for shallow- and deep-infaunal tiers, but rebounded during the early Smithian (Fig. 5.2a). Ecological categories such as facultatively motile suspension feeders and freely motile, non-specialized deposit feeders within the shallow infaunal tier recovered in the late Griesbachian and remained at high levels throughout the rest of the early Triassic (Figs. 5.1a and 5.2a). In contrast, freely motile, non-specialized deposit feeders and facultatively motile suspension feeders occupied very few ecological categories during the Griesbachian–Dienerian, only rebounding in the early Smithian (Figs. 5.1a and 5.2a).

No significant losses in ecosystem engineering were observed across the P-Tr extinction (Fig. 5.1b). Categories of ecosystem engineering across all tiers showed a slight decline at the P-Tr boundary, followed by moderate fluctuations throughout the early Triassic (Fig. 5.1b). However, the abundance of ichnogenera associated with all ecosystem engineering styles in every tier decreased sharply across the P-Tr boundary (Fig. 5.2b). Notably, shallow-tier sediment conveyors that interacted with sediment through backfilling, and deep-tier excavators and regenerators were most abundant during the late Changhsingian (Fig. 5.2b). The abundance of ecosystem engineering modes rebounded in the late Griesbachian and stabilized into a plateau across all tiers by the Smithian (Figs. 5.1b and 5.2b).

The time series of ichnofauna from the studied sections reveal that all four metrics exhibit a broad "W"-shaped pattern of variation from the Changhsingian to the Smithian. This pattern is characterized by a dramatic decline across the P-Tr extinction, followed by a marked increase in the late Griesbachian, and then a sharp decrease across the Griesbachian–Dienerian boundary (Fig. 5.3a). Ichnodiversity, ichnodisparity, and ecospace utilization increased during the early Smithian, forming plateaus, while ecosystem engineering remained at high levels. The two lowest values, occurring in the early Griesbachian and Dienerian, suggest a two-phase depletion pattern across all metrics. However, the decline in ecosystem engineering occupation was less pronounced across the P-Tr extinction compared to the other metrics (Fig. 5.3a).

Range-through data can help mitigate the effects of an incomplete body fossil record, as species cannot re-evolve and must therefore be present between their first and last appearances (Minter et al. 2017). However, trace fossils differ from body fossils in this respect: a variety of organisms can produce the same ichnotaxon by exhibiting the same behavior, and the same organism can create different ichnotaxa by performing different behaviors. Consequently, ichnotaxa are typically broadly defined and long-ranging, meaning their absence is generally interpreted as a true absence rather than a result of poor sampling (Minter et al. 2017). For this reason, raw counts of trace fossils were used in our analysis to identify patterns in behavioral strategies and ecological styles of trace-making organisms (Table 5.7). Additionally, to account for sampling bias, we conducted rarefaction analysis using ichnogenus richness versus occurrences of ichnofossils from the five environmental settings. The results indicate that sampling is sufficient across all intervals (Fig. 5.4a).

To facilitate comparison with ichnofossils, we compiled a dataset of all known marine invertebrate genera using range-through data spanning the late Permian to the Spathian from South China, organized into half-substage bins (Table 5.6). This dataset was primarily derived from an existing database, with revisions to some genera and further subdivisions of the Smithian and Spathian into early and late intervals (i.e., early Smithian, late Smithian, early Spathian, and late Spathian), as well as additional categorization by environmental settings (Song et al. 2018). We employed the ecospace model for marine animals developed by Bambach et al. (2007) and Bush et al. (2007) as the foundation for the quantitative analysis of the ecology of all known benthic marine genera from the late Permian to early Triassic. To simplify the model and enhance its suitability for comparison with trace fossils, certain parameters—such as tiering, motility level, and feeding strategy—were combined (Table 5.6). Additionally, rarefaction analysis was conducted to assess sampling bias, using generic richness versus occurrences of body fossils from the late Permian to early Triassic (Fig. 5.4b). The results confirm that sampling is sufficient across all time bins.

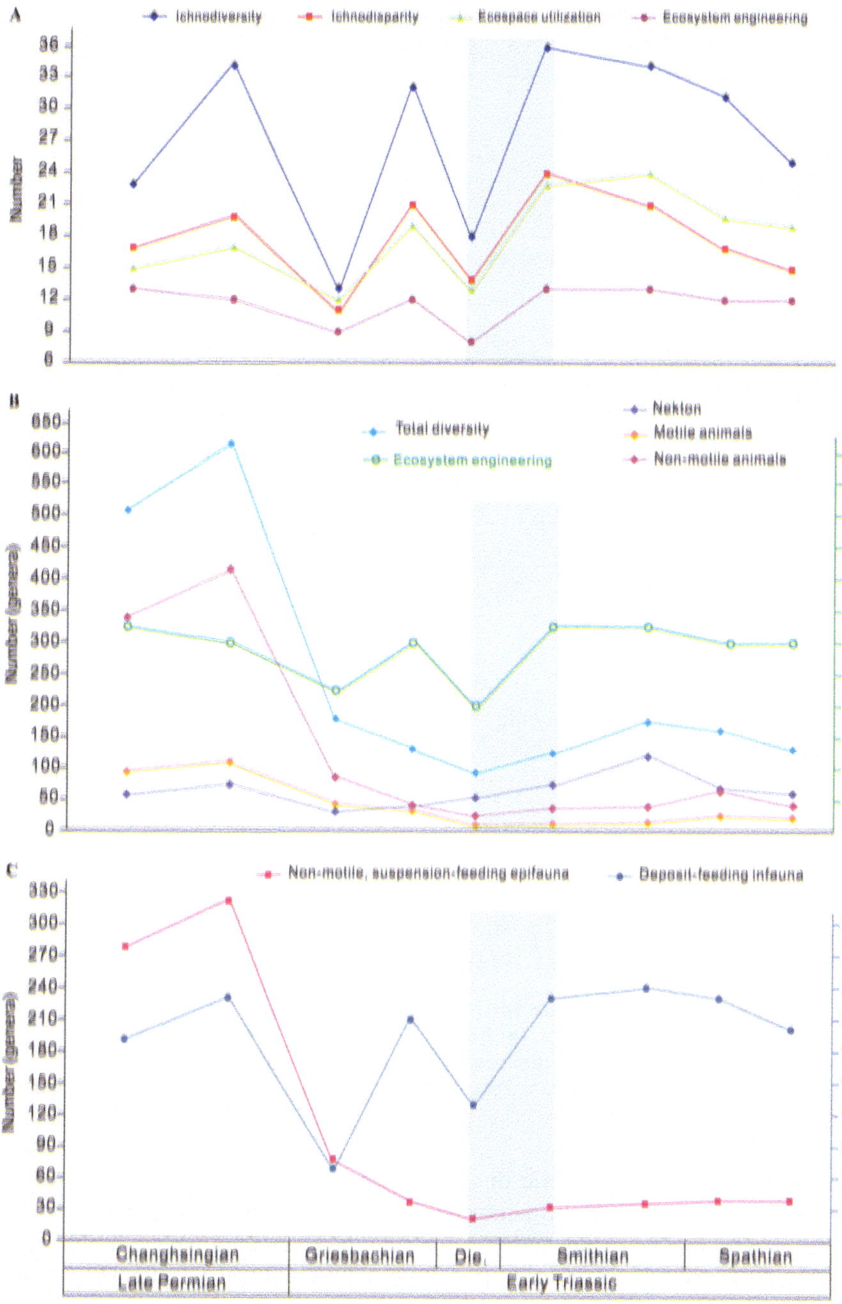

Fig. 5.3 Behavioral and ecologic richness of infauna and generic richness of different functional groups of skeletonized animals through the latest Permian to end of the early Triassic. (**a**) Variations of ichnodiversity, ichnodisparity, ecospace occupation and ecosystem engineering of infauna (trace fossils). (**b**) Generic richness variations of total, nekton, motile, non-motile animals (body fossils). Note that the ecosystem engineering curve of infauna is also cited for comparison. (**c**) Generic richness variations of non-motile, suspension-feeding epifauna, and ichnogeneric richness of deposit-feeding infauna

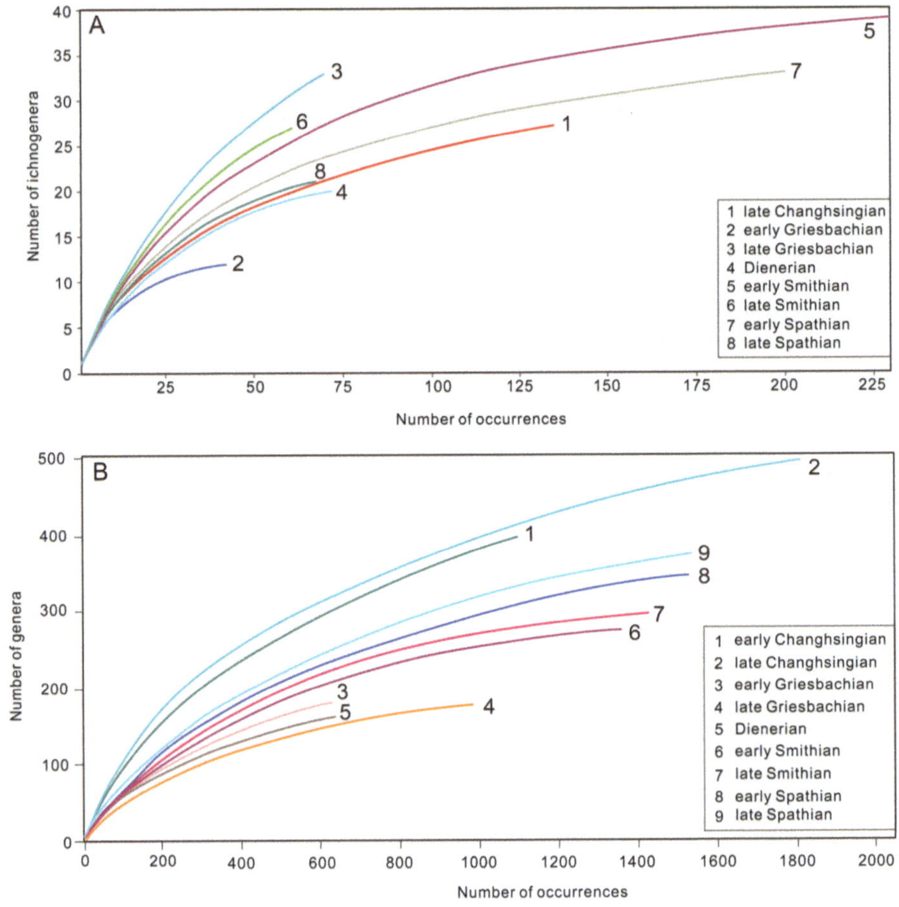

Fig. 5.4 Rarefaction analysis of trace fossil records and body-fossil richness through the late Permian to early Triassic from China. (**a**) Rarefaction curves of ichnogenus richness vs. occurrences of ichnofossils from all environments through the late Permian to early Triassic. (**b**) Rarefaction curves of generic richness vs. occurrences of body fossils through the late Permian to early Triassic

The trajectory of total generic diversity among skeletonized animals shows a sharp decline across the P-Tr extinction, followed by a continued decrease from the Griesbachian to the Dienerian, and then a rise in the early Smithian, culminating in a small peak in the late Smithian across these environmental settings (Fig. 5.3b). Total generic diversity was heavily influenced by the high diversity of non-motile animals during the Permian period. In contrast, the Triassic recovery of generic diversity is primarily driven first by the diversification of nekton and then by the gradual recovery of both motile and non-motile animals (Fig. 5.3b). Non-motile, suspension-feeding animals experienced a sharp decline across the P-Tr extinction, continued to decrease into the Dienerian, and then saw a slight increase during the Smithian.

5.2 Implications for the Biotic Recovery in the Aftermath of the Permian-Triassic Mass Extinction

Ichnodiversity and ichnodisparity reflect the range of infaunal behavioral strategies and the variability of basic morphological plans in biogenic structures, respectively. Additionally, ichnodisparity captures innovations in body plans, locomotory systems, and behavioral programs of infaunal organisms (Mángano and Buatois 2014). The marked declines in these two metrics suggest that both architectural designs and major behavioral innovations among burrowing animals were significantly impacted during the P-Tr mass extinction (Fig. 5.3a). In comparison, more gentle declines in ecological diversities of burrowing animals (Fig. 5.3a and Tables 5.1, 5.2, and 5.3) suggest that ecologically Diverse groups of survivors were able to recolonize and occupy vacant ecospace, when conditions ameliorated during the early Triassic. The decoupled behavioral and ecological diversities suggest that some burrowing animals played a key ecologic role in benthic communities as ecosystem engineers immediately after the mass extinction.

The four ichnological metrics enable us to determine when infaunal communities had fully recovered and re-established themselves following the P-Tr extinction. The initial rebound of behavioral and ecologic diversity occurred in the late Griesbachian, followed by a loss in the Dienerian (Fig. 5.3a). Freely motile, non-specialized deposit-feeders and potential predators, such as arthropods, colonized the surficial-tier ecospace (Fig. 5.1a) and produced horizontal trackways and scratch imprints, and they acted as biodiffusors moving sediment particles over short distances. After the extinction, infaunal ecologic dynamics (ecospace utilization, ecosystem engineering) diversity usually exceeded pre-extinction levels (Fig. 5.3a), but their abundance was still very low (Fig. 5.2a and Tables 5.4 and 5.5). This indicates that the post-extinction ecospace was mainly occupied by opportunists or survivors in some niches when conditions improved. When living conditions were harsh again, the behavioral and ecological diversities of these opportunists declined to some extent. Thus, during the Griesbachian–Dienerian, the infaunal ecospace functioned much like a ship manned by a skeleton crew: each role was occupied, but only by a few individual taxa or behavioral programs. By the early Smithian, however, there was not only a full recovery of behavioral and ecological diversity but also a near-identical restoration of the abundance of these ecological styles (Figs. 5.1, 5.2, 5.3, and 5.4 and Tables 5.1, 5.2, 5.3, 5.4, and 5.5). This ecological structure was characterized by the emergence of a wide repertoire of behavioral strategies, reflecting the interactions of newly developed, distinctive body plans with the substrate, increased complexity in trophic structures, a reorganized occupation of infaunal ecospace, and more complex ecosystem engineering (Fig. 5.1). As a result, the infaunal ecological structure was firmly established by the early Smithian, approximately three million years after the Permian-Triassic (P-Tr) extinction (Fig. 5.3a).

Although the modes of life of benthic marine invertebrate genera (based on body fossils) show no noticeable loss in functional diversity, and early Triassic marine

Table 5.1 Ichnodisparity in each environment through the latest Permian to end of the early Triassic in half-substage time bins

	Brackish								Carbonate platform								Carbonate ramp								Lower shoreface								Shelf-slope-basin								
Architectural design (ichnodisparity)	L. Changhsingian	E. Griesbachian	L. Griesbachian	Dienerian	E. Smithian	L. Smithian	E. Spathian	L. Spathian	L. Changhsingian	E. Griesbachian	L. Griesbachian	Dienerian	E. Smithian	L. Smithian	E. Spathian	L. Spathian	L. Changhsingian	E. Griesbachian	L. Griesbachian	Dienerian	E. Smithian	L. Smithian	E. Spathian	L. Spathian	L. Changhsingian	E. Griesbachian	L. Griesbachian	Dienerian	E. Smithian	L. Smithian	E. Spathian	L. Spathian	E. Changhsingian	L. Changhsingian	E. Griesbachian	L. Griesbachian	Dienerian	E. Smithian	L. Smithian	E. Spathian	L. Spathian
Bilaterally symmetrical short, scratched impressions and burrows	0	0	0	0	0	0	0	0	0	0	0	0	0	0	0	0	0	0	0	0	0	0	0	0	0	0	0	0	0	0	0	0	1	0	0	1	0	0	0	0	0
Bilobate trails and paired grooves	0	0	1	0	1	1	0	0	0	0	0	0	0	0	0	0	0	0	0	0	0	0	0	0	0	0	0	0	0	0	0	0	1	0	0	1	0	0	0	0	0
Burrows with complex vertically oriented spreiten	0	0	0	0	0	0	0	0	0	0	0	0	1	1	1	0	0	0	0	0	1	1	1	1	0	0	1	1	0	0	0	0	1	0	0	0	0	0	0	0	0
Burrows with helicoidal spreiten	0	0	0	0	0	0	0	0	1	0	0	0	0	0	0	0	1	0	0	0	0	0	0	0	0	0	0	0	0	0	0	0	1	1	0	0	0	1	1	0	0
Burrows with horizontal spreiten	0	0	0	0	1	1	0	1	0	0	0	0	0	0	1	0	1	0	0	0	0	1	1	1	1	1	1	1	1	0	0	0	0	1	0	0	0	0	0	0	0
Burrows with shaft or bunch with downwards radiating probes	0	0	0	0	0	0	0	0	0	0	0	0	0	0	0	0	1	0	0	0	1	1	0	0	0	0	0	0	0	0	0	0	1	1	0	1	0	0	0	0	0
Chevronate trails	0	0	1	0	0	0	0	0	0	0	0	0	0	0	0	0	0	0	0	0	0	0	0	0	1	0	0	0	0	0	0	0	1	1	0	0	0	0	0	0	0
Complex actively filled horizontal structures	0	0	0	0	0	0	0	0	0	0	0	0	1	1	0	0	0	0	0	0	0	1	1	1	0	0	0	0	0	0	0	0	1	1	0	0	1	0	0	0	0
Dumbbell and arrow-shaped burrows	0	0	0	0	0	0	0	0	0	0	0	1	1	0	0	0	0	0	0	0	0	0	0	0	0	0	1	0	0	0	0	0	0	0	0	0	0	0	0	0	0

Guided meandering graphoglyptids	0	0	0	0	0	0	0	0	0	0	0	0	0	0	0	0	0	0	0	0	0	0	0	0	0	0	0	0	0	0	0	0	0	1	0	0	0	1	0	0	0
Horizontal burrows with horizontal to vertical branches	0	0	1	0	0	0	1	0	0	0	0	1	1	1	1	0	0	0	0	0	1	1	1	1	1	0	1	0	0	0	0	0	1	1	0	1	0	1	1	1	0
Horizontal burrows with simple vertically oriented spreiten	0	0	0	1	1	1	0	0	0	0	0	0	0	1	0	0	0	0	0	0	0	1	0	0	0	0	0	0	0	0	0	0	1	1	0	0	0	0	0	0	0
Isolated and serial oval to almond-shaped burrows	0	0	1	0	1	1	0	1	0	0	0	0	0	0	1	0	0	0	0	0	0	1	0	0	0	1	0	0	0	0	0	0	1	0	0	0	0	0	0	0	0
Maze and boxwork burrows	0	0	0	0	0	0	1	1	1	0	0	0	1	1	1	1	1	0	0	0	1	1	1	1	1	1	0	1	1	0	0	0	1	1	1	0	0	1	1	0	0
Passively filled horizontal to oblique burrows	0	0	0	1	1	1	1	0	0	0	0	0	1	1	1	1	1	0	0	1	1	1	1	1	0	1	1	1	1	0	0	0	1	1	1	1	1	1	1	1	0
Pentameral-shaped imprints and burrows	0	0	1	0	1	1	0	1	0	0	0	0	0	0	1	0	0	0	0	0	0	0	0	0	0	0	0	0	0	0	0	0	0	0	0	0	0	0	0	0	0
Radial graphoglyptids	0	0	0	0	0	0	0	0	0	0	0	0	0	0	0	0	0	0	0	0	0	0	0	0	0	0	0	0	0	0	0	0	0	0	0	1	0	1	0	0	0
Radial to rosetted structures	0	0	0	0	0	0	0	0	0	0	0	0	0	1	0	0	0	0	0	0	0	1	1	1	0	0	0	0	0	0	0	0	0	0	0	1	0	0	0	0	0
Regular to irregular network graphoglyptids	0	0	0	0	0	0	0	0	0	0	0	0	0	0	0	0	0	0	0	0	0	0	0	0	0	0	0	0	0	0	0	0	0	1	1	1	0	1	1	1	0
Simple actively filled (meniscate) horizontal to oblique structures	0	0	0	0	0	0	1	0	0	0	0	0	0	0	0	0	1	0	0	0	0	0	0	0	0	0	0	0	0	0	0	0	1	1	0	0	1	1	1	0	0

(continued)

Table 5.1 (continued)

	Brackish								Carbonate platform								Carbonate ramp								Lower shoreface								Shelf-slope-basin								
Architectural design (ichnodisparity)	L. Changhsingian	E. Griesbachian	L. Griesbachian	Dienerian	E. Smithian	L. Smithian	E. Spathian	L. Spathian	L. Changhsingian	E. Griesbachian	L. Griesbachian	Dienerian	E. Smithian	L. Smithian	E. Spathian	L. Spathian	L. Changhsingian	E. Griesbachian	L. Griesbachian	Dienerian	E. Smithian	L. Smithian	E. Spathian	L. Spathian	L. Changhsingian	E. Griesbachian	L. Griesbachian	Dienerian	E. Smithian	L. Smithian	E. Spathian	L. Spathian	E. Changhsingian	L. Changhsingian	E. Griesbachian	L. Griesbachian	Dienerian	E. Smithian	L. Smithian	E. Spathian	L. Spathian
Simple horizontal trails	0	0	1	0	1	0	1	1	0	0	0	0	1	1	1	0	0	1	1	0	1	1	1	1	0	0	0	1	0	0	0	0	1	1	1	1	1	1	1	0	0
Simple to complex burrows with terminal chambers	0	0	0	0	0	0	0	0	0	0	0	0	0	0	0	0	0	0	0	0	0	0	0	0	0	0	0	0	0	0	0	0	0	0	0	0	0	0	0	0	0
Simple, actively filled (massive) horizontal to oblique structures	0	0	1	1	1	1	0	1	1	1	1	1	1	1	1	1	1	1	1	1	1	1	1	1	1	1	1	1	0	0	0	0	1	1	1	1	1	0	0	0	0
Spiral graphoglyptids	0	0	0	0	0	0	0	0	0	0	0	0	0	0	0	0	0	0	0	0	0	0	0	0	0	0	0	0	0	0	0	0	0	0	1	0	0	0	1	0	0
Trackways and scratch imprints	0	0	1	0	1	0	0	0	0	0	0	0	0	0	1	0	0	0	0	0	1	1	0	0	0	0	0	0	0	0	0	0	1	0	0	0	0	1	0	0	0
Uniramous meandering graphoglyptids	0	0	0	0	0	0	0	0	0	0	0	0	0	0	0	0	0	0	0	0	0	0	0	0	0	0	0	0	0	0	0	0	0	0	0	1	0	0	0	0	0
Vertical concentrically filled burrows	0	0	0	1	1	1	0	0	0	0	0	0	0	0	0	0	0	0	0	0	0	0	0	0	1	0	0	0	0	0	0	0	0	0	1	0	0	0	0	0	0
Vertical single U- and Y-shaped burrows	0	0	1	1	1	1	0	0	1	0	0	0	0	1	1	0	1	0	0	0	1	1	1	1	0	1	1	0	0	0	0	0	1	1	0	0	0	0	0	0	0
Vertical unbranched burrows	0	0	1	1	1	1	0	0	0	0	0	0	1	1	1	0	1	0	0	0	1	1	1	1	1	1	1	1	1	0	0	0	0	0	0	1	1	1	1	1	0
Vertical plug-shaped burrows	0	0	0	0	0	0	0	0	0	0	0	0	0	0	0	0	0	0	0	0	0	0	0	0	0	0	0	0	0	0	0	0	0	0	0	0	0	1	0	0	0

Table 5.2 Ecospace utilization in each environment through the latest Permian to end of the early Triassic in half-substage time bins

		Brackish								Carbonate platform								Carbonate ramp								Lower shoreface								Shelf-slope-basin								
NO.	Mode of life (ecospace utilization)	L. Changhsingian	E. Griesbachian	L. Griesbachian	Dienerian	E. Smithian	L. Smithian	E. Spathian	L. Spathian	L. Changhsingian	E. Griesbachian	L. Griesbachian	Dienerian	E. Smithian	L. Smithian	E. Spathian	L. Spathian	L. Changhsingian	E. Griesbachian	L. Griesbachian	Dienerian	E. Smithian	L. Smithian	E. Spathian	L. Spathian	L. Changhsingian	E. Griesbachian	L. Griesbachian	Dienerian	E. Smithian	L. Smithian	E. Spathian	L. Spathian	E. Changhsingian	L. Changhsingian	E. Griesbachian	L. Griesbachian	Dienerian	E. Smithian	L. Smithian	E. Spathian	L. Spathian
1	Surficial + Free + Deposit (non-specialized)	0	0	1	0	1	1	0	1	0	0	0	0	0	0	1	0	0	0	0	0	1	1	0	0	0	0	0	0	0	0	0	0	1	0	0	1	0	0	0	0	0
2	Surficial + Free + Predator	0	0	1	0	1	1	0	1	0	0	0	0	0	0	1	0	0	0	0	0	1	1	0	0	0	0	0	0	0	0	0	0	1	0	0	1	0	0	0	0	0
3	Semi-infaunal + Free + Deposit (non-specialized)	0	0	1	1	1	1	1	1	0	0	0	0	1	1	1	0	0	1	1	0	1	1	1	1	0	1	0	1	0	0	0	0	1	1	1	1	1	1	1	0	0
4	Semi-infaunal + Free + Predator	0	0	1	0	0	0	0	0	0	0	0	0	0	0	0	0	0	0	0	0	0	0	0	0	0	0	0	0	0	0	0	0	0	0	0	0	0	0	0	0	0
5	Semi-infaunal + Facultative + Suspension	0	0	1	0	0	0	0	0	0	0	0	0	0	1	1	0	0	0	0	0	1	1	1	1	0	0	0	0	0	0	0	0	0	0	0	0	0	0	0	0	0
6	Semi-infaunal + Facultative + Predator	0	0	1	0	0	0	0	0	0	0	0	0	0	1	1	0	0	0	0	0	1	1	1	1	0	0	0	0	0	0	0	0	0	0	0	0	0	0	0	0	0
7	Semi-infaunal + Free + Deposit (specialized)	0	0	1	0	0	0	1	0	0	0	0	1	1	1	1	0	0	0	0	0	1	1	1	1	1	0	1	0	0	0	0	0	1	1	0	1	0	1	1	1	0
8	Semi-infaunal + Free + Farming	0	0	0	0	0	0	0	0	0	0	0	0	0	0	0	0	0	0	0	0	0	0	0	0	0	0	0	0	0	0	0	0	0	1	1	1	0	1	1	1	0
9	Shallow + Facultative + Suspension	0	0	0	1	1	1	1	0	0	0	0	1	1	1	1	1	1	0	0	1	1	1	1	1	0	1	1	1	1	0	0	0	1	1	1	1	1	1	1	1	0
10	Shallow + Facultative + Predator	0	0	0	1	1	1	1	0	0	0	0	1	1	1	1	1	1	0	0	1	1	1	1	1	0	1	1	1	1	0	0	0	1	1	1	1	1	1	1	1	0
11	Shallow + Free + Deposit (non-specialized)	0	0	1	1	1	1	1	1	1	1	1	1	1	1	1	1	1	1	1	1	1	1	1	1	1	1	1	1	0	0	0	0	1	1	1	1	1	1	1	0	0
12	Shallow + Free + Deposit (specialized)	0	0	0	0	0	0	0	0	0	0	0	0	1	1	0	0	0	0	0	0	0	1	1	1	0	0	0	0	0	0	0	0	1	1	0	1	1	0	0	0	0
13	Shallow + Free + Predator	0	0	0	0	0	0	1	0	0	0	0	0	0	0	0	0	1	0	0	0	0	0	0	0	0	0	0	0	0	0	0	0	1	1	0	0	1	1	1	0	0

(continued)

Table 5.2 (continued)

		Brackish								Carbonate platform								Carbonate ramp								Lower shoreface								Shelf-slope-basin								
NO.	Mode of life (ecospace utilization)	L. Changhsingian	E. Griesbachian	L. Griesbachian	Dienerian	E. Smithian	L. Smithian	E. Spathian	L. Spathian	L. Changhsingian	E. Griesbachian	L. Griesbachian	Dienerian	E. Smithian	L. Smithian	E. Spathian	L. Spathian	L. Changhsingian	E. Griesbachian	L. Griesbachian	Dienerian	E. Smithian	L. Smithian	E. Spathian	L. Spathian	L. Changhsingian	E. Griesbachian	L. Griesbachian	Dienerian	E. Smithian	L. Smithian	E. Spathian	L. Spathian	E. Changhsingian	L. Changhsingian	E. Griesbachian	L. Griesbachian	Dienerian	E. Smithian	L. Smithian	E. Spathian	L. Spathian
14	Shallow + Facultative + Deposit (non-specialized)	0	0	0	0	1	1	0	1	0	0	0	0	0	0	1	0	0	0	0	0	0	0	1	1	1	1	1	1	1	0	0	0	0	0	0	0	0	0	0	0	0
15	Shallow + Free + Farming	0	0	0	0	0	0	0	0	0	0	0	0	0	0	0	0	0	0	0	0	0	0	0	0	0	0	0	0	0	0	0	0	0	0	1	0	0	1	1	0	0
16	Intermediate + Facultative + Suspension	0	0	1	0	1	1	0	0	0	0	0	0	0	1	0	0	0	0	0	0	1	1	1	1	1	1	0	0	0	0	0	0	1	1	0	0	0	0	0	0	0
17	Intermediate + Free + Deposit (non-specialized)	0	0	0	1	1	1	0	0	0	0	0	0	1	0	1	0	0	0	0	0	0	1	0	1	0	0	0	0	0	0	0	0	0	0	0	0	0	0	0	0	0
18	Intermediate + Free + Deposit (specialized)	0	0	0	0	0	0	0	0	0	0	0	0	0	0	0	0	1	0	0	0	0	1	0	0	0	0	0	0	0	0	0	0	0	1	0	0	0	0	0	0	0
19	Intermediate + Free + Predator	0	0	0	0	0	0	0	0	0	0	0	0	1	0	1	0	0	0	0	0	0	1	0	1	0	0	0	0	0	0	0	0	0	0	0	0	0	0	0	0	0
20	Deep + Facultative + Suspension	0	0	0	1	1	1	0	0	0	0	0	0	1	0	1	0	1	0	0	0	0	1	1	1	1	1	1	1	1	0	0	0	0	1	0	1	1	1	1	1	0
21	Deep + Facultative + Predator	0	0	0	1	1	1	0	0	0	0	0	0	1	0	1	0	1	0	0	0	0	1	1	1	1	1	1	1	1	0	0	0	0	1	0	1	1	1	1	1	0
22	Deep + Facultative + Deposit (specialized)	0	0	0	0	0	0	0	0	0	0	0	0	0	0	0	0	0	0	0	0	0	0	0	0	0	0	0	0	0	0	0	0	1	0	0	0	0	0	0	0	0
23	Deep + Free + Deposit (non-specialized)	0	0	0	0	0	0	1	1	1	0	0	0	1	1	1	1	1	0	0	0	1	1	1	1	1	1	0	1	1	0	0	0	1	1	1	0	0	1	1	0	0
24	Deep + Free + Predator	0	0	0	0	0	0	1	1	1	0	0	0	1	1	1	1	1	0	0	0	1	1	1	1	1	1	0	1	1	0	0	0	1	1	1	0	0	1	1	0	0
25	Deep + Facultative + Farming	0	0	0	0	0	0	0	0	1	0	0	0	0	0	0	0	1	0	0	0	0	0	0	0	0	0	0	0	0	0	0	0	1	1	0	0	0	1	1	0	0
26	Deep + Facultative + Chemosymbiosis	0	0	0	0	0	0	0	0	0	0	0	0	0	0	0	0	1	0	0	0	1	1	0	0	0	0	0	0	0	0	0	0	1	1	0	1	0	0	0	0	0

Table 5.3 Ecosystem engineering in each environment through the latest Permian to end of the early Triassic in half-substage time bins

		Brackish								Carbonate platform								Carbonate ramp								Lower shoreface								Shelf-slope-basin								
No.	Impacts upon sediment (ecosystem engineering)	L. Changhsingian	E. Griesbachian	L. Griesbachian	Dienerian	E. Smithian	L. Smithian	E. Spathian	L. Spathian	L. Changhsingian	E. Griesbachian	L. Griesbachian	Dienerian	E. Smithian	L. Smithian	E. Spathian	L. Spathian	L. Changhsingian	E. Griesbachian	L. Griesbachian	Dienerian	E. Smithian	L. Smithian	E. Spathian	L. Spathian	L. Changhsingian	E. Griesbachian	L. Griesbachian	Dienerian	E. Smithian	L. Smithian	E. Spathian	L. Spathian	E. Changhsingian	L. Changhsingian	E. Griesbachian	L. Griesbachian	Dienerian	E. Smithian	L. Smithian	E. Spathian	L. Spathian
1	Surficial + Compression + Biodiffusion	0	0	1	0	1	1	0	1	0	0	0	0	0	0	1	0	0	0	0	0	1	1	0	0	0	0	0	0	0	0	0	0	1	0	0	1	0	0	0	0	0
2	Semi-infaunal + Compression + Gallery Biodiffusion	0	0	1	0	0	0	0	0	0	0	0	0	1	1	1	0	0	0	0	0	1	1	1	1	0	0	1	0	0	0	0	0	1	1	1	1	0	1	1	1	0
3	Semi-infaunal + Backfill + Conveyor	0	0	0	1	1	1	1	1	0	0	0	1	1	1	1	0	0	1	1	0	1	1	1	1	1	0	1	0	0	0	0	0	1	1	1	1	1	1	1	1	0
4	Semi-infaunal + Excavation + Regenerator	0	0	1	0	0	0	0	0	0	0	0	0	0	0	0	0	0	0	0	0	0	0	0	0	0	0	0	0	0	0	0	0	1	0	0	1	0	0	0	0	0
5	Semi-infaunal + Compression + Biodiffusion	0	0	1	0	1	1	0	1	0	0	0	0	0	1	1	0	0	0	0	0	1	1	1	1	0	1	0	1	0	0	0	0	1	0	0	0	0	0	0	0	0
6	Shallow + Compression + Gallery Biodiffusion	0	0	0	1	1	1	1	0	0	0	0	1	1	1	1	1	1	0	0	1	1	1	1	1	0	1	1	1	1	0	0	0	1	1	1	1	1	1	1	1	0

(continued)

Table 5.3 (continued)

No.	Impacts upon sediment (ecosystem engineering)	Brackish								Carbonate platform								Carbonate ramp								Lower shoreface								Shelf-slope-basin								
		L. Changhsingian	E. Griesbachian	L. Griesbachian	Dienerian	E. Smithian	L. Smithian	E. Spathian	L. Spathian	L. Changhsingian	E. Griesbachian	L. Griesbachian	Dienerian	E. Smithian	L. Smithian	E. Spathian	L. Spathian	L. Changhsingian	E. Griesbachian	L. Griesbachian	Dienerian	E. Smithian	L. Smithian	E. Spathian	L. Spathian	L. Changhsingian	E. Griesbachian	L. Griesbachian	Dienerian	E. Smithian	L. Smithian	E. Spathian	L. Spathian	E. Changhsingian	L. Changhsingian	E. Griesbachian	L. Griesbachian	Dienerian	E. Smithian	L. Smithian	E. Spathian	L. Spathian
7	Shallow + Backfill + Conveyor	0	0	1	1	1	1	1	1	1	1	1	1	1	1	1	1	1	1	1	1	1	1	1	1	1	1	1	1	0	0	0	0	1	1	1	1	1	1	1	0	0
8	Shallow + Intrusion + Biodiffusion	0	0	1	0	0	0	0	0	0	0	0	0	0	0	0	0	0	0	0	0	0	0	0	0	1	0	0	0	0	0	0	0	1	1	0	0	0	0	0	0	0
9	Shallow + Excavation + Gallery Biodiffusion	0	0	0	0	1	1	0	1	0	0	0	0	0	0	1	0	0	0	0	0	0	0	1	1	1	1	1	1	1	0	0	0	0	0	0	0	0	0	0	0	0
10	Intermediate + Compression + Gallery Biodiffusion	0	0	1	0	1	1	0	0	0	0	0	0	0	1	0	0	0	0	0	0	1	1	1	1	1	1	0	0	0	0	0	0	1	1	0	0	0	0	0	0	0
11	Intermediate + Backfill + Conveyor	0	0	0	1	1	1	0	0	0	0	0	0	0	0	0	0	1	0	0	0	0	1	0	0	0	0	0	0	0	0	0	0	0	1	0	0	0	0	0	0	0
12	Intermediate + Excavation + Regenerator	0	0	0	0	0	0	0	0	0	0	0	0	1	0	1	0	0	0	0	0	0	1	0	1	0	0	0	0	0	0	0	0	0	0	0	0	0	0	0	0	0

13	Deep + Compression + Gallery Biodiffusive	0	0	0	1	1	1	0	0	0	0	0	0	1	0	1	0	1	0	0	0	1	1	1	1	1	1	1	1	1	0	0	0	1	1	0	1	1	1	1	1	0
14	Deep + Backfill + Conveyor	0	0	0	0	0	0	0	0	0	0	0	0	0	0	0	0	0	0	0	0	0	0	0	0	0	0	0	0	0	0	0	0	1	0	0	0	0	0	0	0	0
15	Deep + Excavation + Regenerator	0	0	0	0	0	0	1	1	1	0	0	0	1	1	1	1	1	0	0	0	1	1	1	1	1	1	0	1	1	0	0	0	1	1	1	0	0	1	1	0	0
16	Deep + Excavation + Conveyor	0	0	0	0	0	0	0	0	1	0	0	0	0	0	0	0	1	0	0	0	0	0	0	0	0	0	0	0	0	0	0	0	1	1	0	0	0	1	1	0	0

Table 5.4 Abundance of ichnogenera in each mode of life (ecospace utilization) in all environments in half-substage time bins

No.	Ecospace utilization	Early Changhsingian	Late Changhsingian	Early Griesbachian	Late Griesbachian	Dienerian	Early Smithian	Late Smithian	Early Spathian	Late Spathian
1	Surficial + Free + Deposit (non-specialized)	2	0	0	6	0	4	3	3	3
2	Surficial + Free + Predator	2	0	0	6	0	4	3	2	3
3	Semi-infaunal + Free + Deposit (non-specialized)	5	4	4	8	5	13	13	7	7
4	Semi-infaunal + Free + Predator	0	0	0	1	0	0	0	0	0
5	Semi-infaunal + Facultative + Suspension	0	0	0	1	0	1	2	2	1
6	Semi-infaunal + Facultative + Predator	0	0	0	1	0	1	2	2	1
7	Semi-infaunal + Free + Deposit (specialized)	2	2	0	4	1	5	5	5	1
8	Semi-infaunal + Free + Farming	0	2	1	4		3	1	1	0
9	Shallow + Facultative + Suspension	1	3	2	4	6	8	4	5	3
10	Shallow + Facultative + Predator	1	3	2	3	5	6	4	5	3
11	Shallow + Free + Deposit (non-specialized)	5	11	4	7	7	6	6	8	7
12	Shallow + Free + Deposit (specialized)	2	1	0	1	1	1	4	2	2

13	Shallow + Free + Predator	1	3	0	0	1	1	1	3	2
14	Shallow + Facultative + Deposit (non-specialized)	0	1	1	1	0	2	1	2	2
15	Shallow + Free + Farming	0	0	1	0	0	1	1	0	0
16	Intermediate + Facultative + Suspension	1	2	1	1	0	2	3	2	2
17	Intermediate + Free + Deposit (non-specialized)	0	0	0	0	1	2	3	1	1
18	Intermediate + Free + Deposit (specialized)	0	2	0	0	0	0	1	0	0
19	Intermediate + Free + Predator	0	0	0	0	0	1	1	1	1
20	Deep + Facultative + Suspension	0	4	2	3	4	6	4	8	3
21	Deep + Facultative + Predator	0	4	1	2	3	5	3	6	2
22	Deep + Facultative + Deposit (specialized)	1	0	0	0	0	0	0	0	0
23	Deep + Free + Deposit (non-specialized)	2	8	2	0	1	4	3	6	5
24	Deep + Free + Predator	2	8	2	0	1	4	3	6	5
25	Deep + Facultative + Farming	1	3	0	0	0	1	1	0	0
26	Deep + Facultative + Chemosymbiosis	1	2	0	1	0	1	1	0	0

Table 5.5 Abundance of ichnogenera in each impact upon sediment (ecosystem engineering) in all environments in half-substage time bins

No.	Ecosystem engineering	Early Changhsingian	Late Changhsingian	Early Griesbachian	Late Griesbachian	Dienerian	Early Smithian	Late Smithian	Early Spathian	Late Spathian
1	Surficial + Compression + Biodiffusion	2	0	0	7	0	4	3	3	3
2	Semi-infaunal + Compression + Gallery Biodiffusion	1	2	1	7	0	5	3	3	1
3	Semi-infaunal + Backfill + Conveyor	4	5	3	6	5	12	11	8	6
4	Semi-infaunal + Excavation + Regenerator	1	0	0	2	0	0	0	0	0
5	Semi-infaunal + Compression + Biodiffusion	1	1	1	2	1	5	7	4	2
6	Shallow + Compression + Gallery Biodiffusion	2	4	3	5	6	9	7	6	4
7	Shallow + Backfill + Conveyor	5	8	4	6	8	7	8	9	8
8	Shallow + Intrusion + Biodiffusion	1	3	0	1	0	0	0	0	0
9	Shallow + Excavation + Gallery Biodiffusion	0	1	1	1	1	2	1	2	2
10	Intermediate + Compression + Gallery Biodiffusion	1	2	1	1	0	2	3	2	2
11	Intermediate + Backfill + Conveyor	0	2	0	0	1	1	3	0	0

12	Intermediate + Excavation + Regenerator	0	0	0	0	0	1	1	1	1
13	Deep + Compression + Gallery Biodiffusive	2	7	2	4	4	7	5	8	3
14	Deep + Backfill + Conveyor	1	0	0	0	0	0	0	1	1
15	Deep + Excavation + Regenerator	1	7	2	0	1	4	3	5	4
16	Deep + Excavation + Conveyor	1	3	0	0	0	1	1	0	0

Table 5.6 Generic richness of different functional groups of skeletonized animals and deposit-feeding burrowing animals through the latest Permian to end of the early Triassic in half-substage time bins

Mode of life (ecospace utilization) body fossils	E. Changhsingian	L. Changhsingian	E. Griesbachian	L. Griesbachian	Dienerian	E. Smithian	L. Smithian	E. Spathian	L. Spathian
Pelagic + Free + Predatory	56	77	33	41	52	75	118	67	60
Pelagic + Non-motile + Other				1					
Pelagic + Non-motile + Suspension	47	65						9	
Erect + Facultative + suspension			1						
Erect + non-motile + suspension	30	35	2						
Surface + Free + suspension	24	27	5	6		1	1	3	1
Surface + Free + suspension/ grazing				1		1	1	1	
Surface + Free + Deposit Feeding	42	51	21	13	5	6	7	16	16
Surface + Free + Deposit Feeding/ Predatory			2						1

Surface + Free + Grazing	17	17	12	8	2	2	2	4	4
Surface + Free + Predator	4	2	4	1					1
Surface + Facultative + Suspension	2	3	1	2	1	0	2	1	0
Surface + Non-motile + Suspension	137	164	48	30	17	21	24	25	27
Semi-infaunal + Free + Suspension	2	4	0	1	0	0	0	0	0
Semi-infaunal + Non-motile + suspension	107	125	29	7	5	11	11	14	12
Shallow infaunal + Free + Suspension	0	1	0	0	1	0	0	0	0
Shallow infaunal + Free + Mining	6	6	3	3	1	0	0	1	0
Shallow infaunal + Free + Predatory	0	0	0	0	0	1	1	0	0
Shallow infaunal + Free + Other	1	1	0	0	0	0	0	0	0
Shallow infaunal + Facultative + suspension	6	8	7	10	9	3	5	7	3

(continued)

Table 5.6 (continued)

Mode of life (ecospace utilization) body fossils	E. Changhsingian	L. Changhsingian	E. Griesbachian	L. Griesbachian	Dienerian	E. Smithian	L. Smithian	E. Spathian	L. Spathian
Shallow infaunal + Non-motile + Suspension	19	24	11	6	2	5	5	9	5
Deep infaunal + Facultative + Suspension	4	4	0	0	1	0	0	0	0
Deep infaunal + Facultative + Other	0	0	0	1	1	0	0	1	1
Numbers represent the number of genera									
Nekton	56	77	33	42	52	75	118	67	60
Motile	96	109	47	33	9	11	12	25	23
Facultative	12	15	9	13	12	3	7	9	4
Non-motile	340	413	90	43	24	37	40	57	44
Total number of genera	504	614	179	131	97	126	177	158	131
Numbers represent the number of genera									
Non-motile, suspension epifauna	167	199	50	30	17	21	24	25	27
Non-motile, suspension semi-infauna	107	125	29	7	5	11	11	14	12

Total	274	324	79	37	22	32	35	39	39
Deposit-feeding infauna (trace fossils)	19	23	7	21	13	23	24	23	20

Numbers represent the number of ichnogenera

Table 5.7 Occurrence of ichnogenera in all environments in half-substage time bins

Late Changhsingian	
Ichnogenus	Occurrence
Arenicolites	1
Balanoglossites	2
Chondrites	14
Cruziana	1
Diplichnites	1
Halopoa	1
Helminthoidichnites	3
Helminthopsis	3
Helminthorhaphe	3
Lockeia	1
Nereites	3
Ophiomorpha	2
Palaeophycus	11
Phycodes	2
Phycosiphon	5
Planolites	24
Protopaleodictyon	2
Protovirgularia	6
Rhizocorallium	2
Rosselia	2
Rusophycus	1
Siphonichnus	1
Skolithos	6
Spongeliomorpha	1
Taenidium	4
Thalassinoides	21
Zoophycos	12
Early Griesbachian	
Arenicolites	3
Cylindrichnus	1
Diplocraterion	5
Helminthopsis	2
Lockeia	2
Palaeophycus	3
Paleodictyon	2
Planolites	14
Rhizocorallium	1
Skolithos	3
Spirorhaphe	2
Thalassinoides	4

(continued)

Table 5.7 (continued)

Late Changhsingian	
Ichnogenus	Occurrence
	Late Griesbachian
Acanthorhaphe	3
Arenicolites	1
Asteriacites	2
Biformites	2
Bifungites	1
Capodistria	3
Chondrites	1
Circulichnis	3
Cruziana	1
Dendrotichnium	3
Dendrotichnium	1
Didymaulichnus	1
Diplichnites	1
Diplocraterion	1
Gordia	1
Gyrochorte	2
Helminthoidichnites	5
Helminthopsis	1
Kouphichnium	2
Laevicyclus	1
Lorenzinia	3
Monomorphichnus	1
Palaeophycus	4
Phycodes	1
Planolites	8
Protovirgularia	2
Ptychoplasma	1
Rhizocorallium	1
Rusophycus	1
Skolithos	3
Treptichnus	5
Ubinia	3
Undichna	1
Dienerian	
Arenicolites	1
Arenituba	2
Bifungites	2
Cylindrichnus	2
Diplocraterion	5
Gordia	1

(continued)

Table 5.7 (continued)

Late Changhsingian	
Ichnogenus	Occurrence
Gyrochorte	4
Helminthoidichnites	2
Helminthopsis	2
Nereites	2
Palaeophycus	12
Phycodes	2
Planolites	17
Rhizocorallium	1
Siphonichnus	4
Skolithos	4
Taenidium	3
Teichichnus	3
Thalassinoides	1
Early Smithian	
Archaeonassa	3
Arenicolites	8
Arenituba	1
Asteriacites	2
Bergaueria	1
Biformites	2
Bifungites	3
Chondrites	2
Circulichnis	1
Cochlichnus	3
Cosmorhaphe	1
Cylindrichnus	3
Diplocraterion	4
Diplopodichnus	2
Glockerichnus	3
Gordia	7
Gyrochorte	6
Helminthoidichnites	5
Helminthopsis	11
Kouphichnium	3
Laevicyclus	4
Lockeia	2
Palaeophycus	36
Paleodictyon	4
Phycodes	9
Phycosiphon	1
Planolites	23

(continued)

Table 5.7 (continued)

Late Changhsingian	
Ichnogenus	Occurrence
Psammichnites	5
Ptychoplasma	1
Rhizocorallium	7
Sinusichnus	2
Siphonichnus	5
Skolithos	10
Spirorhaphe	4
Taenidium	3
Teichichnus	4
Thalassinoides	21
Treptichnus	14
Zoophycos	4
Late Smithian	
Archaeonassa	2
Arenicolites	4
Arenituba	2
Circulichnis	1
Cylindrichnus	2
Diplocraterion	2
Gordia	3
Gyrochorte	3
Helminthoidichnites	1
Kouphichnium	1
Laevicyclus	1
Lockeia	3
Palaeophycus	10
Paleodictyon	1
Phycodes	2
Phycosiphon	1
Planolites	5
Psammichnites	1
Rhizocorallium	1
Sinusichnus	2
Siphonichnus	1
Skolithos	2
Taenidium	1
Teichichnus	3
Thalassinoides	4
Treptichnus	1
Zoophycos	1
Early Spathian	

(continued)

Table 5.7 (continued)

Late Changhsingian	
Ichnogenus	Occurrence
Archaeonassa	1
Arenicolites	9
Arenituba	3
Asteriacites	1
Balanoglossites	1
Beaconites	7
Biformites	2
Camborygma	1
Circulichnis	3
Diplichnites	1
Diplocraterion	8
Gordia	6
Gyrochorte	9
Helminthoidichnites	6
Helminthopsis	4
Kouphichnium	1
Laevicyclus	2
Lockeia	3
Macanopsis	1
Monomorphichnus	1
Palaeophycus	30
Paleodictyon	2
Phycodes	6
Planolites	24
Psammichnites	3
Rhizocorallium	10
Scoyenia	5
Sinusichnus	4
Skolithos	20
Spongeliomorpha	2
Taenidium	5
Thalassinoides	17
Treptichnus	2
Late Spathian	
Arenicolites	4
Arenituba	1
Asteriacites	1
Beaconites	3
Circulichnis	2
Gordia	5
Gyrochorte	2

(continued)

Table 5.7 (continued)

Late Changhsingian	
Ichnogenus	Occurrence
Helminthoidichnites	2
Helminthopsis	3
Laevicyclus	1
Palaeophycus	8
Planolites	6
Psammichnites	1
Ptychoplasma	1
Rhizocorallium	9
Scoyenia	2
Sinusichnus	2
Skolithos	4
Spongeliomorpha	1
Thalassinoides	7
Treptichnus	3

ecosystems were not ecologically depauperate (Foster and Twitchett 2014), all functional groups of benthic fauna documented from sections across South China exhibit significant declines in generic richness (Fig. 5.3b and Table 5.6). Our data show that the functional diversity and ecosystem engineering activities of infauna reached or exceeded their pre-extinction levels in the Smithian and remained at relatively high levels thereafter. For infaunal communities, motile deposit feeders consistently acted as important components of marine benthic ecosystems (Fig. 5.3c). This functional group is generally coupled with epifaunal non-motile, suspension feeders in generic diversity before the P-Tr extinction, but decoupling between the two is conspicuous after the extinction (Fig. 5.3c). This decoupling is likely attributed to large-scale trophic amensalism, where the biological bulldozing activities of motile, deposit-feeding infauna may have disrupted fragile ecological habitats on the seafloor and destabilized the habitats of non-motile, suspension-feeding epifauna during the early Triassic (Fig. 5.3c). This phenomenon may have contributed, at least in part, to the so-called suppressed recovery (Song et al. 2018) of non-motile, suspension-feeding animals during the early Triassic.

As ecosystem engineers, bioturbators can have both negative and positive effects on biodiversity. Bioturbation enhances the complexity of geochemical gradients in sediments by altering the rate of redox cycling through biomixing and bioirrigation (Meysman et al. 2006; Erwin 2008). Biomixing and bioirrigation often exert opposing influences on sedimentary biogeochemistry. While biomixing tends to increase oxygen consumption, bioirrigation typically oxygenates the sediment (van de Velde and Meysman 2016; van de Velde et al. 2018, 2021). Ecologically, this distinction is critical, as biomixers and bioirrigators can have contrasting impacts on biodiversity and the structure of benthic and infaunal communities. Historically, increased levels of biomixing have been linked to ocean anoxia and global warming (van de Velde et al. 2018), while increased bioirrigation promotes productivity by recycling nutrients such as nitrogen (Cribb and Bottjer 2020) and phosphorus (Tarhan et al. 2021). However, our data demonstrate a positive interaction between ecosystem engineers and biodiversity (Fig. 5.3b), in which the increased infaunal ecosystem engineering proxies are coupled with slow increases in the total generic richness of skeletonized animals and all functional groups at that time (Fig. 5.3b). This positive effect may be a result of the amelioration of environmental conditions in the aftermath of the extinction.

Thus, a robust infaunal ecological structure emerged by the late early Triassic, preceding the full restoration of epifauna-dominated ecosystems in the Middle Triassic. Pioneering ecosystem engineers likely facilitated seafloor colonization and occupied vacant ecospace, potentially creating positive feedback mechanisms that supported the initial recovery of skeletonized animals during this interval. Phylogenetic studies of coeval fishes and marine reptiles further reveal an explosive diversification event in the early Triassic (Chen and Benton 2012; Benton et al. 2013). These innovations and radiations reflect opportunistic ecospace occupation following the mass extinction (Hull 2015; Erwin 2021). The reworking of substrates in the late early Triassic may have partly driven these evolutionary breakthroughs (Wood and Erwin 2017). This study highlights how intrinsic ecosystem engineering by burrowing organisms catalyzed the early recovery of marine epifauna and ecosystems after one of Earth's most severe biotic crises.

References

Bambach RK, Bush AM, Erwin DH (2007) Autecology and the filling of ecospace: key metazoan radiations. Palaeontology 50:1–22

Benton MJ, Zhang QY, Hu SX, Chen ZQ, Wen W, Liu J, Huang JY, Zhou CY, Xie T, Tong JN, Choo B (2013) Exceptional vertebrate biotas from the Triassic of China, and the expansion of marine ecosystems after the Permo–Triassic mass extinction. Earth Sci Rev 123:199–243

Bush AM, Bambach RK, Daley GM (2007) Changes in theoretical ecospace utilization in marine fossil assemblages between the mid-Paleozoic and late Cenozoic. Paleobiology 33:76–97

Chen ZQ, Benton MJ (2012) The timing and pattern of biotic recovery following the end-Permian mass extinction. Nat Geosci 5:375–383

Cribb AT, Bottjer DJ (2020) Complex marine bioturbation ecosystem engineering behaviors persisted in the wake of the end-Permian mass extinction. Sci Rep 10:203

Erwin DH (2008) Macroevolution of ecosystem engineering, niche construction and diversity. Trends Ecol Evol 23:304–310

Erwin DH (2021) A conceptual framework of evolutionary novelty and innovation. Biol Rev 96:1–15

Foster WJ, Twitchett RJ (2014) Functional diversity of marine ecosystems after the Late Permian mass extinction event. Nat Geosci 7:233–238

Hull PM (2015) Life in the aftermath of mass extinctions. Curr Biol 25:R941–R952

Mángano MG, Buatois LA (2014) Decoupling of body-plan diversification and ecological structuring during the Ediacaran–Cambrian transition: evolutionary and geobiological feedbacks. Proc R Soc B 281:e20140038

Meysman FJR, Middelburg JJ, Heip CHR (2006) Bioturbation: a fresh look at Darwin's last idea. Trends Ecol Evol 21:688–695

Minter NJ, Buatois LA, Mángano MG, Davies NS, Gibling MR, Macnaughton RB, Labandeira CC (2017) Early bursts of diversification defined the faunal colonization of land. Nat Ecol Evol 1:1–10

Song HJ, Wignall PB, Dunhill AM (2018) Decoupled taxonomic and ecological recoveries from the Permo-Triassic extinction. Sci Adv 4:eaat5091

Tarhan LG, Zhao M, Planavsky NJ (2021) Bioturbation feedbacks on the phosphorus cycle. Earth Planet Sci Lett 566:116961

van de Velde S, Meysman FJR (2016) The influence of bioturbation on iron and Sulphur cycling in marine sediments: a model analysis. Aquat Geochem 22:469–504

van de Velde S, Mills BJW, Meysman FJR, Lenton TM, Poulton SW (2018) Early Palaeozoic Ocean anoxia and global warming driven by the evolution of shallow burrowing. Nat Commun 9:2554

van de Velde SJ, James RK, Callebaut I, Hidalgo-Martinez S, Meysman FJR (2021) Bioturbation has a limited effect on phosphorus burial in salt marsh sediments. Biogeosciences 18:1451–1461

Wood R, Erwin DH (2017) Innovation not recovery: dynamic redox promotes metazoan radiations. Biol Rev 93:863–873

MIX
Papier aus verantwortungsvollen Quellen
Paper from responsible sources
FSC® C105338

If you have any concerns about our products,
you can contact us on
ProductSafety@springernature.com

In case Publisher is established outside the EU,
the EU authorized representative is:
Springer Nature Customer Service Center GmbH
Europaplatz 3, 69115 Heidelberg, Germany

Printed by Libri Plureos GmbH
in Hamburg, Germany